THE TRIPLE PACKAGE

ALSO BY

AMY CHUA

Battle Hymn of the Tiger Mother

Day of Empire:
How Hyperpowers Rise to Global Dominance—
and Why They Fall

World on Fire:
How Exporting Free Market Democracy
Breeds Ethnic Hatred and Global Instability

ALSO BY

JED RUBENFELD

The Death Instinct

The Interpretation of Murder

Freedom and Time:
A Theory of Constitutional Self-Government

Revolution by Judiciary:
The Structure of American Constitutional Law

THE TRIPLE PACKAGE

What Really Determines Success

AMY CHUA

JED RUBENFELD

B L O O M S B U R Y

LONDON · NEW DELHI · NEW YORK · SYDNEY

First published in Great Britain 2014

Copyright © 2014 by Amy Chua and Jed Rubenfeld

The moral rights of the authors have been asserted

No part of this book may be used or reproduced in any manner whatsoever
without written permission from the publisher except in the case of brief
quotations embedded in critical articles or reviews

Bloomsbury Publishing Plc
50 Bedford Square
London
WC1B 3DP

www.bloomsbury.com

Bloomsbury Publishing, London, New Delhi, New York and Sydney

A CIP catalogue record for this book is available from the British Library

ISBN 978 1 4088 5223 1 (hardback edition)
ISBN 978 1 4088 5224 8 (trade paperback edition)

10 9 8 7 6 5 4 3 2 1

Printed and bound in Great Britain by CPI Group (UK) Ltd, Croydon CR0 4YY

To
SOSO
and
LULU

CONTENTS

INTRODUCTION *1*

CHAPTER 1 ★ **THE TRIPLE PACKAGE** *5*

CHAPTER 2 ★ **WHO'S SUCCESSFUL IN AMERICA?** *29*

CHAPTER 3 ★ **THE SUPERIORITY COMPLEX** *59*

CHAPTER 4 ★ **INSECURITY** *85*

CHAPTER 5 ★ **IMPULSE CONTROL** *117*

CHAPTER 6 ★ **THE UNDERSIDE OF THE TRIPLE PACKAGE** *145*

CHAPTER 7 ★ **IQ, INSTITUTIONS, AND UPWARD MOBILITY** *167*

CHAPTER 8 ★ **AMERICA** *199*

ACKNOWLEDGMENTS *227*

NOTES *231*

INDEX *309*

THE TRIPLE PACKAGE

INTRODUCTION

IT IS ONE OF HUMANITY's enduring mysteries why some individuals rise from unpromising origins to great heights, when so many others, facing similar obstacles and with seemingly similar capabilities, don't rise at all.

This book is about that age-old question. We wrote it hoping that readers might come away with a better understanding of the world we live in—a world in which certain individuals and groups do strikingly better than others in terms of wealth, position, and other conventional measures of success.

The paradoxical premise of this book is that successful people tend to feel simultaneously inadequate and superior. Certain groups tend to make their members feel this way more than others; groups that do so are disproportionately successful. This unlikely combination of qualities is part of a potent cultural package that generates drive: a need to prove oneself that makes people systematically sacrifice present gratification in pursuit of future attainment. Groups that instill this kind of drive in their members have a special advantage in America, because contemporary American culture teaches a contrary message—a message of self-acceptance and living in the moment.

THIS BOOK BRINGS together two very different bodies of work and expertise. One of the two authors has written for almost twenty years about successful ethnic minorities all over the world, from Southeast Asia to Africa to the former Soviet Union. The other has written extensively on how the desire to live in the present has come increasingly to dominate modern Western culture, especially in America, undermining the country's ability to live for the future. America was not always this way; in fact, as we'll discuss, the United States was born a Triple Package country.

THAT CERTAIN GROUPS DO much better in America than others—as measured by income, occupational status, test scores, and so on—is difficult to talk about. In large part this is because the topic feels racially charged. The irony is that the facts actually debunk racial stereotypes. There are black and Hispanic subgroups in the United States far outperforming many white and Asian subgroups. Moreover, there's a demonstrable arc to group success—in immigrant groups, it typically dissipates by the third generation—puncturing the notion of innate group differences and undermining the whole concept of "model minorities."

This book offers a new way to look at success—its hidden spurs, its inner dynamics, its costs. These costs can be high, even crippling. But when properly understood and harnessed, the package of three cultural traits described in this book becomes a source of empowerment unconfined by any particular definition of success. As we'll show, the Triple Package can be a ladder to accomplishment of any kind, including that which is measured not by gain to oneself, but by service to others.

Ultimately, the Triple Package is accessible to anyone. It's a set of values and beliefs, habits and practices, that individuals from any background can make a part of their lives or their children's lives, enabling them to pursue success as they define it.

CHAPTER 1

THE TRIPLE PACKAGE

A SEEMINGLY UN-AMERICAN FACT about America today is that certain groups starkly outperform others. Some of America's most successful groups won't surprise you; others might.

What do the current or recent CFOs or CEOs of American Express, Black & Decker, Citigroup, Dell, Fisher-Price, Deloitte, Jet-Blue, Marriott International, Sears, Roebuck, Huntsman, Skullcandy, Sam's Club, and Madison Square Garden have in common? They are all members of the Church of Jesus Christ of Latter-day Saints. In 1980, it was hard to find a Mormon on Wall Street. Today, Mormons are dominant players in America's corporate boardrooms, investment firms, and business schools.

Mormons aren't the only ones to rise out of nowhere. The death of upward mobility in America has been widely reported in recent years. If you're an American born after 1960, we're told, how well you do is heavily dependent on how well your parents did. Overall this may be true, but the American Dream—including the old-fashioned rags-to-riches version—is very much alive for certain groups, particularly immigrants.*

After 1959, hundreds of thousands of Cubans fled to Miami, most arriving destitute. Initially facing hostility—NO DOGS, NO CUBANS signs on rental buildings were common—they crammed into small apartments and became dishwashers, janitors, and tomato pickers. These Cuban exiles, together with their children, helped transform sleepy Miami into one of America's most vibrant business centers. By 1990, the percentage of U.S.-born Cuban Americans with household incomes over $50,000 was double that of Anglo-Americans. Although less than 4 percent of the U.S. Hispanic population, Cuban Americans in 2002 accounted for five of the top ten wealthiest Hispanics in the United States, and today are two and a half times more likely than Hispanic Americans overall to be making over $200,000 a year.

In 2004, two Harvard professors created an uproar when they pointed out that a majority of Harvard's black students—possibly up to two-thirds—were immigrants or their children (as opposed to blacks whose families had been in America for many generations). Immigrants from many West Indian and African countries—such as Jamaica, Haiti, Ghana, Ethiopia, and Liberia—are climbing America's higher education ladder, but the most prominent are Nigerians. A mere 0.7 percent of the U.S. black population, Nigerian Americans, most of them raised by hardworking, often struggling immigrant parents, account for at least ten times that percentage of black students at America's most elite universities and professional schools.

---

* Although rarely mentioned in media reports, the studies said to show the demise of upward mobility in America largely exclude immigrants and their children. Indeed, the Pew Foundation study most often cited as proof of the death of upward mobility in the United States expressly cautions that its findings do not apply to "immigrant families," for whom "the American dream is alive and well."

Predictably, this academic success has translated into economic success. Nigerian Americans are already markedly overrepresented at Wall Street investment banks and blue-chip law firms.

Of course, who's "successful" in America depends on how you define success. For some, the most successful life may be the one spent doing the most good. For others, it could be the life most devoted to God. For Socrates, wealth and prizes were but the false semblances of success; life had to be examined to be worth living. But for reasons that could fill the pages of an entire library—or be sterilely reduced to the laws of supply and demand—goodness, religiosity, and self-awareness are not what modern economies reward. Oliver Wendell Holmes Sr. referred to success "in its vulgar sense" as "the gaining of money and position" (and then went on to give advice about how to attain it). Later we'll turn to the costs and narrowness of such success, but the stubborn question remains: Why do some groups rise in this "vulgar sense" while others don't? And why do some groups simply outperform the rest?

Indian Americans have the highest income of any Census-tracked ethnic group, almost twice the national average. Chinese, Iranian, and Lebanese Americans are not far behind. Asians are now so overrepresented at Ivy League schools that they're being called the "new Jews," and many believe that tacit quotas are being applied against them. It's important to emphasize that even the children of poor and poorly educated East Asian immigrants—Chinese seamstresses, Korean grocers, Vietnamese refugees—outperform other racial minority groups and their white American counterparts.

Meanwhile, the actual Jews continue to rack up Nobel Prizes, Pulitzer Prizes, Tony Awards, and hedge-fund billions at a rate wildly disproportionate to their numbers. In a nationwide study of young to middle-age adults, median American household net worth in 2004 was found to be $99,500; among Jewish respondents, it was $443,000.

In 2009, although just 1.7 percent of the adult population, Jews accounted for twenty of *Forbes*'s top fifty richest Americans and over a third of the top four hundred.

Groups can also fall precipitously in their fortunes. In the early 1900s, when Max Weber wrote his classic *The Protestant Ethic and the Spirit of Capitalism*, Protestants still dominated the American economy. Today, American Protestants are below average in wealth, and being raised in an Evangelical or fundamentalist Protestant family is correlated with downward economic mobility.

THIS BOOK IS ABOUT the rise and fall of groups. Its thesis is that when three distinct forces come together in a group's culture, they propel that group to disproportionate success. Unfortunately, there's a darker side to the story as well. The same forces that boost success also carry deep pathologies, and this book is about those pathologies too.

It turns out that for all their diversity, America's overachieving groups are linked together by three cultural commonalities, each one of which violates a core tenet of modern American thinking. For lack of a less terrible name, we'll call these three cultural forces, taken together, the Triple Package. Its elements are:

1. A SUPERIORITY COMPLEX. This element of the Triple Package is the easiest to define: a deeply internalized belief in your group's specialness, exceptionality, or superiority. This belief can derive from widely varying sources. It can be religious, as in the case of Mormons. It can be rooted in a story about the magnificence of your people's history and civilization, as in the case of Chinese or Persians. It can be based on identity-defining social distinctions that most Americans have never even heard of, such as descending from the "priestly" Brahman caste, in the case of some Indian Americans, or belonging to the famously entrepreneurial Igbo people, in the case of many Nigerian immigrants. Or it could be a mix. At their first Passover Seders, Jewish children hear that Jews are the "chosen" people; later they will be taught that Jews are a moral people, a people of law and intellect, a people of survivors.

A crucial point about the Superiority Complex is that it is antithetical to mainstream liberal thinking, which teaches us to refrain from judging any individual or any life to be better than another. Everyone is equal to everyone else. And if individual superiority judgments are frowned on, group superiority judgments are anathema. Group superiority is the stuff of racism, colonialism, imperialism, Nazism. Yet every one of America's extremely successful groups fosters a belief in its own superiority.

2. INSECURITY. As we will use the term, insecurity is a species of discontent—an anxious uncertainty about your worth or place in society, a feeling or worry that you or what you've done or what you have is in some fundamental way not good enough. Insecurity can take many different forms: a sense of being looked down on; a perception of peril; feelings of inadequacy; a fear of losing what one has. Everyone is probably insecure in one way or another, but some groups are more prone to it than others. To be an immigrant is almost by definition to be insecure—an experience of deep economic and social anxiety, not knowing whether you can earn a living or give your children a decent life.

That insecurity should be a critical lever of success is another anathema, flouting the entire orthodoxy of contemporary popular and therapeutic psychology. Feelings of inadequacy are a diagnostically recognized symptom of personality disorder. If you're insecure—if you feel that in some fundamental way you're not good enough—then you lack self-esteem, and if you lack self-esteem, you aren't on your way to a successful life. On the contrary, you should probably be in therapy. And the greatest anathema of all would be parents working to instill insecurity in their children. Yet insecurity runs deep in every one of America's most successful groups, and these groups not only suffer from insecurity; they tend, consciously or unconsciously, to promote it.

Note that there's a deep tension between insecurity and a superiority complex. It's odd to think of people being simultaneously insecure but also convinced of their divine election or superiority. Yet this tense, unstable combination, as we'll discuss shortly, is precisely what gives the Triple Package its potency.

3. IMPULSE CONTROL. As we'll use the term, impulse control refers to the ability to resist temptation, especially the temptation to give up in the face of hardship or quit instead of persevering at a difficult task. No society could exist without impulse control; as Freud speculated, civilization may begin with the suppression of primal sexual and aggressive instincts. Nevertheless, against the background of a relatively permissive America, some groups decidedly place greater emphasis on impulse control than others.

Impulse control, too, runs powerfully against the grain of contemporary culture. The term "impulse control" conjures up all kinds of negative connotations: "control freaks," people who are "too controlled" or "too controlling," people who can't be "impulsive" and enjoy life. People who control their impulses don't live in the present, and living in the present is an imperative of modernity. Learning to live in the here and now is the lesson of countless books and feel-good movies, the key to overcoming inhibition and repression. Impulse control is for adults, not for the young, and modern culture is above all a youth culture.

As we increasingly disrespect old age and try to erase its very marks from our faces, we correspondingly romanticize childhood, imagining it as a time of what ought to be unfettered happiness, and we grow ever more fearful of spoiling that happiness through excessive restraints, demands, hardships, or discipline. By contrast, every one of America's most successful groups takes a very different view of childhood and of impulse control in general, inculcating habits of discipline from an early age—or at least they did so when they were on the rise.

Because all three elements of the Triple Package run so counter to modern American culture, it makes sense that America's successful groups are all outsiders in one way or another. It also makes sense that so many immigrant communities are pockets of exceptional upward mobility in an increasingly stratified American economy. Paradoxically, in modern America, a group has an edge if it doesn't buy into— or hasn't yet bought into—mainstream, post-1960s, liberal American principles.

WHY IS THE TRIPLE PACKAGE so powerful an engine of group success? It begins with drive, generated by the unlikely convergence of the first two Triple Package forces.

You might think that a superiority complex and insecurity would be mutually exclusive, that it would be hard for these two qualities to coexist in a single individual. And that's exactly the point: it *is* an odd and unstable combination. But this fusion of superiority and insecurity lies at the heart of every Triple Package culture, and together they tend to produce a goading chip on the shoulder, a need to prove oneself or be recognized.

Jews have had a chip of just this kind for basically all of recorded history—or at least since Hadrian, after crushing the Bar Kokhba revolt, barred Jews from entering Jerusalem and erected a large marble pig outside the city gates. In America, the millions of poor Eastern European Jews who arrived in the early twentieth century, viewed as filthy and degenerate not only by Gentiles but also by America's German Jews as well, seemed, as one author puts it, "almost to wear a collective chip on the shoulder." The same could be said of the next generation: men like Alfred Kazin, Norman Mailer, Delmore Schwartz, Saul Bellow, Clement Greenberg, Norman Podhoretz, and so many of the New York intellectuals who grew up excluded from anti-Semitic bastions of education and culture but went on to become famous writers and critics.

This chip on the shoulder, this "I'll show them" mentality, is a Triple Package specialty, the volatile product of a superiority complex colliding with a society in which that superiority is not acknowledged. It's remarkable how common this dynamic is among immigrant groups: a minority, armed with enormous ethnocentric pride, suddenly finds itself disrespected and spurned in the United States. The result can border on resentment—and resentment, as Nietzsche taught, is one of the world's great motivators.

A particularly intense variant involves status collapse. Typically, immigrants to the United States are moving up the economic ladder, achieving a better standard of living. In some cases, however, they experience the opposite: a steep fall in status, wealth, or prestige—an especially bitter experience when their new society is wholly unaware of the respect they formerly commanded. Several of America's most successful immigrant groups have suffered this additional sting. As we'll discuss later, the Cuban Exiles offer a vivid example, and so too Iranian Americans.

Superiority and insecurity can combine to produce drive in a quite different but equally goading way: by generating a fierce, sometimes tormenting need to prove oneself not to "the world," but within one's own family.

In many Chinese, Korean, and South Asian immigrant families, parents impose exorbitantly high academic expectations on their children ("Why only a 99?"). Implicit in these expectations are both a deep assumption of superiority (we know you can do better than everyone else) and a needling suggestion of present inadequacy (but you haven't done remotely well enough yet). Comparisons to cousin X, who just graduated as valedictorian, or so-and-so's daughter, who just got into Harvard, are common—and this is true in both lower- and higher-income families.

To further pile on, East Asian immigrant parents often convey to their children that their "failing"—for example, by getting a B+— would be a disgrace for the whole family. "In Chinese families," one Taiwanese American mother explained in a recent study, "the child's personal academic achievement is the value and honor of the whole family. . . . If you do good, you bring honor to the family and [do] not lose face. A lot of value is placed on the child to do well for the family. It starts from kindergarten."

The East Asian case may be the most conspicuous, but the phenomenon of extravagant parental expectations, with the same double-message of superiority and inadequacy, is common to many immigrant communities. Sixty years ago, Alfred Kazin wrote, "It was not for myself alone that I was expected to shine, but for [my parents]—to redeem the constant anxiety of their existence." In Kazin's Jewish immigrant neighborhood, "If there were Bs" on a child's tests or papers, "the whole house went into mourning." This dynamic recurs in various forms in almost every Triple Package culture, creating enormous pressure to succeed. The result can be anxiety and misery, but also drive and jaw-dropping accomplishment.

BEFORE GOING ON, because we use terms like "East Asian parents," we need to say a word about cultural generalizations and stereotypes. Throughout this book, we will never make a statement about any group's economic performance or predominant cultural attitudes unless it is backed up by solid evidence, whether empirical, historical, or sociological (see the endnotes for sources). But when there *are* differences between groups, we will come out and say so. It's just a statistical fact, for example, that Mormon teenagers are less likely to drink or have premarital sex than other American teenagers. Of course there will be exceptions, but if the existence of exceptions blinded us to—or censored us from talking about—group differences, we wouldn't be able to understand the world we live in.

Group generalizations turn into invidious stereotypes when they're false, hateful, or assumed to be true of every group member. No group and no culture is monolithic. Even a high-earning, "successful" group like Indian Americans includes more than two hundred thousand people in poverty. Moreover, within every culture there are competing subcultures, and there are always individuals who reject the cultural values they're raised with. But that doesn't make culture less real or powerful. "Let me summarize my feelings toward Asian values," writes author-provocateur Wesley Yang. "Fuck filial piety. Fuck grade-grubbing. Fuck Ivy League mania. Fuck deference to authority. Fuck humility and hard work. Fuck harmonious relations. Fuck sacrificing for the future." Whether a person chooses to embrace or run screaming from his cultural background, it's still there, formative and significant.

· · ·

RETURNING NOW TO how the Triple Package works: superiority and insecurity combine to produce drive (or so this book will try to show), but it takes more than drive to succeed. In fact the metaphor of "drive" is misleading to the extent that it conjures up a car-and-open-highway image, where the only thing a successful person needs is a full tank of gas and a foot on the pedal. To invoke a different set of metaphors, life can also be a battle—against the slings and arrows of outrageous fortune, against a system that keeps slapping you down, against the almost irresistible urge to give up.

Nearly everyone confronts obstacles, adversity, and disappointment at some point in life. Drive is offense, but success requires more than offense. One of the greatest chess players in history reportedly said of another grand master, "[H]e will never become World Champion since he doesn't have the patience to endure worse positions for hours." The Triple Package not only instills drive. It also delivers on defense—with toughness, resilience, the ability to endure, the capacity to absorb a blow and pick yourself up off the ground afterward.

In part the superiority complex itself has this effect, providing a kind of psychological armor of special importance to minorities who repeatedly face hostility and prejudice. Fending off majority ethnocentrism with their own ethnocentrism is a common strategy among successful minorities. Benjamin Disraeli used it against British anti-Semitism. "Yes, I am a Jew," he famously replied to a slur in the House of Commons, "and when the ancestors of the right honorable gentleman were brutal savages in an unknown island, mine were priests in the temple of Solomon."

Much more powerful, however, is an interaction between superiority and impulse control—a belief in achieving superiority *through* impulse control—that can generate a self-fulfilling cycle of greater and greater endurance. A superiority complex built up around impulse control can be very potent. A person with such a complex eagerly demonstrates and exercises his self-control to achieve some difficult goal; the more sacrifice or hardship he can endure, the more superior he feels and the better able to accomplish still more demanding acts of self-restraint in the future, making him feel even more superior. The result is a capacity to endure hardship—a kind of heightened resilience, stamina, or grit.

This dynamic is another Triple Package specialty. America's successful immigrant communities nearly always build impulse control into their superiority complex. They tend to believe (probably because it's true) that they can endure more adversity and work more hours than average Americans are willing to. They see this capacity as a virtue, incorporate it as a point of pride into their self-definition, and then try—frequently with great effectiveness—to inculcate this virtue in their children.

But this phenomenon is not unique to immigrants. Mormons, too, weave into their superiority complex their discipline, abstemiousness, and the hardships they endure on mission (a usually two-year proselytizing stint in an assigned location anywhere from Cleveland to Tonga). Notice that impulse control is again understood here as a virtue—moral, spiritual, and characterological. It's admirable and righteous in itself. When on mission, Mormons must give up dating, movies, magazines, and popular music. They are permitted e-mail once a week, but only from public facilities (not their homes), and can call home only on Christmas and Mother's Day. While other American eighteen-year-olds are enjoying the binge-drinking culture widespread on college campuses, Mormons are working six days a week, ten to fourteen hours a day, dressed in white shirt and tie or neat skirt, knocking on doors, repeatedly being rejected and often ridiculed.

In itself, the capacity to endure hardship has nothing to do with economic gain or conventional success. In principle it can lead to hair shirts or Kafkaesque hunger artists. But when the ability to endure hardship is harnessed to a driving ambition—when grit meets chip—the result is a deferred gratification machine.

Superiority plus insecurity is a formula for drive. Superiority plus impulse control is a formula for hardship endurance. When the Triple Package brings all three elements together in a group's culture, members of that group become disproportionately willing and able to do or accept whatever it takes today in order to make it tomorrow.*

BUT THIS SUCCESS COMES AT A PRICE. Each of the components of the Triple Package has its own distinctive pathologies. Deeply insecure people are often neurotic. Impulse denial can undercut the ability to experience beauty, tranquillity, and spontaneous joy. Belief in the superiority of one's own group is the most dangerous of all, capable of promoting arrogance, prejudice, and worse. Some of history's greatest evils—slavery, apartheid, genocide—were predicated on one group's claim of superiority over others.

But even when it functions relatively benignly as an engine of suc-

---

* Needless to say, no amount of grit or drive will lift a group to economic success in a society that wholly denies them economic opportunity. Most of America's Jewish immigrants probably had the Triple Package before they got to this country, but that didn't do them much good in the *shtetls* of Eastern Europe. As an engine of success, the Triple Package is dependent on institutions rewarding hard work and deferred gratification. An entire nation could have a Triple Package culture yet remain mired in poverty if governed by a kleptocracy. Thus as Daron Acemoglu and James Robinson have argued, institutions rather than culture may best explain many between-country wealth differences. North and South Korea provide an obvious example. (On the other hand, Jared Diamond might be right that geography ultimately explains why some continents or regions developed wealth-creating institutions before others.) But to explain why, *within* a single country, some groups rise from penury to affluence while facing roughly the same economic system—and often discrimination as well—institutions can't be the full answer.

cess, the Triple Package can still be pathological—because of the way it defines success.

Triple Package cultures tend to focus on material, conventional, prestige-oriented success. This is a function of the insecurity that drives them. The "chip on the shoulder," the need to show the world or prove yourself, the simple fear that, as a newcomer who doesn't even speak the language, you may not be able to put food on the table—all these characteristic Triple Package anxieties tend to make people put a premium on income, merit badges, and other forms of external validation.

But material success obviously cannot be equated with a well-lived or successful life. James Truslow Adams, the historian who popularized the term "American Dream," wrote that everyone should have two educations, one to "teach us how to make a living, and the other how to live." Triple Package success, with its emphasis on external measures of achievement, does not provide the latter education. On the contrary, being raised in a high-achieving culture can be a source of oppression and rage for those who don't or choose not to achieve. Deferring gratification can lead to a nothing-is-ever-good-enough mentality, requiring years of therapy to not fix. Hence the joke that making partner in a Wall Street law firm is like winning a pie-eating contest where the prize is—more pie.

At its worst, the Triple Package can misshape lives and break psyches. Children made to believe they are failures or worthless if they don't win every prize may realize in their twenties that they've spent their lives striving for things they never even wanted.

MUCH OF THIS BOOK is about America's most successful groups— their cultural commonalities, their generational trajectory, their pathologies. But we'll also look in detail at some of America's poorer groups. As we'll explain, confirming the basic thesis of this book, these groups lack the Triple Package, but—and this point is so important it needs to be highlighted in advance—the absence of the Triple Package was not the original cause of their poverty. In almost every case, America's persistently low-income groups became poor because of systematic exploitation, discrimination, denial of opportunity, and institutional or macroeconomic factors having nothing to do with their culture.

Moreover, in some cases, a group's lack of the Triple Package was America's doing. Centuries of slavery and denigration can make it difficult, if not impossible, for a group to have a deeply internalized sense of superiority. At the same time, if members of certain groups learn not to trust the system, if they come to believe that discipline and hard work won't really be rewarded—if they don't think that people like them can make it—they will have little reason to engage in impulse control, sacrificing present satisfactions for economic success down the road. Thus the same conditions that cause poverty can also grind the Triple Package out of a culture.

But once that happens, the situation worsens. America's poorest groups may not have fallen into poverty because they lacked the Triple Package, but now that they do lack it, their problems are intensified and harder to overcome. In these circumstances, it takes much more—more grit, more drive, perhaps a more exceptional individual—to break out.

Groups that do achieve Triple Package success in the United States become enmeshed in a process of creative destruction that will change them irrevocably. America's own cultural antibodies invariably attack these groups, encouraging their members to break free from their cultures' traditional constraints.

In one possible outcome of this process, the group does so well it simply disappears—as the once-extremely successful American Huguenot community has all but disappeared. A group's demise can be celebrated as a triumph of the American melting pot, or mourned as the loss of a heritage and identity, but the bottom line is the same: a successful group, precisely because of its success, assimilates, intermarries, and Americanizes away. Lebanese Americans (who, along with other Arab Americans, have extremely high out-marriage rates) may be a twentieth-century example of this phenomenon.

Another possible outcome is decline. Triple Package success is intrinsically hard to sustain. Success softens; it erodes insecurity. Meanwhile, modern principles of equality tend to undercut group superiority complexes. And with its freedom-loving, get-it-now culture, America undercuts impulse control too. WASP economic dominance in the United States declined under the weight of all these pressures.

Many American Jews today fear the Huguenot outcome—disappearance through assimilation and intermarriage. Perhaps they should be more concerned about the WASP outcome, in which success is followed by decline. As Ellis Island and the Lower East Side recede into the past, and with a strong presence from Washington to Wall Street to Hollywood, Jews may feel less insecure in the United States today than they have been in any country in a thousand years. Moreover, as we'll discuss later, Jewish culture today appears to be much less oriented around impulse control than it used to be. If so, and if the thesis of this book is correct, continuing Jewish success should not be taken for granted. In fact, mounting evidence today indicates a precipitous drop in performance among younger Jews across numerous academic activities in which American Jews were once dominant.

But Triple Package groups are not condemned to either disappearance or decline. Another possibility is more tantalizing and volatile. As the children of Triple Package groups grow into adults in America, they learn to question what their family's culture has taught them about who they are and how they should live. They begin to internalize American attitudes without, however, being fully Americanized. Instead they are likely to feel like outsiders both within their own culture and in the larger society.

Straddling this cultural edge may make people feel that they don't belong anywhere, that they have no cultural home. But it can also be a source of prodigious vitality and creativity. It can lead people to break free from their group's cultural constraints—rejecting would-be limits on their personality, their sexuality, their careers—while retaining the core traits of the Triple Package. Thus, Triple Package groups can reinvent themselves across generations, and individuals can achieve forms of success, grand or simple, their parents never dreamed of.

THROUGHOUT THIS BOOK we'll be referring to Triple Package cultures and Triple Package groups, but to avoid any misunderstanding, we want to emphasize two important points.

First, a Triple Package culture will not produce the qualities we've described—a sense of superiority, a chip on the shoulder, a capacity to endure hardship, and so on—in all its members. It doesn't have to in order to produce group success; it just has to do better than average. In 1941, baseball great Ted Williams got a hit only 4 times out of every 10 at-bats, but because the average Major Leaguer gets a hit about 2.6 times out of 10, Williams achieved a Hall of Fame feat that hasn't been equaled since. Similarly, a culture that produced four high achievers out of ten would attain wildly disproportionate success if the surrounding average was, say, one out of twenty.

Second, and conversely, an individual can possess every one of the

Triple Package qualities without being raised in a Triple Package culture. Steve Jobs had a legendarily high opinion of his own powers; long before he was famous, a former girlfriend believed he had narcissistic personality disorder. His self-control and meticulous attention to detail were equally famous. At the same time, according to one of his closest friends, "Steve always had a kind of chip on his shoulder. At some deep level, there was an insecurity that Steve had to go out and prove himself. I think being an orphan drove Steve in ways that most of us can never understand."

Possibly Jobs was born with these Triple Package traits, or perhaps, as his friend speculated, being an orphan played a role. In any given family, no matter what the background, an especially strong parent or even grandparent can instill children with a sense of exceptionality, high expectations, and discipline, creating a kind of miniature Triple Package culture inside the home. Individuals can also develop these qualities on their own. Being raised in a Triple Package culture doesn't guarantee you anything unique or inaccessible to others; it simply increases your odds.

EVERY ONE OF THE PREMISES underlying the theory of the Triple Package is supported by a well-substantiated and relatively uncontroversial body of empirical evidence. Later chapters will elaborate, but we'll briefly summarize here.

The capacity of group superiority complexes to enhance success is borne out by repeatedly confirmed findings of stereotype threat and stereotype boost, both in laboratory experiments and field work. Basically, belonging to a group you believe is superior at something—whether academic work or sports—psychologically primes you to perform better at that activity. Moreover, sociologists specializing in immigrant communities have found that certain groups turn a sense of cultural pride and distinctive heritage into an "ethnic armor" directly contributing to higher levels of educational achievement.

That insecurity can spur accomplishment is corroborated by a recent groundswell of studies showing that a personal feeling of not being good enough—or not having done well enough—is associated with better outcomes. This conclusion is also supported by two of the leading twentieth-century studies of individuals who have risen to eminence, including one conducted by Howard Gardner, most famous for his theory of multiple intelligences. Both studies found that insecurity, particularly stemming from childhood, figured prominently as a surprisingly common driver of success. Gardner quotes Winston Churchill:

> the twinge of adversity, the spur of slights and taunts in early years are needed to evoke that ruthless fixity of purpose and tenacious mother-wit without which great actions are seldom accomplished.

Lastly, an entire subfield of experimental psychology today is devoted to phenomena variously called "effortful control," "self-regulation," "time discounting," "ego strength," or (more appealingly) "willpower" and "grit." These concepts are all connected to impulse control, as we're using the term: the capacity to resist temptation, especially the temptation to give up in the face of hardship. The results of these studies—beginning with the well-known "marshmallow test"—are conclusive and bracing. Kids with more impulse control go on to get better grades; spend less time in prison; have fewer teenage pregnancies; get better jobs; and have higher incomes. In several studies, willpower and grit proved to be better predictors of grades and future success than did IQ or SAT scores.

. . .

BEFORE CLOSING THIS CHAPTER, we need to say a word about something we didn't include in the Triple Package: education. It's often said that Jewish and Asian Americans do well in the United States because they come from "education cultures." Given that the Triple Package is essentially a cultural explanation of group success, why isn't education one of its core elements?

Because, to begin with, there are some flat-out exceptions to the rule that successful groups emphasize learning. The immensely successful but highly insular Syrian Jewish enclave in Brooklyn does not stress education or intellectualism; indeed, higher education at prestigious universities is often disfavored. Instead this community prioritizes business, tradition, "taking over the family company," and keeping younger generations within the fold. Because of its insularity, most people probably have never even heard of America's Syrian Jewish community, but it's been thriving for generations, economically as well as culturally, and elite education has decidedly not been part of its formula.

Of course it's true that most successful groups in America do emphasize education. They also tend to save and work hard. The question is why. The worst move to make at this point—the kind of move that gives cultural theories a bad name—is to take these behaviors, turn them into adjectives, impute them to culture, and offer them up as "explanations." Why do the Chinese save at such higher-than-average rates? Because they come from a "thrifty" culture. It's the same with education. Why do parents from so many successful groups harp on education? Because they hail from an "education culture."

In fact, many of America's rising groups, although they stress academics today, do not have longstanding "education cultures." For example, although early Mormon pioneers founded many schools and colleges in the American West, an important current of Mormon culture for much of the twentieth century remained relatively closed to intellectual and scientific inquiry, emphasizing "the authority of scripture over human reason." In 1967, future Church president Spencer Kimball urged the faculty of Brigham Young University to remember that Mormons are "men of God first and men of letters second, and men of science third . . . men of rectitude rather than academic competence."

Even when we consider cultures supposedly steeped in centuries-old scholarly traditions, the conclusion that they focus on education today *because* of those traditions can be much too facile. Jews, for example, are sometimes said to have the quintessential "learning culture." Yet many of the Ellis Island Jewish immigrants were barely schooled, having lived most of their lives in *shtetls* or ghettos in extreme poverty. Perhaps these unintellectual butchers and tailors transmitted to their children the great Jewish "learning tradition" through synagogues, Passover rituals, or the respect they accorded rabbis. Or perhaps not.

Nathan Glazer says that his immigrant parents and many of their generation knew nothing of Jewish learning. The influential social psychologist Stanley Schachter made a similar point:

> I went to Yale much against my father's wishes. He couldn't have cared less about higher education and wanted me to go to a one-year laundry college (no kidding) out in the Midwest and join him then in the family business. I never have understood what this intellectually driven Jewish immigrant business is all about. It wasn't true of my family, and I know very few families for which it was true.

Indeed, the Jewish subgroup arguably most dedicated to and organized around the old tradition of Talmudic study is the ultra-Orthodox Satmar community of Kiryas Joel, in Orange County, New York, which is one of the poorest groups in the entire nation.

What is it about certain groups that makes their members, however poor or "uncultured," seize on education as a route to upward mobility? It's simply not illuminating to say that these groups come from "hardworking cultures" or "education cultures." That's one step away from saying that successful groups are successful because they do what it takes to be successful—and two steps from saying that unsuccessful groups are unsuccessful because they come from "indolent cultures" and don't do what it takes to be successful.

In short, education—like hard work—is not an *independent*, but a *dependent* variable. It's not the explanatory factor; it's a behavior to be explained. Successful groups in America emphasize education for their children because it's the surest ladder to success. The challenge is to delve deeper and discover the cultural roots of this behavior—to identify the fundamental cultural forces that underlie it.

FOR ALL ITS VAST DIVERSITY, America has an overarching culture of its own—a very strong one. That's why we hear so much about America's worldwide "cultural hegemony" or how "globalization is Americanization." Which raises the question of whether American culture is a Triple Package culture.

Certainly it used to be. In fact, America was for a long time the quintessential Triple Package nation, convinced of its exceptional destiny, infused with a work ethic inherited from the Puritans, seized with a notorious chip on the collective shoulder vis-à-vis aristocratic Europe, and instilling a brand-new kind of insecurity in its citizens— a sense that every man must prove himself through material success, that a man who doesn't succeed economically is a failure. Tocqueville observed all this when he described Americans' "longing to rise."

But as we'll discuss at length later in this book, America has changed, especially in the past fifty years. Today, American culture—whether high or low, blue state or red, blue collar or ivory tower—is much more ambivalent about, and undermining of, everything the Triple Package stands for. The overwhelming message taught in American schools, public and private, is that no group is superior to any other. In America, embracing yourself as you are—feeling secure about yourself—is supposed to be the key to a successful life. People who don't live in the present are missing out on happiness and life itself. Whatever kernels of truth may underlie these propositions, the irony is this: America still rewards people who don't buy into them with wealth, prestige, and power.

In other words, there is a disconnect today between the story Americans tell themselves about how to think and how to live—and the reality of what the American economy rewards. Triple Package groups are taking advantage of that disconnect.

CHAPTER 2

WHO'S SUCCESSFUL
IN AMERICA?

IN THIS CHAPTER WE'LL be taking a look at America's most successful groups as measured by income, academic accomplishment, corporate leadership, professional attainment, and other conventional metrics. But first we should clarify the kind of groups we're looking at.

There are infinite ways to slice up the U.S. population. Countless economic mobility studies break down American wealth by race—typically white, black, Asian, and Hispanic. A recent countertrend focuses on class and class rigidity instead, dividing the population into quintiles, rich and poor, 99 percent and 1 percent. But gigantic umbrella terms like "race" and "class" obscure as much as they reveal.

The reality, uncomfortable as it may be to talk about, is that some religious, ethnic, and national-origin groups are starkly more successful than others. Without looking squarely at such groups, it's impossible to understand economic mobility in America and what the levers of success in this country really are.

A distinctive feature of many—but by no means all—religious, ethnic, and national-origin groups is that they are "cultural groups": their members tend to be raised with, identify themselves by, and pass down certain culturally specific values and beliefs, habits and practices.* Needless to say, religion, ethnicity, and national origin are cultural starting points, not end points. Cultural subdivisions within these categories—for example, fundamentalist versus non-fundamentalist, first-generation immigrant versus third-generation—can have dramatic effects on group success, and we'll be highlighting these finer distinctions throughout.

IF THERE'S ONE GROUP in the U.S. today that's hitting it out of the park with conventional success, it's Mormons.

Just fifty years ago, Mormons were often regarded as a fringe group; many Americans had barely heard of the Church of Jesus Christ of Latter-day Saints. (The term "Mormon" is not part of the official Church name and comes from the Book of Mormon, a new work of scripture that the church's founder, Joseph Smith, said he translated from golden plates received from an angel.) Concentrated in Utah and neighboring states, Mormons were a largely isolated and insulated community, resisting many developments in modern America. As late as 1978, the LDS Church expressly discriminated against

---

* To say there's no generally accepted definition of culture would be an understatement. A 2006 compilation found more than three hundred definitions in the literature. The way we use the term in this book is probably consistent with most of those definitions, although we start with certain baseline premises: that a group's culture is highly dynamic (not static), capable of changing radically even in a generation; that culture is usually many-sided (not monolithic), weaving together numerous and even conflicting strands; that people can change their cultural conditions (so that a family, for example, can have a distinctive culture shaped by parental decisions); and that there is no principle of one-person-one-culture (meaning that individuals can be and in America usually are defined by more than one culture). For more on how we use "culture" in this book, see the endnote.

blacks, refusing to ordain them into the priesthood. In 1980, Mormons were still a rarity on Wall Street and in Washington.

Three decades later, it's hard not to notice the Mormons' explosive success. Overwhelmingly, Mormon success has been of the most mainstream, conventional, apple-pie variety. You don't find a lot of Mormons breaking the mold or dropping out of college to form their own high-tech start-ups. (Omniture cofounder Josh James is a notable exception.) What you mostly find is corporate, financial, and political success, which makes perfect sense given the nature of the Mormon chip on the shoulder. Long regarded as a polygamous, almost crackpot sect, Mormons seem determined to prove they're more American than other Americans—with a particular penchant for presidential runs.

Whereas Protestants make up about 51 percent of the U.S. population, America's 5 to 6 million Mormons represent just 1.7 percent. Yet a stunning number have risen to the top of America's corporate and political spheres.

Most famous of course is Mitt Romney, who, before serving as governor of Massachusetts for four years, was CEO of Bain Capital (and now has an estimated net worth of $230 million). Jon Huntsman Jr., former U.S. Ambassador to China and for a while Romney's rival for the 2012 Republican nomination, is also Mormon, as is majority leader of the U.S. Senate, Harry Reid. Other leading Mormon politicians include Senator Orrin Hatch (who lost his bid for the 2000 Republican nomination to George W. Bush), Congressman Morris Udall (who lost his bid for the 1976 Democratic nomination to Jimmy Carter), and Mitt's father, former Michigan governor George Romney (who lost his bid for the 1968 Republican nomination to Richard Nixon).

In the business world, prominent Mormons include David Neeleman, founder and former CEO of JetBlue; J. W. Marriott, chairman and son of the founders of Marriott International; Thomas Grimm, CEO of Sam's Club; Dave Checketts, the former CEO of Madison Square Garden and former president of the New York Knicks who now heads up the sports and entertainment firm SCP Worldwide; Kevin Rollins, the former CEO of Dell; Gary Crittenden, the former CFO of Citigroup, American Express, and Sears, Roebuck; Gary Baughman, former CEO of Fisher-Price; Kim Clark, former dean of Harvard Business School; Alison Davis-Blake, the first female dean of the University of Michigan's Ross School of Business; Stephen Covey, author of *The Seven Habits of Highly Effective People*, which has sold more than 25 million copies; and Clayton Christensen, author of *The Innovator's Dilemma* (which Intel CEO Andy Grove said was the most important book he'd read in ten years), who was recently the subject of a *New Yorker* profile titled "When Giants Fail: What Business Has Learned from Clayton Christensen."

And that's just the tip of the iceberg. Mormons have risen to the top of American Motors, Lufthansa, Deloitte, Kodak, Black & Decker, SkyWest Airlines, Lord & Taylor, Skullcandy, and PricewaterhouseCoopers. Jon Huntsman Sr. became a billionaire on the *Forbes* 400 list after founding one of America's most successful chemical companies. Alan Ashton cofounded WordPerfect Corporation, making him in the 1990s one of the four hundred richest people in America. Edwin Catmull, raised in a traditional Mormon Salt Lake City family, became a pioneer of three-dimensional computer animation in the 1980s; today he's the president of Walt Disney Animation Studios and its subsidiary, the twenty-six-time Academy Award–winning Pixar Studios.

Mormons have achieved fame outside the corporate world as well. They are reportedly overrepresented in the CIA and foreign service (apparently because of their missionary-trained language skills and clean habits), and "out of nowhere," Brigham Young University's video animation program has become a main line into the country's major animation and special effects studios. Stephenie Meyer, author of the blockbuster *Twilight* novels, is Mormon, as is talk-radio host Glenn Beck, *Napoleon Dynamite* star Jon Heder, and all-time *Jeopardy!* record-holder Ken Jennings (seventy-four consecutive wins).

To be sure, a list of superstars, however impressive, doesn't by itself prove disproportionate success, and it's worth noting that Mormons are not (yet) overrepresented among CEOs of Fortune 500 companies. But here's one way to look at the startling rise of Mormons from relative obscurity into America's business elite. The Fortune 500 list has been published since 1955. Before 1970, there appear to have been no Mormon senior executives in any Fortune 500 company. Since 1990, there have been fourteen, including twelve CEOs, one president, and one CFO.

Here's another data point. In February 2012, Goldman Sachs announced the addition of 300 more employees to the 1,300 already working in the firm's third largest metropolitan center of operations (after New York/New Jersey and London). Where is this 1,600-employee location? In Salt Lake City, Utah. By reputation, the University of Pennsylvania's Wharton business school is one of the nation's best and most prestigious. In 2010, Wharton placed thirty-one of its graduates with Goldman—exactly the same number as did Brigham Young University's less well-known Marriott School of Management.

Getting a statistical fix on Mormon income and wealth is notoriously difficult. The country's leading researcher on the correlation of faith and money in the United States, Lisa Keister, says that the sample sizes studied so far are too small to support definitive conclusions (although judging by available information, she surmises that Mormon wealth is probably higher than average). Survey data paint a picture of Mormons as solidly middle-class. They are somewhat more likely to make $50,000–$100,000 than Americans generally (38 percent of Mormons versus 30 percent of the general population), somewhat less likely to make under $30,000 (26 percent versus 31 percent), and no more likely to make over $100,000 (in fact slightly less: 16 percent versus 18 percent).

But these numbers are hard to interpret. First, they represent *household* income, and Mormon women are encouraged to be full-time mothers; the percentage of Mormon women who describe themselves as housewives is double that of non-Mormons. While this gives LDS men some advantages (Mormon journalist Jeff Benedict calls the "stay-at-home" wives of nine famous Mormon CEOs the "secret" to their success), it also means that LDS men have to earn considerably more than non-Mormon men in order to keep on a par with or above overall American household income.

More important, these figures lump all Mormons together, which can be highly misleading. Mormonism is spreading rapidly around the world; one fourth of America's Mormons are converts. While some of these converts are famous—Mr. Beck being an example—most are relatively poor, which brings down overall Mormon income. Non-convert Mormons are significantly more likely to make at least $50,000 a year than Americans overall (58 percent as compared with a national figure of 45 percent).

Not all Mormon households, of course, are sending their young men to Goldman Sachs. Small fundamentalist Mormon communities still exist, which tend to be insular, polygamous, and relatively poor. (These groups are excommunicated from, and not considered Mormon by, the official LDS Church.) In Colorado City, Arizona, the "prophet" Rulon Jeffs remained the leader of one of these groups into his nineties; as of 2003, "Uncle Rulon" had married some seventy-five women and fathered at least sixty-five children. Although they view the United States as a satanic force, the fundamentalist residents of

Colorado City are happy to accept welfare, and a third are on food stamps. But from a cultural point of view, fundamentalist Mormons are radically different from the vast majority of present-day Mormons, and as we'll discuss later, their economic backwardness confirms the efficacy of the Triple Package (which fundamentalist Mormons lack).

The real testament to Mormons' extraordinary capacity to earn and amass wealth, however, is the LDS Church itself. The church keeps its finances a closely guarded secret, both from the outside world and from its members; the actual numbers are known only to the Apostles, the church's highest leadership council, made up of twelve men (women are excluded). It's clear, however, that church assets are vast and spread all over America—a huge conglomerate of for-profit and nonprofit enterprises.

The amount of American land owned by the Mormon church is larger than the state of Delaware. According to a former president of the Mormon Social Science Association, the church owns ten times more Florida real estate than the Walt Disney company, including "a $1 billion for-profit cattle and citrus ranch"—the largest non-government subsidized cattle operation in the United States. The LDS Church is one of America's largest producers of nuts and one of its largest potato growers. Beneficial Life is a church-owned insurance company, with assets of $1.6 billion. Church holdings also include at least twenty-five radio stations, commercial real estate, shopping centers, and a theme park in Hawaii containing replica Polynesian villages, which, with a million annual visitors, is one of that state's leading tourist attractions. Brigham Young University's endowment alone is worth nearly $1 billion.

The entire Church of England, with its grand history, had assets of about $6.9 billion as of 2008. In 2002, counting all the money raised by all the parishes in the United States, the total annual revenue of the U.S. Catholic Church, with its 75 million members, was estimated at $7.5 billion. By comparison, the LDS Church, with less than a tenth of the membership, is believed to have owned $25–$30 billion in assets as of 1997—and this is said to be a "very conservative" estimate—with present revenues of $5–$6 billion a year. As one study puts it, "Per capita, no other religion comes close to such figures."

THE GOIZUETA BUSINESS SCHOOL is part of Emory University in Atlanta, Georgia. You'll find the Goizueta name in a lot of places in Atlanta—the Atlanta Ballet, the Atlanta History Center, Georgia Tech—because Roberto Goizueta was chairman and CEO of Coca-Cola from 1981 until his death in 1997 and one of Atlanta's leading philanthropists. Goizueta, who in 1960 defected from Cuba with his family and little more than forty dollars in his pocket, was the most successful CEO in Coca-Cola's history, launching Diet Coke, globalizing the Coca-Cola brand, quadrupling its profits, and taking its market capitalization from $4 billion to $180 billion, an increase of 4,400 percent.

Between 1959 and 1973, hundreds of thousands of anti-Castro Cubans—the "Cuban Exiles"—fled to the United States, most of them settling in Miami.* As mentioned earlier, theirs was a classic case of Triple Package status collapse. In Cuba, most of the Exiles

---

* There have been four major waves of post-Castro Cuban immigration: roughly 200,000 initial exiles between 1959 and 1962 (the so-called Golden Exiles); roughly 260,000 between 1965 and 1973 (when persons from America were permitted to go to Cuba and return with relatives); more than 125,000 by boat in 1980 (the Marielitos, so called because of the port, Mariel, from which they left Cuba); and about 20,000 per year since 1994 (beginning with the "rafter" crisis). Following Professor Susan Eckstein, we will refer to the pre-1980 émigrés as "Cuban Exiles" and those who began emigrating in the 1990s as "New Cubans." In 2010, the total population of Cuban Americans was approximately 1,760,000.

had been middle and upper class, including many judges, professionals, engineers, academics, and white-collar employees of large corporations. Some came from families with summer homes and art collections, at the pinnacle of a highly stratified society. To the humiliating sting of becoming menial workers, looked down on and discriminated against as a racial or ethnic underclass, was added the resentment they felt against Castro, who had not only defeated them, but taken everything they had. Their "disgrace," as one Cuban historian writes, "became the psychological impetus that fueled their efforts to prevail in the economic arena."

And they did prevail. In the 1950s, Miami was still largely a resort and retiree town, servicing winter visitors and refugees from the cold Northeast. Today, the greater Miami area is home to more than 1,100 multinational corporations. The city is a global economic hub with the eleventh-highest gross metropolitan product in the country and a thriving business life in which Cuban Americans play a central role.

Some Cuban Exiles had a head start when they arrived in America. Alfonso Fanjul, a sugar magnate whose lands were confiscated in 1959, had enough money to restart his business on 4,000 acres he bought in Florida; today the Fanjul Corporation (still controlled by the Fanjul family) is the world's largest sugar producer, with Domino Sugar among its many holdings. Unlike the Fanjuls, the Bacardis (along with a number of other elite businessmen in Cuba) had jubilantly supported the Revolution, even throwing Castro an extravagant welcome luncheon at one of their breweries near Havana—which Castro did not attend. But within a year Castro seized all of the Bacardis' Cuban assets. Before leaving, the family destroyed their prize possession—a unique yeast culture used in the fermentation of rum—getting just enough of it out of Cuba to continue their rum dynasty in Puerto Rico and Mexico.

Most of the Exiles, however, came to the United States the way Roberto Goizueta did: with practically nothing. As of 1961, émigrés leaving Cuba were prohibited from taking more than five dollars and were required to relinquish the rest of their property to the state. Like Goizueta, some worked their way up to dizzying heights in the business world. Gedalio Grinberg fled to Miami in 1960; in 1983 he bought the Movado watch company, and today his family members remain the controlling shareholders (and his son the CEO) of the Movado Group, with its $500 million in annual net sales. Carlos Gutierrez arrived in Miami at the age of six with his family, knowing no English; he would become CEO of Kellogg and was the United States secretary of commerce from 2005 to 2009. In 1962, Ralph de la Vega, the ten-year-old son of a wholesale grocer in Cuba, arrived in Miami with no family, no English, and no money; in his memoir, *Obstacles Welcome*, he writes of living on U.S. government-supplied "Spam-like meat" that came in containers "the size of paint cans," and cleaning bathrooms to put himself through college. In 2008, de la Vega became CEO of AT&T Mobility, with 65,000 employees and annual wireless revenues of more than $50 billion.

It's often pointed out that the Exiles, poor as they were when they arrived in Miami, had been relatively privileged in Cuba. Some had attended the finest Havana private schools; some knew English; some had vacationed in the United States. But it's simply not true that these immigrants stepped off a plane right into America's middle and professional classes. Most were not only penniless when they arrived, but "had no idea . . . where to take their families, much less where to find a job." For the overwhelming majority, the story of their success was one of arduous toil, swallowed pride, and sacrifice for their children. Former executives parked cars; judges washed dishes; doctors delivered newspapers. Women who had never held jobs before worked as seamstresses, hotel maids, or shrimp sorters at warehouses by the

Miami River—work so painful they called it *la Siberia*. As one émigré put it, "I was determined that my children would be middle class even if I had to have two jobs—which I did for fourteen years."

These descriptions may sound like overly uplifting American immigrant success stories, but as Cuba expert Susan Eckstein puts it, "Cubans more than most Latin American immigrants have lived the American dream." By 1997, a list of the eighty wealthiest Hispanic American multimillionaires contained thirty-two Cubans, even though Cuban Americans constituted only about 5 percent of the U.S. Hispanic population. As of 2004, over 30 percent of U.S.-born Cuban Americans held managerial and professional positions, as compared with 18 percent of Hispanics as a whole. According to a 2002 analysis, over 90 percent of Cuban American students between the ages of eighteen and twenty-four were enrolled full-time in college, reportedly the highest rate of any ethnic group in the country. Only forty years after the Exiles had arrived in the United States, the total revenue of Cuban American businesses exceeded that of the entire island of Cuba.

About a third of Miami's population, Cuban Americans also dominate Miami politics. Both the mayor of Miami-Dade County (one of the most powerful positions in the state) and the mayor of the city of Miami are Cuban American, the latest in a series of Cuban Americans to have held those offices. Nationally, all three Latinos elected in 2012 to the U.S. Senate—Ted Cruz, Marco Rubio, and Robert Menendez—were Cuban American. Cuban actors and entertainers are household names, including Andy García (whose family moved to Miami after the failed Bay of Pigs Invasion and built a million-dollar perfume company), Cameron Diaz, Eva Mendes, and Gloria Estefan.

In trying to quantify Cuban American success, it's important to distinguish the Exiles from later waves of Cuban immigrants who began arriving in 1980. These more recent immigrants came from much poorer backgrounds, including the 125,000 Marielitos and tens of thousands of *balseros*, or rafters, who floated across ninety miles of water on makeshift vessels. The Cuban Exile community is mostly white, whereas a substantial fraction of the Marielitos and post-1990 New Cubans were black or of mixed race.

The Cuban American "economic miracle" has been primarily a phenomenon of the Exiles and their children. Cuba's pre-Castro social stratifications have been replicated in the United States. The Marielitos and New Cubans frequently received a cold shoulder from the Exiles. They are not prominent in business and are largely absent from Miami's power elite.

Thus the Cuban American story is a complex one. Statistically, the New Cubans have fared no better—indeed on some dimensions worse—than other Hispanics. Only 13 percent of the Marielitos have college degrees—about the same percentage as other Hispanics. Some 70 to 80 percent of New Cubans speak English "less than very well." Their poverty rates (about 15 percent of adults under the age of sixty-five; almost 40 percent over the age of sixty-five) are roughly double that of the Exiles, and their overall median income is below that of other Hispanics.

By contrast, the Exiles have distanced themselves from America's other Hispanic groups in both their self-identification and their bank accounts. Not only did the Exile community far outperform other Hispanic groups in the U.S.; by 1990, nearly 37 percent of the Exiles' U.S.-born children earned more than $50,000 a year, whereas only 18 percent of Anglo-Americans did.

The majority of the nonwhite Marielitos chose to resettle outside Miami. Many live in New York City and Los Angeles, joining the large populations of other Hispanics in those cities. By contrast, the

Exiles live in a world of upward mobility, thriving business success, exclusive Miami private clubs, outsize political influence, and rising representation on Ivy League campuses—a group that can justifiably boast of having joined America's elite.

THE DISPROPORTIONATE SUCCESS OF certain West Indian and African immigrant groups, as compared with non-immigrant American blacks, has been the subject of intense debate for some time. Although immigrants make up only 8 percent of America's black population,* their overrepresentation at America's best universities and on Wall Street is well-known. Hard numbers, however, are surprisingly hard to come by.

Ivy League universities, for example, every year announce the "African American" percentage of their incoming class, but they will not disclose how many of these students are from immigrant families or provide national-origin information about them. Investment banks are in general extremely secretive about who works for them. They don't make public any information about the racial composition of their employees. As a result, researchers are obliged to rely on sample surveys and are sometimes consigned to poring over lists of surnames trying to decide which ones "sound African."

The most comprehensive and reliable survey on the black immigrant presence at American universities, conducted by Princeton sociologist Douglas Massey in 1999, indicated that 41 percent of black Ivy League freshmen had at least one foreign-born parent. In 2004, Harvard professors Henry Louis Gates Jr. and Lani Guinier asserted

---

* The 2010 Census showed 3.2 million black immigrants in the U.S. out of a total black population of 38.4 million.

that by their reckoning, a majority of Harvard's black students were immigrants or their children. Today, if Yale Law School is any measure, the overrepresentation of first- and second-generation black immigrants at top U.S. schools may be even higher.

In the first-year class of 2011–2012 at Yale Law, 18 students out of total class of 205 were members of the Black Law Students Association. Of these 18 students, only 2 were African American (a term we'll use—unideally—to refer to blacks in the United States who are neither immigrants nor the children of immigrants) on both sides. Of the other 16, 3 were Nigerian, 2 Ethiopian, 1 Liberian, 1 Haitian, 1 half-Ethiopian and half-Jewish, 1 half-Haitian and half-Korean, 1 half-Jamaican and half–Puerto Rican, 1 half-Panamanian and half–African American, and 1 half-Swedish and half–African American. In all, 12 of the 18 had at least one foreign-born parent.

Among black immigrants, Nigerians are the largest and most successful group. In 2010, there were some 260,000 Nigerians in the U.S., a mere 0.7 percent of the black American population. Yet in 2013, 20 to 25 percent of the 120 black students at Harvard Business School were Nigerian. As early as 1999, Nigerians were overrepresented among black students at elite American colleges and universities by a factor of about ten.

As Nigerians graduate from these schools, they have predictably flourished. Nigerians have done particularly well in medicine. Overall, Nigerian Americans probably make up around 10 percent of the nation's black physicians. And medicine may not even be the real Nigerian forte. By comparison with other blacks in the United States, according to a PhD dissertation on high-achieving second-generation Nigerian Americans, "Nigerians dominate" investment banking. Or as one African American analyst at Goldman Sachs recently joked, "If my only life experiences were at Goldman, my impression would be that Nigeria must have a billion people, because most of the black people I met were Nigerian." In addition, Nigerians appear to be over-represented at America's top law firms by a factor of at least seven, as compared to their percentage of the U.S. black population as a whole.

The success of Nigerians—as well as Ghanaians and certain other African immigrant communities—continues a trend observed decades ago in America's black West Indian immigrants. Several of the country's most famous 1960s civil rights and black power activists, including Marcus Garvey, Malcolm X, Shirley Chisholm (America's first black congresswoman), and Stokely Carmichael, were West Indian by birth or descent. In 1999, only Jamaica sent more blacks to America's selective colleges than did Nigeria. In 2001, the son of two Jamaican immigrants, Colin Powell, became America's first black secretary of state. And the man who promoted Powell to general in 1978 was Clifford Alexander, the country's first black secretary of the Army; Mr. Alexander's father was a Jamaican immigrant. For decades, at least in America's northern states, West Indians and their children "have been disproportionately represented in the business, professional, and political elites."

It's important to emphasize that West Indian and African immigrant success in the United States is not at all uniform and in any event is not comparable to that of, say, Indian or Jewish Americans, whose incomes are the highest of any groups in the country. Several African immigrant groups, particularly those from Somalia, Sudan, or other countries with large refugee populations, are among the nation's poorest. And even the most successful groups are not (yet) topping the income charts, although they are doing better than the national average.

For example, in terms of income, Nigerian Americans dramatically outperform black Americans and also outperform Americans in general. Almost 25 percent of Nigerian households make over $100,000 a year; only 10.6 percent of black American households overall do. Five percent of Nigerian American households earn over $200,000 a year; the figure is only about 1.3 percent for black America overall. The median Nigerian American household income is $58,000 a year; the national median is $51,000. In 2010, Nigerian men working full-time earned a median income of $50,000, while the figure for all U.S.-born men was $46,000.

Given that Nigerian Americans obtain college and advanced degrees at one of the highest rates in the country, the fact that they only somewhat out-earn the national average is noteworthy. A number of factors almost certainly contribute. For one thing, the Nigerian community includes a significant number of new arrivals and their children. Some of these new arrivals are very poor (pulling down the overall income statistics); many are in the middle of their university or postgraduate careers. Thus the big payoffs from Nigerian educational attainment are probably still to come. But a larger factor may well be discrimination. African immigrants frequently testify to experiences of racial discrimination, and data repeatedly show that high-skilled immigrants from Africa end up in unskilled jobs far more than any other immigrants (except those from Latin America) and other Americans overall.

For all these reasons, however, Nigerian American success—including their extraordinary performance relative to the overall black population—is of special importance to this book. Conservatives like Dinesh D'Souza have argued that African Americans' poverty is the result not of discrimination, but of a "dysfunctional" culture that includes a "conspiratorial paranoia about racism" and a "celebration of the criminal and the outlaw as authentically black." Some liberals assert, on the contrary, that African American poverty is entirely the product of racism, whether overt, covert, or "structural." Both sides of this argument are mistaken.

As we'll discuss later, the success of Nigerian Americans and certain other black immigrants—who face many of the same institutional obstacles and prejudices as African Americans—*is* significantly due to cultural forces. While many factors contribute to the lower overall socioeconomic status of African Americans, the Triple Package is an important part of the story behind black immigrant success. The lesson to take, however, is not that native-born American blacks have only themselves to blame for their economic position. The lesson is that the United States did everything it could for centuries to grind the Triple Package out of African American culture—and is still doing so today.

IN JUNE 2012, the Pew Research Center released a report called *The Rise of Asian Americans*, describing Asians as "the highest-income, best-educated and fastest-growing racial group in the United States."

We'll use "Asian American" the same way the Census Bureau (as well as Pew) does, covering all U.S. residents "having origins in any of the original peoples of the Far East, Southeast Asia, or the Indian subcontinent." But for our purposes, "Asian American" isn't a very useful classification. It embraces vastly disparate groups with entirely different cultures, including two of the most successful groups in the nation (Indian and Chinese Americans) and several of the poorest (such as Hmong, Laotian, and Cambodian Americans).*

Among the most successful Asian American groups, one well-

---

* The six largest U.S. Asian groups are Chinese (approximately 4 million as of 2010), Filipino (3.4 million), Indian (3.2 million), Vietnamese (1.7 million), Korean (1.7 million), and Japanese (1.3 million). Other, less numerous groups include Bangladeshis, Burmese, Cambodians, Hmong, Indonesians, Laotians, Pakistanis, and Thais. Two of these groups—Cambodian and Hmong Americans—have per capita incomes below $12,000, lower than any major racial or ethnic group in the country, with more than a third of their children living below the federal poverty line. Altogether, Asian Americans make up 5–6 percent of the U.S. population.

known phenomenon is the breathtaking accomplishment of their youth, who top list after list of prestigious awards and competitions. For example, over the last five years, twenty-three of the fifty top prizewinners of the Intel Science Talent search—a nationwide high school competition that George H. W. Bush called the "Super Bowl of science"—were Asian Americans, overwhelmingly of Indian, Chinese, and to a lesser extent Korean heritage.

The résumés of these Intel winners are terrifying to even your run-of-the-mill Tiger Mom. Take Amy Chyao of Richardson, Texas, a 2012 finalist. By attaching a nitric oxide donor to titanium dioxide nanotubes, Amy "synthesized a nanoparticle," as the *New York Times* reported, which "essentially is a remotely triggered bomb that attacks cancer cells," offering a potential noninvasive alternative to chemotherapy for deep tumors. The pretty seventeen-year-old is the co-author of two articles in peer-reviewed journals, has perfect ACT scores, was first in her high school class of 1,473, is an accomplished cellist, and founded a nonprofit organization to teach immigrant children spelling. Dozens of similar bios can be found among the Intel winners: for example, the saxophone-playing Chinese American teen who developed "a treatment for phenylketonuria" while serving as editor-in-chief of the school newspaper, or the Indian American teen who "gave the first non-trivial analytic lower bound for odd perfect numbers" and was also crowned homecoming king.

Asians and Asian Americans constitute 30–50 percent of the student bodies at the country's leading music programs. At the Juilliard School of Music, they make up more than half the students; the two largest groups are Chinese and Korean violinists and pianists. The quadrennial International Tchaikovsky Competition, perhaps the most prestigious music competition worldwide, has a junior section awarding prizes to young musicians in violin, piano, and cello. Since 1992, precisely four Americans have won first prizes in any instrument; their names were Emily Shie, Jennifer Koh, Sirena Huang, and Noah Lee.

Indians, meanwhile, have cornered the market in spelling bees. In 1999, the winner of the nationwide Scripps National Spelling Bee was Nupur Lala, a fourteen-year-old from Tampa, Florida, with an Indian mother and a Pakistani Indian father. After the documentary *Spellbound* made Nupur famous—and ESPN began broadcasting the Bee finals live—America's Indian community got serious about competitive spelling. In 2012, the champion and first two runners-up were all Indian Americans; it was the fifth consecutive year an Indian American took first prize.

And lest it be thought that Asian Americans dominate only in fierce but narrow single-discipline endeavors, consider the United States Presidential Scholars, honoring one young man and woman from each of the fifty states, the District of Columbia, and Puerto Rico, as well as thirty-seven others, not only on the basis of academic success, but also artistic excellence, community service, leadership, and commitment to high ideals. In 2012, of the 141 well-rounded winners of this award, approximately 48 were Asian American; in 2011, it was 52. Again, the overwhelming majority of these were Chinese and Indian.

Unsurprisingly, these hypersuccessful Asian American teens are outperforming other groups on standardized tests and in admissions to elite universities. Overall, Asian American SAT scores are 143 points (out of 2400) above average, including a 63-point edge over whites, with the gap apparently increasing. About 5 percent of the U.S. college-age population, Asians make up 19 percent of the undergraduates at Harvard, 16 percent at Yale, 19 percent at Princeton, and 19 percent at Stanford. Indeed, if admissions were based solely on National Merit Scholarship and SAT scores, these percentages would be even higher, and some believe that there is an implicit "anti-Asian admissions bias" in the Ivy League. At CalTech, said to base admissions solely on academic criteria, nearly 40 percent of the students are Asian.

Turning to adults, Indian Americans have the highest median household income of any Census-tracked ethnic group in the United States: $90,500 per year (the figure for the entire U.S. population is $51,200). An extraordinary 44 percent of Indian American households make over $100,000 a year—again, the highest rate of any ethnic group—with 12 percent over $200,000. (The nationwide figures are 21 percent over $100,000 and 4 percent over $200,000.) Meanwhile, Chinese Americans are not too far behind, with exceptionally high median household income and strongly disproportionate representation in the higher income brackets. In fact, Taiwanese Americans have the second highest household income of any ethnic group in the country, just below Indians; their median individual income is even higher than Indians'.

Indian and Chinese American success has tended so far to be relatively conventional and prestige-oriented, starting with National Merit Scholarships, valedictorian titles, and brand-name schools. It's no coincidence that East Asian American musical virtuosos typically play classical music, as opposed to jazz or rock. Their professional successes have similarly been concentrated in fields perceived as practical, stable, "respectable," and "impressive." Clichéd as it sounds, there are in fact a disproportionate number of Asian Americans in engineering and the sciences, although career aspirations are beginning to change among the young. Since 1965, Indian Americans have won three Nobel Prizes, and Chinese Americans have won six.

Although many Chinese and Indian American families place a special premium on scholarliness, both groups have achieved remarkable business success. Yahoo cofounder Jerry Yang, Zappos founder Tony Hsieh, YouTube cofounder Steve Chen, Guitar Hero cofounder

Kai Huang, and venture capitalist Alfred Lin are all Chinese American. The Indian American list is even more impressive. The Bose Corporation was founded by Amar Gopal Bose, son of a Bengali immigrant, whose 2011 net worth was estimated at $1 billion. Indra Nooyi, a first-generation Indian immigrant, is the CEO of PepsiCo. Other companies with Indian Americans as their present or recent CEOs or presidents include: Sun Microsystems, MasterCard, United Airlines, Motorola, Adobe Systems, Citigroup, Citibank, HSBC North America, McKinsey, and US Airways.

Indian Americans also appear to be bigger high-finance risk takers than other Asian Americans. Another billionaire, Vinod Khosla, came to America from India for graduate school, cofounded Sun Microsystems, and became a partner at the legendary venture capital firm Kleiner Perkins. Khosla is just one of dozens of Indian American Silicon Valley entrepreneurs. In fact, Indians have founded more Silicon Valley start-ups than any other immigrant group—more than the four next immigrant groups (British, Taiwanese, Chinese, and Japanese) combined.

To a greater extent than any other Asian group, Indian Americans are prominent in politics, the public sphere, and the media. Bobby Jindal—who got into both Harvard Medical School and Yale Law School but opted to attend Oxford as a Rhodes Scholar—is the governor of Louisiana. Nikki Haley, the daughter of immigrant Sikh parents, is the governor of South Carolina. Featured on the cover of *Time* magazine as the man "busting Wall Street," Preet Bharara is the U.S. Attorney for the Southern District of New York. With their own TV shows, former *Newsweek International* editor Fareed Zakaria and Emmy-winning Dr. Sanjay Gupta are household names. Other Indian American public intellectuals include *New Yorker* contributor Dr. Atul Gawande and Pulitzer Prize winner Dr. Siddhartha Mukherjee (at an awards ceremony in Boston 2012, the former introduced the latter by joking, "I thought I had the Indian-Rhodes-Scholar-Harvard-Medical-School thing cornered").

By contrast, Chinese American success has been more restricted in scope. There are not many (or possibly any) prominent Chinese American talking heads, and Chinese Americans are conspicuously underrepresented at the top levels of corporate America. As of 2013, there were five Indian American CEOs of Fortune 500 companies, but no Chinese Americans.

In part, discrimination and stereotyping may explain the shortage of Chinese Americans in leadership positions. But as several Asian activists have argued, part of the problem may also lie in a tendency within East Asian culture to encourage deference to authority, while discouraging the self-promotion and quarterback mentality said to be necessary to leadership success in America.

We'll have much more to say about all this later, but the bottom line is that, for all their differences, both Indian and Chinese Americans are paradigmatic Triple Package groups. In both cases, it's been Triple Package insecurity—not only economic, but also social and racial—that has driven these groups toward narrowly defined, prestige-oriented avenues of success. And it's a Triple Package dynamic that will put pressure on these constraints going forward.

The Triple Package has an arc. Basically, Chinese and Indian Americans are today where Jewish Americans were two or three generations ago. Over 65 percent of Chinese Americans and 87 percent of Indian American adults are foreign-born; over 90 percent of both these groups are either immigrants or their children. (By contrast, only about 15 percent of American Jewish adults are foreign-born.) Chinese and Indian parents are hardly the first immigrant parents to want their children to be doctors, not poets. But as we'll discuss, the inner dynamic of the Triple Package, as it interacts with American society, will lead young Chinese and Indian Americans increasingly to balk and bridle at the conformity their parents try to impose on them.

TURNING, THEN, TO THE JEWS: it's generally much harder to talk about Jewish wealth than that of any other group. Exaggerated or even patently false claims of Jewish economic "control" have in the past led to discrimination, ghettoization, and some of the worst atrocities in history. As a result, while one can relatively freely explore the phenomenon of, say, a 3 percent Chinese minority owning 70 percent of a country's corporate wealth (as was true of Indonesia in the 1990s), it is far more difficult to ascertain or even discuss the extent of Jewish economic influence. The U.S. Census Bureau used to compile data on religion, but this was largely discontinued after World War II, and today the Census is barred by law from asking mandatory questions about individuals' faith.

But the fact is that Jews are the quintessential successful minority. Jews do not appear to have been particularly economically successful in antiquity—but that's about the last time they weren't, at least when left alone to pursue their livelihoods. Today, about 43 percent of the world's Jews (5.9 million) live in Israel, and about 40 percent (5.4 million) in the United States, making Jewish Americans about 1.7 percent of the U.S. adult population.

American Jews are disproportionately successful by pretty much any economic measure, except maybe something like "fortunes amassed through golf." (Of the ninety players in the 2010 Masters, roughly zero were Jewish.) Four of the country's top ten highest-paid CEOs in 2011 were Jewish: Disney's Robert Iger; Coach's Lew Frankfort; Polo's Ralph Lauren; and Verizon's Ivan Seidenberg. (For purposes of comparison, no Mormons or Asian Americans made the top ten.) But these CEOs are wage slaves compared with the ten top-earning hedge fund managers, four of whom in 2011 were also Jewish: Jim Simons (2011 earnings: $2.1 billion); Carl Icahn ($2 billion); Steve Cohen ($600 million); and Bruce Kovner ($210 million).

In 2009, a reporter counted the Jews on the Forbes 400: 20 of the top 50 were Jews, as were 139 overall. Of the people on the 2011 *Forbes* list who made their fortunes in real estate, 60 to 70 percent were Jewish. Almost twenty of the forty billionaires who have pledged to leave half their estates to charity are Jewish, including New York City mayor Michael Bloomberg, media mogul Barry Diller, and Diller's fashion-designer wife Diane von Fürstenberg (formerly Diane, Princess of Fürstenberg, but born Diane Halfin to a Jewish Romanian father and a Jewish Greek Holocaust survivor mother).

Jewish success extends well beyond commerce and finance. The Antoinette Perry Award for Excellence in Theatre, better known as the Tony, was first given out in 1947. Since then, 69 percent of the Tony-winning composers and 70 percent of the lyricists have been Jewish. Many of America's best-loved comics—not Charlie Chaplin, contrary to popular belief, but the Marx Brothers, the Three Stooges, Jack Benny, Jerry Lewis, Bette Midler, Woody Allen, Jerry Seinfeld, and more recently Andy Samberg, Jon Stewart, Adam Sandler, and Sarah Silverman—have been Jewish. Winners of the Academy Award for best director: 37 percent Jewish. Pulitzer Prize winners for nonfiction: 51 percent Jewish (for fiction: 13 percent).

Certain professions seem to be particularly attractive to Jews. In 2007, Reuters reported that 29 percent of America's psychiatrists were Jewish, as were 13 percent of its physicians. As early as the 1930s, Jews already filled New York City's law offices; by one estimate, 65 percent of New York's lawyers were Jewish at that time.

Today, three of the nine justices on the United State Supreme Court are Jewish. Architecture superstars Richard Meier, Frank Gehry, Daniel Libeskind, and Louis Kahn are Jewish. Advice-giving is another favorite Jewish occupation. Ann Landers was born Esther Pauline Friedman; Abigail Van Buren ("Dear Abby") was her identical twin sister, Pauline Esther Friedman; Miss Manners was born Judith Perlman.

Because the Census does not track religion, there's no authoritative measure of overall Jewish income. Poverty certainly exists among American Jews and may be rising; according to a recently released United Jewish Association report, nearly one in four Jewish households in New York City is poor. Nevertheless, numerous studies confirm that, overall, Jewish income is higher than that of any religious group or Census-tracked ethnic or national-origin group in the country. Jewish median household income in 2010 was probably in the range of $97,000 to $98,000, almost twice the national median. (Indian Americans, the highest-earning Census-tracked group, had a 2010 median household income of about $90,500.) Research dating back decades indicates that Jewish income exceeds that of Protestants by up to 246 percent and of white Catholics by up to 243 percent. According to Pew, 46 percent of Jewish households in 2006 were making at least $100,000 a year (again, the highest rate for any religious or ethnic group in the country). For Reform Jews, 55 percent had incomes of $100,000 or more in 2006; for the nation as a whole, the figure was 18 percent.

Jewish intellectuals have been influential in American public life on all sides of the political spectrum. Both the 1960s New Left and the neoconservative movement a few decades later were heavily influenced by Jewish thinkers (in some cases, the same thinkers). Three of the leading opinion writers for the *New York Times* today are Jewish:

Thomas Friedman, David Brooks, and Paul Krugman. The extraordinary Jewish contribution to American academia, from the sciences to the liberal arts, has been well documented. "[I]t is impossible," concludes one author, "to find an academic domain in which Jews have not played a disproportionate role."

Jews are also awkwardly prominent in Hollywood, a fact that many Jews prefer not to highlight, an exception being journalist Joel Stein, who in 2008 noticed that:

> When the studio chiefs took out a full-page ad in the *Los Angeles Times* a few weeks ago . . . , the open letter was signed by: News Corp. President Peter Chernin (Jewish), Paramount Pictures Chairman Brad Grey (Jewish), Walt Disney Co. Chief Executive Robert Iger (Jewish), Sony Pictures Chairman Michael Lynton (surprise, Dutch Jew), Warner Bros. Chairman Barry Meyer (Jewish), CBS Corp. Chief Executive Leslie Moonves (so Jewish his great uncle was the first prime minister of Israel), MGM Chairman Harry Sloan (Jewish) and NBC Universal Chief Executive Jeff Zucker (mega-Jewish). If either of the Weinstein brothers had signed, this group would have not only the power to shut down all film production but to form a minyan.

Worldwide, though Jews make up only about 0.2 percent of the total population, a fifth of all Nobel laureates have been Jewish. As of 2007, of the forty-two U.S. winners of the Nobel Prize for economics, twenty had been Jewish, and several more have won since, including Mr. Krugman. Roughly 2 percent of the U.S. population, Jews account for an estimated 36 percent of all Nobel Prizes ever awarded to Americans.

But perhaps the most important contribution made by Jewish scientists in America came in the 1940s, after anti-Semitic persecution had led to a flight of incalculable scientific talent out of Europe. America became home to some of the greatest physicists of all time, including Edward Teller (known as the "father of the hydrogen bomb"), John von Neumann (a founder of game theory as well as a principal member of the Manhattan Project), and of course Albert Einstein. Together with Robert Oppenheimer, a native-born Jewish American, these men played a critical role in enabling the United States to win the race for the atom bomb.

IRANIAN AND LEBANESE AMERICANS are also among the nation's most successful national-origin groups in terms of median household income. In later chapters, we'll discuss these groups in more detail. Here is a snapshot.

Iranian immigration to America does not track European immigration. From the mid-1800s to the early 1900s, the number of known Iranian immigrants in the United States was 130. Only in the latter part of the twentieth century did they begin to arrive in significant numbers, mostly as students before the 1979 Revolution, followed by a flood of exiles and refugees, including royalists as well as persecuted minorities. In all, according to the Census, their population is about 450,000.

Iranian immigrants in this country include numerous ethnic and religious groups: Shiite Muslims of course, but also Armenians, Assyrian Christians, Jews, Kurds, Turks, Zoroastrians, and Bahá'ís. As a whole, they have tended to keep a relatively low profile: in the 1980s, because of American hostility in the wake of the hostage crisis, and later because of concern about being associated with Muslim terrorists. An exception is the Iranian community in Los Angeles, by far the country's largest and currently the subject of a reality TV show, *Shahs of Sunset*. Featuring a circle of relatively affluent Iranian

Los Angelenos, the show has been criticized for presenting them stereotypically as "vulgar, materialistic show-offs."

One thing *Shahs* does not focus on is Iranian American academic success. A study based on 2000 Census data found that more than a quarter of Iranian Americans over twenty-five had a graduate degree, reportedly making them "the most highly educated ethnic group in the United States," with five times the percentage of doctorates as the national population. As of 2010, over 17 percent lived in houses with a value greater than $1,000,000 (compared with a national figure of 2.3 percent), possibly the highest percentage of any Census-reported ancestry group. Their median household income of $68,000 (compared with a national average of $51,000) is higher than that of Chinese Americans. One in three earn over $100,000 a year; the national figure is one in five.

The famously entrepreneurial Lebanese have one of the most successful diasporas in the world. Today, as Beirut continues to suffer from violence and sectarian strife, a Christian Lebanese from Mexico—Carlos Slim, net worth $73 billion—is the world's richest man. Although tiny in numbers, Lebanese minorities are disproportionately successful throughout Latin America, West Africa, and the Caribbean.

Lebanese immigrants arrived in the United States in two main waves. The roughly 100,000 who came between 1881 and 1925 were predominantly Christian and poor but enterprising. Along with German Jews, they rose quickly as "pack peddlers," going door-to-door with suitcases weighing up to two hundred pounds, selling everything from ribbons and jewelry to children's clothes. Many accumulated small fortunes, eventually opening grocery stores and banks; a number even became millionaires. By contrast, the Lebanese who arrived in the United States after 1967, many of them fleeing the Lebanese civil war or the Arab-Israeli conflict, were generally more educated (contributing to Lebanon's brain drain) and included Muslims and Druze as well as Christians.

Today, the Lebanese American population numbers about 497,000, approximately the same size as the Iranian American population. The Lebanese have income numbers very comparable to Iranian Americans. Among ethnic groups in the United States, Lebanese are close to the top of the charts in terms of median household income, percentage earning over $100,000, and percentage earning over $200,000.

The demographic profile, however, of Lebanese Americans is quite different. Whereas 65 percent of Iranian Americans are foreign-born, only 23 percent of Lebanese Americans are. In fact, most Americans who identify themselves as having Lebanese ancestry are probably only a half or quarter Lebanese. This is true, for example, of former White House chief of staff John H. Sununu (at most half-Lebanese) and his son, the former New Hampshire senator John E. Sununu (at most quarter-Lebanese), who was defeated in his 2008 bid for reelection by Jeanne Shaheen, wife of attorney Bill Shaheen (three-quarters Lebanese).

THE GROUPS DISCUSSED in this chapter are by no means the only successful ones in America. Many other ethnic and national-origin groups are disproportionately successful too. We could, for example, have included Japanese Americans, who significantly out-earn the national average, or Greek Americans, whose remarkable economic ascent in the second half of the twentieth century is well documented. Both these groups have median incomes not far below the groups we focus on.

In an endnote to this paragraph, we offer more detail about how we chose the eight groups we focus on. But in a nutshell, five of our groups—Jewish, Indian, Chinese, Iranian, and Lebanese Americans—were by standard metrics arguably the five most successful in the country as of 2010. Two are stark outperformers among a larger class of statistical underperformers; the success of Nigerians (as well as certain other African immigrants) is exceptional given the relatively poor outcomes of other black Americans, as is the success of the Cuban Exile community given the relatively poor outcomes of other Hispanic Americans. Finally, Mormons may be the most economically dynamic group in the country, rising in a few decades from average or even below-average status to extraordinary levels of corporate success.

The rest of this book attempts to explain the common key to these groups' disproportionate success.

CHAPTER 3

THE SUPERIORITY COMPLEX

THE DESIRE TO SEE one's group as superior may be one of those rare universals in human culture. William Graham Sumner, who coined the term *ethnocentrism* in 1906, believed that every group suffered from this vice, and anthropologists today confirm that the impulse to paint "one's own group (the in-group) as virtuous and superior . . . and out-groups as contemptible and inferior" is a "syndrome" found in human societies virtually everywhere.

So do *all* groups have a superiority complex?* The answer is no. Some have had their superiority narrative ground into the dirt for so long it's hard to reclaim (typically this is done by other groups to entrench their own superior status). Some celebrate the idea of their

---

* Note that our use of the term "superiority complex" is not the same as Alfred Adler's. For Adler, who is often (although almost certainly incorrectly) credited with coining the term, a "superiority complex" was in every case a kind of self-deceiving defense mechanism masking a more basic "inferiority complex." In our usage, groups or individuals can be said to have a superiority complex only when they have a genuine, deeply internalized belief in their own exceptionality. For sources, see the endnote.

middle-ness, neither high nor low. Some have so internalized modern postulates of equality that they frown on or even censor notions of group superiority. Some may lay claim to a sense of specialness, but in a way that exalts their victimization (for example, most downtrodden people ever), often revealing a lack of belief in their superiority.

At the very least, in some groups a superiority complex runs deeper and stronger than it does in others. Such is the case with every one of America's most successful groups.

JEWS MAY HAVE INVENTED the idea of a "chosen people." Certainly the Jewish claim to chosenness is the most famous; for three thousand years, it's been a source of inspiration, derision, and imitation. Colonial Americans believed that New England was the New Israel; before that, Oliver Cromwell had pronounced himself an "Israelite"; and long before that, the New Testament told Christians that they were now the "chosen" ones. Every new "Israelite" heralding a new chosen people reminds the world of the Jewish claim to that status and acknowledges its priority, if only in time.

The Jewish understanding of *why* God chose them has always been a little mysterious. It wasn't because they were already a great and flourishing people: "The Lord," Moses tells the Jews, "did not set his love upon you, nor choose you, because you were more in number than any people; for you were the fewest of all people." And it certainly wasn't because of their purity of heart. "Not for thy righteousness," says Moses after having led the Israelites through the desert for forty years. At various points the Old Testament refers to the Jews as "corrupt," "warped," "foolish," "perverse," "unfaithful," and "a nation without sense."

Nevertheless, in the face of mammoth historical evidence to the contrary, and despite all the self-questioning without which the Jewish conception of chosenness would not be Jewish, Jews maintained for millennia the idea that they were God's chosen people. Wherever Jews settled, whatever their hardships, Jewish children were raised hearing that proposition in synagogue, celebrated in the home on Sabbath evenings, and by the entire community on holy days. ("Blessed art Thou, Lord our God, King of the universe, who hast chosen us above all people and exalted us above all nations.")

The utter brazenness of the idea—that there is only one God, maker of all the universe, and He has chosen *us* (not *you*, or anyone else) as His people, His "peculiar treasure" (Exodus 19:5)—remains to this day a special burr for many. The reason Jews are a "contaminated" people, said the Portuguese novelist José Saramago, honored with a Nobel Prize in 1998, lies in their "monstrous and rooted 'certitude' that . . . there exists a people chosen by God." Beloved Greek songwriter Mikis Theodorakis, who scored the film *Zorba the Greek* (and composed the Palestinian national anthem), has said that Jews "are at the root of evil"; in an interview with an Israeli reporter, he emphasized "the feeling that you are the children of God. That you are the chosen."

But do modern Jews still believe in their "chosenness," and does chosenness imply superiority?

The great Jewish philosophers of the modern era have long been uncomfortable with these ideas. Spinoza believed that, fundamentally, "God has not chosen one nation before another"—and was excommunicated at the age of twenty-three. Walking the tightrope of Jewish assimilation in anti-Semitic eighteenth-century Germany, the philosopher Moses Mendelssohn (grandfather of the composer Felix) held that God had revealed "legislation" to the Jews, not a "religion"; Jews had no special claim to salvation or any eternal religious truths.

The early generations of Jewish Americans felt especially conflicted about the idea of the Jews as a separate, divinely chosen people. "To abandon the claim to chosenness," as Arnold Eisen puts it, would have been "to discard the raison d'être that had sustained Jewish identity and Jewish faith through the ages, while to make the claim was to question or perhaps even to threaten America's precious offer of acceptance." Reconstructionist Judaism, founded by an American rabbi in the 1920s, renounced chosenness as incompatible with equality and democracy. Reform Judaism speaks of a Jewish "mission" to be "witnesses to God's presence," deemphasizing if not rejecting the idea of chosenness.

But even as the notion of chosenness waned, Jews rarely gave up the idea of their exceptionality. "There is no doubt," wrote Freud of the Jews, "that they have a particularly high opinion of themselves, that they regard themselves . . . as superior to other peoples." As Jewish Americans rose in prominence in the early twentieth century, they grew less afraid to express this sense of exceptionality. In 1915 Louis Brandeis, soon to be a justice of the United States Supreme Court, made the following extraordinary statements at a speech in New York City:

> And what people in the world has . . . a nobler past? Does any possess common ideas better worth expressing? . . . Of all the peoples in the world those of two tiny states stand preeminent as contributors to our present civilization, the Greeks and the Jews. The Jews gave to the world its three greatest religions, reverence for law, and the highest conception of morality. . . . Our conception of law is embodied in the American constitution.

For Brandeis, the persecution Jews had suffered through the ages was a point in their favor. "Persecution," he added, had "broadened" the Jews' "sympathies," training "them in patient endurance, in self-control, and in sacrifice. . . . It deepened the passion for righteousness."

The idea of Jewish historical exceptionality made the concept of chosenness practically irrelevant. "The point is not whether we feel or do not feel that we are chosen," declared Martin Buber, the Austrian-born Jewish philosopher. "The point is that our role in history is actually unique." This feeling of historical uniqueness, seemingly proven by a three-thousand-year record of survival and accomplishment, beats close to the heart of Jewish culture everywhere.

But uniqueness is a double-edged sword. Michael Chabon has described the "foundational ambiguity" in Jewish exceptionalism: simultaneously a "treasure" and a "curse," a "blessing" and a "burden," a "setting apart that may presage redemption or extermination. To be chosen has been, all too often in our history, to be culled." Yet even the Nazi genocide became a kind of twisted emblem of Jewish superiority for some Jewish leaders—particularly in America, starting around the 1970s—who engaged in what many other Jews, such as the historian Peter Novick, viewed as a "perverse sacralization" of Auschwitz, competing with other groups over "'who suffered most,'" and behaving as if they were "almost proud of the Holocaust."

Most American Jews today would politely applaud Britain's former chief rabbi Lord Sacks, who in 2001 condemned the notions of both "Jewish superiority" and "Jewish inferiority" as "two sides of the same coin." But they might take greater satisfaction, if only in private, from articles like "Jewish Genius," Charles Murray's 2007 catalog of the "extravagant" Jewish overrepresentation in "the top ranks of the arts, sciences, law, medicine, finance, entrepreneurship, and the media."

In 2009, 70 percent of Israeli Jews said they still believed that Jews were God's chosen people. The figure cannot be nearly so high for American Jews, given that less than half say they believe in God, but even for those without theological faith, the sense of Jewish exceptionality is often part of their upbringing. Some are taught to locate this exceptionality in a Jewish commitment to justice, law, or morality; others, in a Jewish insistence on questioning; others, in Jewish intellectual achievements.

Or, according to the novelist Philip Roth, there may be no articulated basis for it whatsoever. Roth said that American Jews inherit from their parents "no body of law, no body of learning and no language, and finally, no Lord," but rather "a kind of psychology," a "psychology without content" that could be "translated into three words: 'Jews are better.'" He added: "There was a sense of specialness, and from then on it was up to you to invent your specialness; to invent, as it were, your betterness." Whether rooted in divine election, history, intellect, morality, or "a psychology without content," the Jewish sense of being somehow exceptional has lasted three thousand years and is unlikely to disappear anytime soon.

MORMON SUPERIORITY IS, like Jewish superiority, historically founded on the idea of chosenness—only without the angst.

Mormons aren't subtle about claiming the Jewish mantle. Mormon leaders call their community "Israel." Utah's Salt Lake Valley is "Zion"; Mormon prophecy calls for the building of a "New Jerusalem." Mormons had their Moses in Brigham Young; they had their exodus when, following Young, they made their arduous trek across the American wilderness away from the enemies who persecuted them and who had murdered their prophet, Joseph Smith. Every Mormon child is taught about the "extermination order" issued against them in 1838 by Missouri governor Lilburn Boggs.

But in terms of its beliefs, origins, and spirit, Mormonism has much less to do with Judaism than with American manifest destiny. Often described as "a religious genius," Smith essentially made a church out of the idea of America's providential place in the world, rewriting Christian scripture to fit that idea. To Smith were revealed previously unsuspected truths: that the Garden of Eden had actually been located in Jackson County, Missouri; that Adam and Eve had lived in America; that a band of Israelites had voyaged to the Western Hemisphere around 600 BC; that Christ himself, after the Resurrection, came to America; that great wars between Christ's followers and his enemies had been fought all over the American continent; that the true Christian gospel was preserved in America, recorded on plates of gold buried in upstate New York; and that Christ will return to America in the Second Coming. "The whole of America is Zion," Smith proclaimed, and he was to be its prophet and leader.

Thus was Mormonism infused from its inception with American exceptionalism—with America's belief in itself, with what Mormon scholar Matthew Bowman calls America's "confident amateurism"— even as Mormons found themselves rejected, ostracized, and attacked by their fellow Americans. As Mormons crossed the country in their covered wagons, braving rattlesnakes, fever, hunger, and the mile-wide Mississippi, they added a pioneer spirit to their already "quintessentially American religion." On their westward trek they fed not on manna or matzo, but on hardtack and salted bacon.

Finding deliverance in the isolated Salt Lake Valley only strengthened Mormons' belief in their divine election. America had turned its back on true religion; all Christendom had fallen into a "Great Apostasy." Theirs alone was Christ's true church on earth. To them alone had God entrusted his priesthood, his truth, and the task of redeeming mankind in anticipation of the imminent Second Coming. This self-understanding—as God's end-of-days special emissaries—has remained central to Mormon identity.

Mormons believe they can and do receive direct divine communications and revelations. Every Mormon male above the age of twelve can receive the "priesthood," enabling him to "access the power of

God: to heal, cast out demons, bless, and dedicate." Mormons go on "missions," to spread the word to the rest of mankind. And they're good at it. All over the world Mormon temples are rising up—which, once dedicated, non-Mormons may not enter.

Although some Christians argue that Mormonism is not a Christian faith, Mormons firmly believe they are followers of Christ, praying to him, hallowing his name, and believing in the Atonement and Resurrection. But in addition to supplementing the Bible with its own scripture, Mormonism departs on key theological points from most Christian denominations (as many Christian denominations do from one another).

In particular, Mormons reject the doctrine of original sin. "Unlike other Christians," writes Columbia Professor (and Mormon) Claudia Bushman, "they consider themselves free from the original sin that degraded mankind." Indeed, Mormon theology disavows Christianity's usually categorical dualism between the human and the divine, holding rather that God is a corporeal, essentially man-like person, and that man in turn shares God's divine nature. Thus Mormonism teaches, as LDS leader B. H. Roberts put it in 1903, that "it is no robbery to be equal to God." Even today, Mormons commonly believe that if they have proven worthy in this life, "they will inherit godhood of their own" in the next.

In their unostentatious way, Mormons are almost preternaturally confident about the afterlife. They see their families as divinely ordained units, destined to be together not only here but hereafter. A special third tier of heaven awaits them—the "celestial kingdom"— where those who have accepted what Mormons consider "the fullness of the gospel" will live for eternity reunited with their earthly families.

But the most important element of contemporary Mormon superiority is moral. In terms of their relation to mainstream American morals and "family values," Mormons have completely turned the historical tables. Throughout much of the nineteenth century, Mormons were openly polygamous. Indeed, early Mormon theology seems to have made the number of wives a man had a mark of his exaltation, bringing him closer to divinity. (Brigham Young had somewhere between twenty-seven and fifty-five wives; Smith had perhaps thirty.) Under intense pressure from United States authorities, the Church finally renounced polygamy in 1890 and disavowed it again in 1904.

So long as Mormons embraced polygamy, Americans could view them as licentious, deviant, immoral. Today, with their abstemiousness, strong families, and clean-cut children, Mormons can view America as licentious and immoral. From the LDS perspective, as Claudia Bushman puts it, Mormonism is "an island of morality in a sea of moral decay."

Thus today, the Mormon sense of exceptionalism is focused less on American manifest destiny and much more on morality and mission. Sit in on a Sunday school class at a Mormon church and you're likely to hear children sing, *I am a child of God, / And He has sent me here*. The hymn, commissioned to teach Mormon children about their unique relationship to God, captures the idea of both divine parentage and divine mission. Every generation of Mormons is taught that they are the ones the world has been waiting for. For six millennia, God has held his "peculiar treasure" back, waiting until now to place them on earth to redeem humanity. As Church president Ezra Taft Benson used to tell young Mormons in the 1970s and 1980s:

For nearly six thousand years, God has held you in reserve to make your appearance in the final days before the Second Coming of the

Lord. . . . God has saved for the final inning some of His strongest children, who will help bear off the Kingdom triumphantly. . . . [F]or you are the generation that must be prepared to meet your God.

IN THE SUMMER OF 2012, the popular Miami blogger, radio host, and YouTube personality Pepe Billete posted the following online entry, titled "I'm Not a Latino, I'm Not a Hispanic, I'm a Cuban American!"

I don't know about you, *pero* every time I hear someone refer to me as "Hispanic" or "Latino," *me dan ganas de meterle una pata por culo a alguin.** My response is always the same . . . I'm a Cuban American, you *comemierda!*†

Oye, the only thing that makes me more proud than being a Miamian is the fact that I am Cuban American.

After a rant about how the term "Hispanic" is never applied to Spaniards, who get to be called "European" or "Spanish," Pepe went on to say:

The fact is that in most of the U.S., when people say "Hispanic" or "Latino," what they really mean is "Mexican." Yes, *pipo*, this is 100% the case. . . . [T]o me, it's . . . pretty fucking offensive. Cubans and Mexicans come from two very different cultures and have very different experiences here in the U.S. . . . [E]very time I turn on the

---

* Idiomatic for, "I feel like kicking someone's ass." *"Alguin"* should be *"alguien."*
† This Cuban expression translates literally as "shit-eater" but is roughly equivalent to "idiot," "asshole," or "dumb-ass."

news and I hear some statistic about "Hispanics," I know *que esos comemierdas* are not talking about my fucking people, *coño!*

To be fair, Pepe is actually a puppet—whose real identity is a mystery—and he scrupulously added that his intent was "not to propagate any idea that Cubans are somehow 'better' than any other group." Nevertheless, the post provoked outrage among many non-Cuban Hispanics and was widely reposted and retweeted throughout the Miami Cuban community. The truth is that Pepe expressed a sentiment probably shared by most Cuban Americans. As a prominent Miami businessman explained:

> Cubans are a different breed. We are not Latin American, we are not Caribbean, we are *Cuban*. We are special and distinct from other Latin American groups. And in contrast to other Hispanics in the U.S., I don't consider myself an immigrant. I am an *exile*. I did not leave Cuba for economic reasons. I left Cuba because of Communism. I left because I had to.

Cubans' sense of their own distinctness long predates Castro's takeover. According to Cuban-born professors Guillermo Grenier and Lisandro Pérez, from "colonial times" through the pre-Castro period, Cubans "believe[d] they occupied a unique and privileged position in the world order," creating a strong feeling "of singularity and self-importance in relation to [their] Latin American and Caribbean neighbors."

It can be a little hard for an outsider to understand exactly why Cubans think their homeland is so exceptional. What Cubans say on this point is sometimes contradictory, sometimes disturbing. For example, Cuba is the "most Spanish" country in Latin America; Cuba is the "most African" country in Latin America; Cuba is "perhaps the only country of Latin America in which the impact of its aboriginal cultures . . . has been virtually erased from the national consciousness"; Cuba was uniquely valuable in the Spanish empire because its deep harbor and perfect location made it a "key to the Americas"; and Cuba was the Latin American country most similar to the United States (prior to the Communist revolution).

Whatever the explanation, Cuban exceptionalism can rival that of the Mormons. In his gripping memoir, historian Carlos Eire—who was one of 14,000 children airlifted by the United States out of Cuba in Operation Pedro Pan in 1962—recalls a boyhood teacher in Havana asking the class what Columbus said when he "set foot on Cuba":

> Looking the teacher straight in the eye, Miguel answered: "Columbus said, 'This is the most beautiful land ever seen by human eyes.'"
>
> "Excellent. I'm glad to see . . . that at least some of you have had a proper upbringing and were told about this at home, before we got to today's lesson. You know, what Columbus said is very, very important. It may be one of the most important things ever said about our island, and one of the most true."
>
> "Yes, Cuba is a paradise," said Ramiro, unbidden. . . . "My dad told me that the Garden of Eden was here in Cuba, and that Adam and Eve were not only the first humans, but also the first Cubans, and that the entire human race is kind of Cuban. And he also said that this is the reason we don't have any poisonous reptiles at all."
>
> "True, Ramiro, Cuba is a paradise, and it might very well have been the original Paradise, the Garden of Eden. How many of you have heard this before?"
>
> A good number of hands went up.
>
> "Yes, Cuba is a paradise. There is no other place on earth

as lovely as Cuba, and that's why you should be so proud of your country."

The Cuban superiority complex was particularly strong in the Exile community (those who fled Cuba between 1959 and 1973). The Exiles largely came from a racial and class background that, in Cuba's highly stratified society, already armed many of them with a sense of entitlement and superiority. They were overwhelmingly white, middle to upper class, and included members of some of Cuba's most prominent families. Like the Bacardi family and Robert Goizueta, Eire came from "the *crème de la crème* of Cuban society." His father was a prominent judge and art collector—in their living room hung a painting of Jesus as a child ascribed to Murillo—whose parents had grown up with slaves in their household.

No matter how penniless, the Cuban Exiles who arrived in Miami never identified themselves with—indeed separated themselves from—America's other, relatively poor Hispanic communities. Most children of Cuban Exiles, writes Cuban-born Professor Miguel de la Torre, "are taught by their parents and the overall Exilic community that they are somehow different from other Hispanics, specifically Puerto Ricans and Mexicans. As they are instilled with pride for their heritage, these children are unconsciously taught never to allow anyone to confuse them with those other Latinos/as. In fact, Cubans learn to regard themselves as equal to, if not more advanced than, the North Americans in intelligence, business acumen, and common sense." The enclave the Exiles created would eventually become the vibrant and prosperous Little Havana, and Miami "the second largest Cuban city in the world."

Which brings us back to Pepe. Fifty years after the first Exiles arrived, the proud sense of *cubanía* they brought with them remains powerful among Cuban Americans today, giving them a distinctive self-definition infamous among other Latinos. In 2010, an article widely circulated on the Internet declared that the Cubans in exile are the "only globally transplanted population in the world which (except for the Jews) in more than a third of a century has not lost its identity." The statement is a wild exaggeration—Cubans and Jews are hardly the only diaspora populations to have retained their identity— but nonetheless telling. Nor is the comparison wholly inapt. Like the biblical Jews, Cuban Americans have created an identity for themselves as an exceptional people expelled from their promised land.

THE IRONY IN THESE CLAIMS to exceptionality is—their unexceptionality. All of America's disproportionately successful groups have a superiority complex; in fact most are famous for it. In Asia, everyone knows about the Chinese superiority complex, which is so deeply ingrained that it held like a rock in the face of centuries of decline and is now surging because of China's meteoric rise. In the Middle East, the Iranians' superiority complex is equally notorious.

We'll leave the details of these and our other remaining groups' superiority complexes to later chapters, where we'll show how they function as part of the Triple Package as a whole. Here, we shift focus to a converse but equally important point: the absence of a superiority complex in one of America's relatively less successful groups. No discussion of superiority and inferiority in the United States could be complete without taking on the fraught subject of race, to which we now turn.

FOR MOST OF ITS HISTORY, America did pretty much everything a country could do to create a narrative of superiority—moral and intellectual, political and economic—for its white population and the opposite for everyone else. White supremacy was an equal-opportunity discriminator, targeting all nonwhites from the Native American tribes to Chinese "coolies" to Mexican laborers. For centuries, however, the central corollary of white superiority was black inferiority. Over and over, African Americans have refuted and fought back against the narrative of inferiority that the United States tried to impose on them, but its legacy persists.

Black America is of course no one thing: "not one or ten or ten thousand things," as President Obama's first inaugural poet Elizabeth Alexander has written. There are black families and communities in the United States occupying every possible socioeconomic position. There are longstanding black "aristocracies" up and down the East Coast, in Chicago, and in California, who do indeed have superiority complexes—quite strong ones, going back generations. But that's not the world in which the majority of African Americans grow up.

Sean "Diddy" Combs—rapper, record producer, fashion tycoon, the second-wealthiest African American and the richest person in hip-hop—says this about growing up black in the United States:

> The thing that really shocked people was when Biggie [fellow rapper The Notorious B.I.G.] admitted that being young and black and living in America makes you feel like you want to kill yourself. . . .
>
> If you study black history, it's just so negative, you know. It's just like, OK, we were slaves, and then we were whipped and sprayed with water hoses, and the civil rights movement, and we're American gangsters. I get motivated for us to be seen in our brilliance. And that's the way I always wanted it to be in my fashion. . . . I'd rather show the kids [wealth] than to constantly see the cutaways on television of us just living in the projects.

Culture is never all-determining. Individuals can defy the most dominant culture and write their own scripts, as Combs himself did. Families and whole communities can create narratives of pride that reject the master narratives of their society, or turn those narratives around, reversing their meaning. As one successful African American lawyer put it:

> I was raised to think that just by the fact of being a black woman in America I would always have to do twice as much for half of the credit. My parents expected a lot of me, and when things seemed unfair I was reminded of the struggles and sacrifices of our ancestors. Simply put: your great-great-grandparents were slaves—what obstacle could there be that you cannot overcome?

In any given family, an unusually strong parent or even grandparent can instill in children every one of the Triple Package traits. But when you're not from a Triple Package group, creating a Triple Package family is much harder. It takes more strength, more inner resources. Over time, exceptional individuals could conceivably change society's master narrative about their group. But today, the fact remains, notwithstanding historic breakthroughs like the presidential election of 2008, that the majority of African Americans typically still do not—to put it mildly—grow up with a group superiority complex.

And this cultural fact, in combination with other intractable problems of race and class in America, helps propel a cycle of poverty, as Nicholas Powers hauntingly describes:

> You are born to a single mother who is one of the ten million black people in poverty. On the television, in casual talk or music you

learn by age five that black equals negative and white, positive. Sub-
consciously you see your skin as a weight, a burden. . . . You hear
stories of family relatives jailed for drugs, who you never met
just being released. . . . Your idols are people who look like you in
videos rapping on how to kill, steal, and buy. You don't talk like the
wealthy. You know where to buy drugs.

You graduate but there are no jobs. . . . You try to make a drug
deal, quick cash you think. . . . You get busted again and again until
you are living inside a cell. . . . [Y]ears later you get out. No one will
hire you. No one can let you stay at their apartment, it's against the
rules. You beg on the train sometimes, but run in shame when you
bump into a relative.

Obviously Powers, like Combs, is speaking at one level only about
inner-city black American communities. But African Americans in
every stratum of American society, including the most successful, repeatedly testify to the internal burdens of being black in the United
States and the "sheer force of will" required to succeed "while being
condescended to (under the best of circumstances)."

But it would be ridiculous to suggest that the lack of a group
superiority complex caused African American poverty. Historically,
the causes of black poverty in the United States barely require repeat-
ing: slavery, violence, the breaking up of families, exclusion, systematic discrimination, and so on. Today, a host of factors contribute to
continuing black poverty, including schools that fail to teach, banks
that refuse to lend, employers who won't hire or promote, and the
fact that a third of young black men in this country are in jail or on
parole. Nor does the absence of a group superiority narrative prevent any given individual African American from succeeding. It
simply creates an additional hurdle, a psychological and cultural

disadvantage, that America's most successful groups don't have to overcome.

Moreover, the theory of the Triple Package allows an additional observation, which is much more difficult to deal with—or even to acknowledge.

At least since *Brown v. Board of Education* and the civil rights acts of the 1960s, America's official racial mantra has been equality. You can criticize America's ideal of equality as unfulfilled—some might even call it hypocritical—but its premises are clear and noble. All individuals are equal; every race is equal; every group is just as good as every other.

But the dirty secret is that the groups enjoying disproportionate success in America do not tell themselves, "We're as good as other people." They tell themselves they're better.

In this paradoxical sense, equality isn't fair to African Americans. Superiority is the one narrative that America has relentlessly denied to or ground out of its black population, not only in the old era of slavery and Jim Crow, but equally in the new era of equality, when everyone must kowtow to the idea that there's no difference between different racial groups. "We're a superior people," is the one belief America has consistently and deliberately tried to deny blacks, from the day the first African captives were bought by American settlers, to the days of affirmative action, white guilt, and mass incarceration.

It's one thing for a group with a longstanding superiority complex to pledge allegiance to the idea of universal equality. After all, a group's silent belief in its own superiority isn't fundamentally altered by this declaration; indeed, as they proclaim the equality of all mankind, members of such a group can pride themselves on their generosity and open-mindedness (showing just how superior they really are). It's quite another thing for a group with a long history of inferiority narratives behind it to be asked to pledge allegiance to the same ideal.

Not coincidentally, most of America's great black colleges and universities—Morehouse, Spelman, Howard, and many others—were founded with the mission of fostering the kind of collective pride that other groups in the United States turn into superiority narratives. There's a huge difference in black students' academic experience when the honors students, student body president, and math and science prizewinners are all African American; placing a special curricular emphasis on the accomplishments of African Americans and the African diaspora, as these colleges usually do, can make a big difference too. Studies strongly suggest that the sense of group pride instilled in students at historically black colleges and universities has contributed to their achieving better academic and economic outcomes. It's hardly a coincidence that so many of America's most influential black figures—including Martin Luther King Jr., Thurgood Marshall, Jesse Jackson, Samuel L. Jackson, Spike Lee, Alice Walker, and Toni Morrison—attended historically black colleges.

In the 1960s, the Black Power, "Black Is Beautiful," Black Panther, Nation of Islam, and Afrocentrism movements sought to reclaim black history and rewrite racial narratives, often turning the tables on white superiority, even claiming the mantle of divine chosenness. Malcolm X, when still a follower of the Nation of Islam, urged blacks to believe

> not only that we're as good as the White man, but better than the White man. . . . That's not saying anything . . . just to be equal with him. Who is he to be equal with? You look at his skin. You can't compare your skin with his skin. Why, your skin looks like gold beside his skin.

Louis Farrakhan would say in 1985, "I declare to the world that the people of God are not those who call themselves Jews, but the people of God who are chosen at this critical time in history is you, the black people of America."

Martin Luther King was surely right when he said that "black supremacy," which he saw in the teachings of the Nation of Islam, would be just "as dangerous as white supremacy." But the fact of the matter is that America's most successful groups are cashing in on superiority stories they still believe in and still pass down to their children. This is an advantage that was denied to African Americans—and continues to be denied to them today.

CAN THE EFFECT of a superiority complex be tested empirically? It has been, and the results dramatically confirm that such complexes lift achievement.

Beginning with the pioneering "stereotype threat" studies conducted by Claude Steele and Joshua Aronson, hundreds of controlled experiments have now shown that people's performance on all kinds of measures is dramatically affected by their belief that they're doing something their group is stereotypically good or bad at. Merely reminding people of a negative group stereotype—sometimes even just by requiring them to identify their race or gender on a questionnaire before a test—can worsen their performance.

Thus black students score lower on standardized test questions when their test instructions remind them about stereotypes concerning differential racial performance on such tests. White male Stanford students specially selected for high math ability scored worse on a difficult math test when told that the researchers were trying to understand "the phenomenal math achievement of Asians." Women chess players lost more online games when reminded that men dominate chess rankings—provided they believed they were playing a man (the effect disappeared when they were falsely told their opponent was another woman).

Researchers have also established the opposite effect: stereotype boost. For example, women performed better at visual rotation when told that the task tested a "perspective-taking" skill that women were expected to perform well on (but worse when told that the task was "spatial" and that women were not expected to excel). In a laboratory-controlled miniature golf experiment—we're not making this up—whites did better when the putting test was described as measuring "sports intelligence" than when told that it measured "natural athletic ability." Asian undergraduates scored significantly better on math questions when their instructions stated that "these types of tests measure individuals' true intellectual ability, which historically have shown differences based on ethnic heritage." Crucially, this effect held only for those Asian students who "strongly identified" with their ethnic heritage; for those who didn't, the instruction made little or no difference.

Outside the laboratory, in-depth studies of Asian and Hispanic American high schoolers in Southern California found that Asian students were profiting from a stereotype lift. In a study including Vietnamese as well as Chinese American students, sociologists Min Zhou and Jennifer Lee found that, even after controlling for socioeconomic status, positive stereotypes and ingrained expectations about superior Asian academic achievement—both internal to the culture of these groups and widespread in society at large—significantly contributed to the exceptional academic outcomes of the children of Asian American immigrants.

Since Steele and Aronson published their early findings in 1995, stereotype threat and boost have been among the most widely studied phenomena in social psychology. Perhaps the most astonishing finding in these studies is the susceptibility of individuals to even a single, one-sentence, subtle suggestion of a group stereotype. Imagine, then, the boost you might derive if belief in your group's superiority were part of the culture you grew up in, instilled by your parents, grandparents, and community from the day you were born.

THE CULTURAL BURDEN BORNE by African Americans—along with their susceptibility to stereotype threat—is thrown into sharp relief by the fact that black immigrants are often free of it, at least when they first arrive. In her gripping memoir *The House at Sugar Beach*, the *New York Times* journalist Helene Cooper, a Liberian American, describes the choice made by her ancestors, free American-born blacks, to leave the United States for Liberia—a choice she sees as having armed her, by sheer fortuity of birthplace, with a worldview different from that of many American blacks:

> Because of that choice, I would not grow up, 150 years later, as an American black girl, weighed down by racial stereotypes about welfare queens. . . . Instead, [they] handed down to me a one-in-a-million lottery ticket: birth into what passed for the landed gentry upper class of Africa's first independent country, Liberia. None of that American post–civil war/civil rights movement baggage to bog me down with any inferiority complex about whether I was as good as white people. No European garbage to have me wondering whether some British colonial master was somehow better than me. Who needs to struggle for equality? Let everybody else try to be equal to me.

Culturally and psychologically, "let everyone else try to be equal to me" is worlds apart from "I'm just as good as other people." It's the expression of a superiority complex, and in a society where negative stereotypes are widespread, the confidence it confers on a minority group can be extremely valuable.

This is certainly true of Nigerian Americans. Among West Africans, the stereotype of Nigerians as "arrogant" is common. But the overwhelming majority of Nigerian Americans are not merely Nigerian. They are Igbo or Yoruba, two peoples renowned—and often resented--throughout Western Africa for being disproportionately successful and ethnocentric. The Yoruba boast an illustrious royal lineage and a once great empire. Upstarts by comparison, the entrepreneurial Igbo are often called the "Jews of West Africa." Chinua Achebe, the late Igbo-Nigerian author and winner of the Man Booker Prize, warned of the "dangers of hubris," "overweening pride," and "showiness" among the Igbo—and of other Nigerians' "resentment" against them.

Because of these superiority narratives, the theory of the Triple Package would predict that black immigrants should be able to fend off negative stereotypes better than African Americans can. Again, empirical evidence confirms this result. In a recent study of more than 1,800 students at twenty-eight American selective colleges, even after controlling for socioeconomic factors, first- and second-generation black immigrant students did not suffer the same stereotype-threat effects that American black students did. And as many similar studies have shown, the more strongly black immigrant students identify with their specific ethnic origins, the better they perform.

Newcomers from Africa and the West Indies frequently point to what they perceive as defeatism among African Americans, identifying this mind-set as an obstacle to black success. In the words of one business-school graduate, born in the United States to two Nigerian parents:

Perception is very important, and I think that is what holds African-Americans back. If you start thinking about or becoming absorbed in the mentality that the whole system is against us, then you cannot succeed. . . . Nigerians do not have this. I feel that Nigerians coming from Nigeria feel they are capable of anything. . . . [T]hey don't feel they can't do chemistry or engineering or anything because they are Black.

Superiority complexes can be invidious, but in a society rife with prejudice they can also provide what sociologists have described as "an ethnic armor" enabling some minorities "to cope psychologically, even in the face of discrimination and exclusion."

Helene Cooper had this kind of armor. She experienced numerous racist episodes after arriving in South Carolina as a fourteen-year-old. What made her proof against them, as she tells her story, was the internalized sense of superiority she brought with her from Liberia, where her family belonged to the elite "Congo people," descendants of the freed American slaves who founded the country, as distinct from the "Country people," a derogatory term for "native" Liberians. On top of that, her family were "Honorables," an even higher distinction. "You could have a Ph.D. from Harvard but if you were a Country man . . . you were still outranked in Liberian society by an Honorable with a two-bit degree from some community college in Memphis, Tennessee." Thus when Cooper's freshman roommate, "a white girl from Seagrove, North Carolina, who didn't want to room with a black girl," transferred out of their room, Cooper "called my father and told him, and we both laughed about it on the phone. I felt no outrage. . . . It was completely incomprehensible to me that she could be that much of an idiot."

The way Cooper warded off the blows of American racism—

fending off one brand of ethnocentrism with another—is surprisingly common among America's disproportionately successful groups. Especially among minorities, this strategy tends to function much more as a defensive shield of self-protection than as a weapon of contempt against others.

To CONCLUDE, every one of America's disproportionately successful groups has a deeply ingrained superiority complex, whether rooted in theology, history, or imported social hierarchies that most Americans know nothing about. If a disproportionately successful group could be found in the United States *without* a superiority complex, that would be a counterexample, undercutting the Triple Package thesis.

But superiority complexes are hard to maintain. As one generation passes to the next, group identity and ethnic pride come under attack. All the forces of assimilation work against it, including the homogenizing pull of American culture. For racial minorities, there will be the additional assaults of prejudice and discrimination. America's ideals of equality will come into play as well, eroding superiority claims. Second- and third-generation Americans may begin (perhaps correctly) to see their parents' superiority complex as bigoted or racist and reject it for that reason. As nature abhors a vacuum, so America abhors a superiority complex—except its own. Yet disproportionate success in the United States comes to groups who, in the face of these pressures, find a way to maintain belief in their own superiority.

Superiority alone, however, is merely complacent. The titled nobility of Victorian England had plenty of superiority but were not famously hardworking; even when in financial straits, they would have found employment or entrepreneurship contemptible. For this reason, important as they are, the stereotype boost experiments capture only a piece of the Triple Package dynamic. Only when superiority comes together with the other elements of the Triple Package does it generate drive, grit, and systematic disproportionate group success.

CHAPTER 4

INSECURITY

WE TURN NOW TO the second component of the Triple Package, *insecurity.*

It's been almost two hundred years since the French nobleman Alexis de Tocqueville noticed a peculiar difference between America and Europe. There were places in the "Old World," he said, where the people, though largely uneducated, poor, and oppressed, "seem serene and often have a jovial disposition." By contrast, in America, where "the freest" men lived "in circumstances the happiest to be found in the world," people were "anxious and on edge." They were "insatiable." They never stopped working—first at one thing, then another; first in one place, then another. Americans suffered, said Tocqueville, from a "secret restlessness."

The anxiety Tocqueville described was not spiritual; nor was it a mere wanderlust, a craving for new experiences; much less was it what a future era would call existential. It was material: Americans wanted more. "All are constantly bent on gaining property, reputation, and power." They "never stop thinking of the good things they have not got," always "looking doggedly" at others who have more than they. This thirst for more prevented them from enjoying what they did possess, distracting them from the happiness they ought to have felt, placing them under a "cloud." Ultimately, Americans' anxiety was connected to their "longing to rise."

In short, Tocqueville was describing a people in the grip of insecurity in precisely the sense we have in mind: a goading anxiety about oneself and one's place in society, which in certain circumstances can become a powerful engine of material striving.

Everyone is probably insecure to some extent. Insecurity may be fundamental to the human condition, an inevitable product of the knowledge of mortality or self-consciousness itself. Perhaps this is why people who are insecure are often described as "self-conscious." But insecurity is not all or nothing. You can be more or less insecure, and you can be insecure about different kinds of things. Nor is insecurity a fixed and stable quantum throughout a person's life. Most people are much more insecure during adolescence, for example.

Above all, you don't need to be a member of any particular group to be insecure. But certainly it isn't true any longer, if it ever was, that all Americans feel the goading, insatiable longing to work and rise that Tocqueville described. Some groups' insecurities differ from others, in both kind and intensity.

The great puzzle for Tocqueville was *why* Americans should feel insecure "in the midst of their prosperity." We'll return to this question later. Here, we want to take a closer look at the particular anxieties of America's successful groups. With striking frequency and remarkable consistency, members of these groups are afflicted with certain distinctive insecurities that—in combination with insecurity's seeming opposite, a superiority complex—are especially likely to fuel a drive toward acquisitive, material, prestige-oriented success.

Among the most powerful sources of these insecurities are *scorn*, *fear*, and *family*. We'll discuss these in turn.

. . .

SCORN IS A LEGENDARY MOTIVATOR. ("Hell hath no fury," as the playwright William Congreve didn't quite put it.) All of America's disproportionately successful groups are strangely united in this respect: each is or has been looked down on in America, treated with derision, disrespect, or suspicion. Every one of them suffers—or at least used to suffer, when on the rise—scorn-based insecurity. And to be scorned socially can create a powerful urge to rise socially. Everything can be borne but contempt, said Voltaire.

Scorn, contempt, and above all resentment: these levers of motivation, so well-known in literature, are wholly uncaptured by the useful but bland terms "human capital" and "social capital." In explaining the Cuban American success story, it's invariably pointed out that the Cuban Exiles brought with them considerable "human capital," much more than most other Hispanic immigrants. Which of course was true: about a third of the first wave of Cuban immigrants (the so-called Golden Exiles) had been elites in Cuba, already trained as professionals and executives. But their Cuban degrees and résumés typically counted for little in the United States, where they were forced to take any work they could find, whether semiskilled or unskilled, as factory hands or domestic servants.

Among the groups with the highest human and social capital in the United States are surely the "blue-blooded" WASPs, who still populate America's finest boarding schools and have old-boy networks going back generations. But as members of this class themselves often observe, a culture of lassitude, of nonstriving, seems to have set in at the upper echelons of WASP society. The "less-advertised corollary" to the Protestant work ethic, writes Tad Friend, "held that if you were born to success, nothing further was required."

(According to one of Friend's cousins, "it was customary for the top executives, most of whom had inherited their wealth, to leave their offices between half past three and four; and Father, having spent the morning reading the newspapers, would join them for backgammon, bridge, billiards, and alcohol.") A culture that "once valued education, ability and striving," adds Peter Sayles, "now looks upon these qualities as optional accoutrements. Intellectualism is also frowned upon within these circles—that's for Jews and nerds." After "generations of affluence," says an heir to the Johnson & Johnson fortune, "[l]ots of people just got lazy."

Laziness was something the Cuban Exiles could not afford, either economically or psychologically. Humiliated by Castro, who had called them "the scum of the Earth and worthless worms," many of the Cuban émigrés felt an almost personal mission to prove Castro and Communism wrong by making good in the land of the free. As one Exile puts it, "prevailing in the economic arena" to "bolster their wounded collective pride" became for Cuban Americans "an ideological quest." At the same time, they encountered in Florida the unexpected contempt of discrimination. "When we first arrived in Miami," another Exile remembers, "there were signs on the doors of many houses for rent that said NO DOGS, NO CUBANS. After reading a sign like that, you can imagine how I felt. I had never been discriminated against."

The Exiles' plummet in status was itself an additional blow and extra goad. A successful Cuban American professor—whose childhood memories include a mansion in Havana with a private amusement park in its backyard—describes what it was like for her father to work as a waiter in Miami. "Many times while at work at a restaurant or hotel my father would run into people who knew him back when he was worth millions. It was very embarrassing for him to work in these menial positions, but the embarrassment just propelled him to work harder."

Capital is never enough for success in a capitalist society; drive is equally essential, and resentment can fuel drive. No emotion is cleverer than resentment, as Nietzsche scholar Robert Solomon wrote. It is "the one dependable emotional motive, constant and obsessive, slow-burning but totally dependable and durable." For groups who arrive in America with a superiority complex, the sudden—sometimes traumatic—experience of disrespect and scorn can be a powerful motivator. Iranian Americans are another case in point.

IN THE UNITED STATES—and the West generally—antiquity means classical Greece and Rome. Ancient Persia is seen, if at all, through a Greek lens. Because the earliest Persian rulers left virtually no written histories of their own empire, most of what we know about Achaemenid Persia comes from a very limited number of Greek sources, including Xenophon's *Anabasis*, Aeschylus's *Persians*, and most important, Herodotus's *Histories*. But the Greeks and Persians were bitter enemies, so Greek authors weren't exactly impartial; imagine Saddam Hussein writing *A History of the United States, 1990–2006*. Greek historians refer to Persians as "the barbarians of Asia," frequently portraying the ancient Persian kings as unctuous and decadent.

To the ire of Iranians in America and around the world, Hollywood recently did the same. In the 2007 blockbuster film *300*, the Spartan king Leonidas (played by Gerard Butler) exudes integrity and heroic masculinity, whereas the Persian king Xerxes is depicted as effeminate, corrupt, and monstrously body-pierced. As one Iranian American blogged, "I just can't get over the humiliation that this stupid movie has brought us."

The film was humiliating to Iranians because they identify so deeply with the glories of ancient Persia. Many Americans may not know it, but Iran *is* Persia (Iran was called Persia in the West before 1935), and Persia once ruled the world.

Founded around 500 BC, Achaemenid Persia was "a superpower like nothing the world had ever seen," governing a three-continent-wide territory larger even than Rome's would be. At its height, Persia ruled up to 42 million people—nearly a third of the world's population at the time. This grand history is taught to every Iranian child and underlies what Middle East experts often refer to as the Persian "superiority complex." "All Iranians," says one Iranian American writer, "learn about their great empire." They are taught that "Iran was the equal, if not the better, of Rome and Athens." As Middle East analyst Kenneth Pollack observes, history is "a source of enormous pride" to Iranians:

> It has given them a widely remarked sense of superiority over all of their neighbors, and, ironically, while Tehran now refers to the United States by the moniker "Global Arrogance," within the Middle East a stereotypical complaint against Iranians is their own arrogant treatment of others.

Iran's superiority complex has powered through the centuries, even in the face of stubborn historical realities. Alexander the Great may have conquered the Achaemenids in 324 BC, but Iranians turned this defeat into a source of cultural pride. As explained by Hooman Majd, grandson of an ayatollah and now a U.S. citizen, Iranians think of Alexander as:

> such a brute and ignoramus that he burned magnificent libraries along with the greatest city in the world, Persepolis, to the ground.

But in a good example of the Persian superiority complex, even this villain is shown to have ultimately had the wisdom to recognize the superiority of the Persians by settling down (until his death) in Persia and marrying a blue-blooded Persian. What could be a better endorsement of the greatest civilization known to man?

The Arab conquest of Persia in the mid-seventh century was another blow to Iranian pride, but once again not insurmountable. For hundreds of years afterward, Iranian literature depicted Arabs as "savage bedouins" who eat nothing but "camel's milk and lizards" and "constantly fight among themselves." With European domination of the Middle East, Iranian intellectuals were able to blame Iran's backwardness on the Arab-Islamic destruction of Persia's glorious civilization. In the novels of Iran's most famous modern author, Sâdeq Hedâyat, Arabs were likened to "locusts and plague"; they were "black, with brutish eyes, dry beards beneath their chins, and ugly voices." For much of the twentieth century, Iranian writers "equated Arab domination of Iran (and hence the advent of Islam) with Iran's political and cultural downfall and glorified the pre-Islamic heritage of Iran."

Even today, a millennium and a half after the Arab conquest, Iranians insist they're not Arab—they speak Farsi, not Arabic—and Iranian condescension toward Arabs remains strong. "Iranians don't like being called Arabs," says one Iranian American, self-critically, in an online Iranian newspaper. "If you call them Arabs by mistake, you might as well be calling them trash."

But centuries of subjugation and upheaval have taken their toll. Alongside the Iranian superiority complex, Middle Eastern experts have long observed a "tremendous sense of insecurity that runs right through the Iranian psyche." This insecurity is particularly acute in relation to the West, which as Robert Graham has written, causes Iran to show two very different faces: "with its immediate neighbours . . . a sense of superiority," while "with the West . . . a sense of not wanting to look inferior." Historian and Yale professor Abbas Amanat, a native Iranian, believes this insecurity fuels Iran's contemporary foreign policy; Iran's pursuit of nuclear weapons, he argues, is "in effect a national pursuit of empowerment" driven by "the mythical and psychological dimensions of defeat and deprivation at the hands of foreigners."

Thus, in addition to a deeply ingrained superiority complex, insecurity, too, was part of the cultural inheritance carried by Iranian immigrants to the United States. In America, this insecurity was exacerbated. Status loss, anxiety, resentment, and even trauma have been dominant themes of the Iranian experience in the United States, beginning with the hostage crisis in 1979.

Like the Cuban Exiles, many Iranians suffered a precipitous status collapse when they fled to this country after the Islamic Revolution in 1979. Professionals, scientists, and once powerful figures suddenly found themselves poor and almost unemployable. Particularly for the men, this was a traumatic loss of stature, which Hollywood has twice captured: in the tragic figure of Colonel Massoud Behrani in *House of Sand and Fog* (played by Ben Kingsley), who puts on a suit every morning so that his wife won't know he works as a trash collector; and in the Iranian shopkeeper in *Crash*, whose medical student daughter barely saves him from committing a terrible, racially motivated crime.

Moreover, Iranian Americans have frequently encountered severe prejudice and animus. When American hostages were seized in the U.S. embassy in Tehran, Iranian flags were burned in public, and demonstrators carried signs saying, GO HOME DUMB IRANIANS, 10 IRANIANS EQUAL A WORM, and GIVE AMERICANS LIBERTY OR GIVE

IRANIANS DEATH. This was a bitter irony for those Iranians who had fled to the United States precisely to escape the Islamic Revolution.

Twenty years later, anti-Iranian hostility surged again in the United States when terrorists brought down the twin towers. One young Iranian American woman working in New York remembers thinking, "God, please don't let them be Muslim or Iranian or Arab." After 9/11, Iranian identity "became a stigma to be hidden," causing feelings of "insecurity and even feelings of self-hatred and shame among second-generation Iranian-Americans." For the last decade, Iranian Americans have had to deal with being lumped together with Arab terrorists and branded part of the "axis of evil." Some Iranian parents tell their children not to volunteer their Iranian ancestry; in a hilarious, self-parodying Internet video, confident Iranians refuse to identify themselves as Iranian, claiming instead to be Italian. In the words of Iranian sociologist Mohsen Mobasher, who came to the United States in 1978, Iranian Americans feel like outcasts both from their home country and in their host country, where they have been "stigmatized and humiliated."

Young Iranians in the United States rankle painfully under this scorn and suspicion, as a recent survey of second-generation Iranian Americans in Northern California confirmed. "When I say 'I'm Iranian,' they say, 'You are [an] Iraqi?'" complained one eighteen-year-old. Others reported being called "Middle Easterners" and "hairy terrorist." Negative portrayals of Iranians in the media are especially grating. Persians are "the kindest people," said one fifteen-year-old girl, "but [the media] depicts us as vicious animals and we are not. They put Iran down so much in the news."

All this has led Iranians in the United States to feel an intense need to distinguish themselves, to acquire visible badges of accomplishment and respect. Study after study portrays the Iranian American community as extraordinarily status-conscious, valuing markers of prestige more even than income. "In Southern California," one Iranian American reported, "every Mercedes you see belongs to an Iranian person. They can live in a little shack yet go out and buy themselves a Mercedes and drive around." This status-consciousness is what television shows like *Shahs of Sunset* exploit, while paying much less attention to the hard work and drive that has allowed Iranian Americans to succeed. Iranian Americans often explicitly describe their motivation to succeed—and their parents' determination for them to succeed—in terms of a need for prestige and respect. As one college student put it:

> The American ideal is to study what you want. . . . [T]he Iranian way is to pick something that is guaranteed to make money or guaranteed to be prestigious[;] study is done as a means to an end, not as an end unto itself. The Iranian parent or parents pick a few professions that their peers' children have done well at. . . . Tara's father wanted . . . her to be a dentist. My father wanted me to become a doctor, a pharmacist or at least a nurse.

America has intensified Iranians' attraction to a Persian identity, which among Iranian Americans today not only separates them from Arabs but also from the Islamic Republic of Iran. In one study, 95 percent of second-generation adolescent Iranian Americans said that Persian culture was "a central element in their sense of self." Over 80 percent called themselves "Persian," while only 2 percent said "Iranian." Over half had taken Farsi language classes; many speak Farsi with their parents. Almost invariably, they were taught that Persian culture is older, richer, and deeper than American culture. Second-generation Iranians may be "confused as to exactly what constitutes"

Persian culture, but they are "certainly sure" that it was "far superior" to American culture.

They are also sure that succeeding is a requirement for Persian Americans. "All Iranians are successful," said one boy. Academic achievement is taken for granted. "If you don't get an A," your parents "get upset with you." This drive to succeed—a classic Triple Package mixture of confidence in their abilities and a need to prove themselves in the United States—is widespread and well internalized. There is "no problem," said an eighteen-year-old, putting in "that extra effort to have the bigger house, it is like a form of sacrifice. Do the sacrifice and be ok with it and become successful. [That is how] I look at it." Or in the words of a college sophomore, "We have to prove it . . . we have to carry the torch and show Americans that we are not terrorists."

DESPITE THE RICHNESS and antiquity of their civilization, Indians in the United States, as in India itself, don't focus so much on magnificent-history narratives, at least by comparison to Persians or Chinese. Instead, Indian superiority complexes tend to be rooted in the highly stratified nature of Indian society, with its bewildering array of caste, regional, ethnic, linguistic, religious, and other distinctions. The great majority of Indian immigrants in America come from the upper echelons of India's social hierarchy.* In the

---

* A few immigrants from East India settled in Salem, Massachusetts, in 1804. The first significant Indian community in the United States was made up of Punjabi Sikhs, mostly men, who came to the West Coast as laborers in the early twentieth century, many of them eventually marrying Mexican women. From 1924 to 1965, Indian immigration was largely barred. The initial wave of post-1965 Indian immigrants was overwhelmingly an elite, educated, professional cohort; as of 1975, 93 percent of immigrating Indians were "professional/technical workers" or their immediate family members. Since

United States, however, they suddenly find themselves outsiders, not fully accepted, often the objects of discrimination. As a result, Indian American sociologists have written about an "ethnic anxiety" widespread in their community, and this anxiety helps explain the extraordinary drive that has made them, by any number of measures, the most successful Census-tracked ethnic group in the country.

Although hard numbers are impossible to find, it's widely agreed that most Indian Americans, apart from Sikhs and Muslims, hail from the three highest traditional Hindu "castes" (or *varna*): Brahmans (priests, scholars), Kshatriyas (warriors, royalty), or Vaishyas (merchants). For centuries, some say millennia (everything about caste is controversial, from its origins to its basic nature), caste was all-important in India. Those at the bottom, the out-caste or "untouchables," were barely considered human beings. Fit only for such unclean occupations as removing sewage, cleaning latrines, handling animal carcasses, and disposing of corpses, they were forbidden to touch members of the upper castes; they could even be "required to place clay pots around their necks to prevent their spit from polluting the ground." One step up were groups such as the low-caste Nadars, who could not wear shoes or use umbrellas in the rain; the "master symbol of their inferiority" was the requirement that their women bare their breasts in public.

For obvious reasons, virtually all of modern India's most famous names have been high-caste. Jawaharlal Nehru, Rabindranath Tagore, and (Nehru's daughter) Indira Gandhi were Brahman; Mohandas Gandhi was born into the Vaishya caste. Even today, the high-caste,

---

1990, more Indian immigrants have come from nonelite educational backgrounds. Indian Americans today are a predominantly immigrant population; an astonishing 87 percent of adults are foreign-born. For sources, see the endnotes.

who represent about a third of the population, still dominate Indian society. "Just try and check how many brahmins there are as Supreme Court judges," commented novelist Arundhati Roy in 2000, "how many brahmins there are who run political parties." Although the Indian Constitution has formally abolished untouchability and prohibits caste-based discrimination, high-caste status remains in India a deeply ingrained source of superiority.

But what Westerners know as "caste" only scratches the surface of Indian social stratification. Many Indian subgroups have deep-seated, often cross-cutting superiority claims. Bengalis pride themselves on being India's intellectuals. (Luminaries Amartya Sen and Siddhartha Mukherjee are both from Bengali Brahman families.) Gujaratis, perhaps the largest Indian group in America, are famous not only in India but all over the world as businessmen; in 2008, a Gujarati website noted triumphantly that two of the top three, and four of the top ten Indian billionaires were Gujaratis. Sikhs, who number about 200,000 in the United States, have their own superiority story as, historically, the armed protectors of Hinduism—a reputation so strong that, according to Ved Mehta, the eldest sons of Hindu families in Punjab used to be raised as Sikhs, so that they could serve as protectors of their families. There are also competing north/south snobberies, in which supposedly fairer-skinned "Aryans" look down on the mostly southern, darker-skinned "Dravidians," who in turn think they're smarter and more academically successful. On top of all this, many Indian immigrants in America are graduates of the prestigious Indian Institutes of Technology, which are like the Ivy League—only far more competitive.

Some of these sources of Indian superiority can be, simultaneously, sources of insecurity as well. At least since the beginning of the twentieth century, perceived "Brahman dominance" has provoked resentment and fueled anti-Brahman movements, including among high-caste, non-Brahman leaders. But the sting of Brahman dominance pales in comparison to the insecurity and resentment generated by centuries of British colonial rule, with its famous condescension, high-handed oppression, cooption of elites, and "white man's burden." When Tagore renounced his knighthood after the Amritsar massacre in 1919, he was only one of many Indians to protest the "glaring" "shame" and "humiliation" inflicted by England on his countrymen, who "suffer a degradation not fit for human beings."

Thus most Indian immigrants to the United States bring with them double or even triple layers of simultaneous superiority and insecurity. In America, they run headlong into a totally different set of social hierarchies, in which their old superiority narratives don't matter even as their insecurity intensifies. The experience of exclusion, scorn, even contempt has been a powerful theme for many Indian Americans, at all social levels.

"I always felt so embarrassed by my name," remembers Pulitzer Prize–winner Nilanjana Sudeshna Lahiri, who now goes by Jhumpa; "you feel like you're causing someone pain just by being who you are." The popular claim that America is a "Christian country," built on "Judeo-Christian values," puts Indians on the outside, and Indian Americans are occasionally teased about "worshipping cows." Academic superstars back home, Indian university students sometimes report a very different reception in America:

> If there's an Indian student who just cooked, and then gone to the office, and he's smelling like curry, professors have actually singled people out and told them, 'Why don't you shower?' And 'why don't you spray some cologne or something before you come to class because you smell like curry all the time' and I found that very funny, but at the same time very demeaning as well.

After 9/11, South Asians of all faiths became targets of American anti-Muslim, anti-Arab suspicion and violence. With their turbans, Sikh men are especially likely to be singled out, pegged as "terrorists" instead of protectors. ("You fucking Arab rag-head, you're all going to die, we're going to kill every one of you," a white man shouted at a Sikh in one post-9/11 episode.) Indian cabdrivers report being spat on and called "Arabs."

Racially, Indians in the United States today are regarded as Asian, although for a long time they were considered Caucasian, but the overriding fact is that they are perceived as nonwhite. (In 1923, the Supreme Court managed to hold that a native Punjabi "of high-caste Hindu stock," although perhaps "Caucasian," was not "white" and therefore ineligible for the privileges of "free white citizens.") Many Indian Americans attest to the continuing prejudice their community faces. To attract more business, Indian American hotel owners typically "whiten" their lobbies—hiring whites as desk clerks. "I think you have to," said one owner. "If you're running an upscale hotel with an Indian at the front desk, you know, unfortunately we still live in a society that, uh, doesn't look upon us kindly at times." Another observed that a European manning the front desk was better than an Indian:

> Even foreign is not a bad thing. I was just at a hotel a few weekends ago, and there was a French lady at the front desk. . . . It gives some, you know, flavor to the place; it's cool to me. . . . You're not going to get that [impression] from an Indian, no way.

Such prejudice is a painful reality for many Indian Americans, but once again, the perverse combination of superiority and insecurity can be a powerful motivator. Turning down plum jobs offered to him in India after graduating from IIT Delhi, Rajat Gupta came to the United States in 1971 to attend Harvard Business School. Gupta, according to journalist Anita Raghavan, was descended from "one of India's oldest bloodlines." His father had been British-educated, a distinction enjoyed by only 0.1 percent of India's population during the Raj. And "[i]n a society where skin color was a defining force, both Rajat and his father, Ashwini, were fair-skinned, a clear advantage that afforded them a natural superiority." Yet Gupta, despite being at the top of his class at Harvard, was passed over by every firm on Wall Street, including (initially) the consulting giant McKinsey. Twenty years later, he would be McKinsey's chief executive.

Sociologist Bandana Purkayastha argues that—following a pattern long familiar in immigrant communities in the United States—many Indian American parents "try hard to succeed as 'model minorities,'" demanding high achievement from their children in order to "distance themselves from those who they see as the 'real' minorities," namely blacks and Hispanics. Other South Asian American parents impose similar demands on their children simply because they believe that nonwhites in the United States have to outperform in order to succeed. Either way, "there is an ongoing pressure on their children to be better, smarter, more high-achieving compared to their white peers."

Second-generation Indian Americans' racial attitudes are extraordinarily complex. Young South Asians who date African Americans or Latinos may well experience intense emotional conflict and frustration with their parents. Some who grow up in relatively affluent white suburbs may be more insulated from American racism, but the color line in the United States is difficult to escape, and most young Indian Americans feel an "ethnic anxiety" of one kind or another. Trying to "whiten" their complexion or appearance is a surprisingly common theme among Indian American adolescents. One twenty-three-year-old Indian American professional recalls:

> I always thought that because I was brown, had more hair—didn't look like an Abercrombie boy—I was disadvantaged in the race for women's affections. Therefore, I had to get their attention in alternate arenas. I settled on academic achievement and success at school politics. . . . In my case, the insecurity as a brown American has been particularly acute post-9/11. The imagery of Abu Ghraib (naked brown male bodies sexually humiliated by white U.S. soldiers) resonated strongly with me.

The superiority complex being passed down to second-generation Indian Americans differs from those their parents brought with them. The cultures of America's immigrant communities are never pure incarnations of age-old traditions; they are inevitably reconfigured through their confrontation with American society. In the United States, regional and caste distinctions have become much less significant for Indians (although it can still be an issue in marriages). Many younger Indian Americans grow up rarely hearing caste mentioned in their families. Such distinctions are generally viewed as discriminatory; moreover, the experience of being "Indian" (or "South Asian") in America has united the Indian American community across caste and regional divides.

Instead, Indian Americans have constructed a new "'superior culture' narrative." Part of this new superiority complex has to do with family. Indian American families are extremely tight-knit, with the lowest out-marriage rate among all Asians. According to a recent Pew study, almost 70 percent of Indians in the United States (more than any other Asian group) believe that families from their country are stronger, and embody better values, than American families. A sense of moral or spiritual superiority is often in play as well. Many Indian Americans are emphasizing their Hinduism in new ways (an identity that excludes the approximately 10 percent of Indian Americans who are Muslim as well as the roughly 18 percent Christian, 5 percent Sikh, and 2 percent Jain). A boom in Hindu temple building is taking place across the United States, even though weekly temple-based worship involves a transformation of Hinduism as practiced in India (traditional Hinduism is decentralized, without a fixed liturgical canon or regular services); these temples also serve as social centers, reinforcing a communal Indian identity.

But the most important piece of Indian Americans' superiority complex today may be the feedback loop from the community's outsize success. It's hard to find an Indian in the United States who doesn't know at some level that Indian Americans are hypersuccessful—more successful than whites. As a result, the Indian American community has developed a strong, if unspoken, belief in their "distinctive and superior family/ethnic culture." This sense of superiority, combined with persistent ethnic anxiety, is a classic Triple Package recipe for drive.

INDIAN AMERICANS ARE not the only Asians who experience racism and discrimination, as a recent Pew study confirmed. In *New York* magazine, Wesley Yang offered this description:

> Sometimes I'll glimpse my reflection in a window and feel astonished by what I see. Jet-black hair. Slanted eyes. A pancake-flat surface of yellow-and-green-toned skin. An expression that is nearly reptilian in its impassivity. . . . Here is what I sometimes suspect my face signifies to other Americans: an invisible person, barely

distinguishable from a mass of faces that resemble it. . . . Not just people "who are good at math" and play the violin, but a mass of stifled, repressed, abused, conformist quasi-robots who simply do not matter, socially or culturally.

In case it's not obvious, there's nothing intrinsically empowering about being the object of discrimination and prejudice. For anyone, including members of groups with a superiority complex, the accumulated weight of having the "wrong" skin color or facial features, the relentless tide of stereotypes and media caricatures, can eventually be too much, crushing the spirit. (Although, for the record, Wesley Yang has plainly not been crushed.) And according to many Asian Americans, "racially gendered stereotypes" have made the situation worse.

In the United States, icons of male leadership—whether in Hollywood films, sports, or corporate boardrooms—are almost never Asian. (This is part of the reason so many Asian Americans were euphoric over Jeremy Lin's superstar run with the New York Knicks.) Cultural attitudes play a role too. "White people," said Columbia law professor Tim Wu in an interview with Yang, "have this instinct that is really important: to give off the impression that they're only going to do the really important work. You're a quarterback. It's a kind of arrogance that Asians are trained not to have." (Self-promotion is not encouraged in Confucian cultures, which view modesty, humility, and self-improvement as core virtues.)

Bullying of East Asian kids at school may be on the rise. A 2004 study of New York City public schools found that attacks on Chinese children, both verbal and physical, far outnumbered those against blacks and Latinos. Reasons for this harassment apparently include not only the physical size differences between Chinese schoolchildren and their peers, but also the perception that Chinese students are high achievers favored by their teachers. In a 2010 incident in Philadelphia, twenty-six Asian students were beaten up by non-Asians; thirteen were sent to a hospital.

Episodes of this kind can have cruel and long-lasting effects. Researchers have found that Chinese American students are not only harassed more, but show higher levels of depressive symptoms. But contrary to the harassers' intentions, the targeting of smaller and more studious kids can also in some cases be a motivating spur. One variant of this spur is the proverbial "Some day you'll all be working for me" mind-set. Another, perhaps unique, is Yul Kwon's.

As a child, Kwon remembers being repeatedly bullied and beaten up at school by kids who called him "chink" or "gook." Kwon says he became reclusive and suffered from anxiety disorders. But he went on to graduate from Stanford University and Yale Law School—and then to win *Survivor* in 2006, partly through physical and strategic prowess that earned him the titles "Ringleader," "Puppetmaster," and "Godfather." Selected as one of *People* magazine's Sexiest Men Alive, Kwon told an interviewer, "However improbable it might be that I would end up in front of a camera, the underlying roots of my insecurities help to explain how I got here."

BEYOND SCORN, *fear* is a second source of goading insecurity active in America's disproportionately successful groups—fear of being unable to survive. The dangers are both real and imagined, the fear both rational and irrational, mixing material worries with the deeper anxieties that come from being an outsider.

All of America's most successful groups today are outsiders in one way or another, and all may suffer this fear to some extent, but Jews are the paradigm case. "Fear of being persecuted and even murdered solely for being a Jew resides in just about every Jew's psyche," asserts Dennis Prager, the nationally syndicated talk-radio host. Not all Jews would agree with Prager, but many would probably endorse author Daniel Smith's only slightly less dramatic expression of the same point:

> As a Jew . . . you are forced to live in a world in which you are—for perplexing, unfathomable reasons—not only the object of a wide spread psychotic rage but also, as the very consequence of that rage, urged and expected to associate all the more strongly with your heritage. Indeed, you are urged and expected to act as a kind of personal repository for nearly 6,000 years of collective memory and as a bearer of an entire people's hopes for surviving. . . . You don't want to be anxious? You don't want to be neurotic? Tough. You were born into anxiety.

A history of persecution can produce a variety of psychological reactions, including despair, paralysis, surrender, even shame. Another reaction, however, is a compulsion to rise, to get hold of money or power and cling to it—to be so successful that you either can't be targeted or at least have the resources to escape. Franz Kafka, who was Jewish, wrote in a 1920 letter to a Catholic friend that the Jews' "insecure position, insecure within themselves, insecure among people," makes "Jews believe they possess only whatever they hold in their hands or grip between their teeth" and feel that "only tangible possessions give them a right to live."

Jews are not the only group to fear for their survival. Mormons were long persecuted. Many Cuban Exiles and African immigrants fleeing oppressive regimes had lost everything before they arrived in this country. And in these communities, too, parents often communicate a sense of life's precariousness to their children. Hence the credo of many immigrant homes: they can take away your home, your business, even your homeland, but never your education—so study harder.

We won't say more about this dynamic here, because we're going to explore it in the next chapter. Instead we turn to a very different kind of insecurity, which is even more common among successful groups, but which (like every other kind of insecurity, it seems) is again especially vivid in Jews. This insecurity stems not from persecution, but from parents—although some might say the two are not mutually exclusive.

IMMIGRANT CHILDREN FREQUENTLY feel a need to redeem their parents' sacrifices, but in Jewish culture, there seems to be an additional turn of the screw: never-ending guilt.

The stereotypical "Jewish mother" is by now an old and worn-out figure, but one of her best-loved features was her power to induce guilt. In a famous 1960 Broadway sketch, Elaine May and Mike Nichols portrayed a Jewish mother telephoning her scientist son, who has just helped the country launch its first rocket ship, to complain that he never calls. ("Is it so hard to pick up a phone?") Suffocating devotion is of course on display, but the biggest laugh comes when the mother communicates a different and quite peculiar message. She tells her son she's been so worried about him that she had to see a doctor so her "nerves" could be "X-rayed."

"Mother," he replies, "I feel awful."

"If I could believe that, I'd be the happiest mother in the world," she says. "That's a mother's prayer."

Why does the mother in this stereotype want her son to "feel awful"? In part because she wants him to worry about her; in part because she wants to be needed. At the same time, however, the message is that a child should always feel he's doing something to make his mother unhappy—or to put it the other way, that nothing he does is ever enough. This message is suggested by a different Jewish mother joke recounted by historian Lawrence J. Epstein:

> A Jewish girl becomes president and says to her mother, "You've got to come to the inauguration, Mom." The mother says, "All right, I'll go, I'll go. What am I going to wear? It's so cold. Why did you have to become president? What kind of job is that? You'll have nothing but tsuris." But she goes to the inauguration, and as her daughter is being sworn in by the chief justice, the mother turns to the senator next to her and says, "You see that girl up there? Her brother's a doctor."

There may be a hint of sexism or female competitiveness here, with the mother favoring her doctor son over her president daughter. But the fundamental lesson applies to sons and daughters alike: not even as president should you feel your parents are happy with you. If you become president, you could have been a doctor; if a doctor, you could have been president.

The "domineering" Jewish mother who figures in these jokes and who came famously to life in Philip Roth's *Portnoy's Complaint* seems to have sprung into existence in the second half of the twentieth century. Previously, the Jewish mother had been a nurturing, matzo-ball mother; according to the cultural critic Martha Ravits, the new incarnation was a "construct developed by male writers in the United States in the 1960s."

Parental dissatisfaction was also the province of Jewish fathers. In the desperately poor immigrant communities of early twentieth-century New York, Jewish men tended to be distant from their children, and "disappointment" was "often the only thing [they] could clearly communicate." As countless Jewish authors would later suggest, these fathers' disappointment with their children, especially their sons, may have been a redirection of their own "self-perception of failure," their own embarrassment at having to take menial or peddling work, their inability to provide a better life for their families. Arthur Miller's *Death of a Salesman* movingly portrays one such father. Isaac Rosenfeld's widely read 1946 novel *Passage from Home* describes a son "forever disappointed in my father, just as, I know, he was disappointed in me."

None of which means that for Jewish parents, the child isn't also the center of the parental universe, the apple of their eye. Roth may have captured it best:

> [W]hat *was* it with these Jewish parents . . . that they were able to make us little Jewish boys believe ourselves to be princes on the one hand, unique as unicorns on the one hand, geniuses and brilliant like nobody has ever been brilliant and beautiful before in the history of childhood—saviors and sheer perfection on the one hand, and such bumbling, incompetent, thoughtless, helpless, selfish, evil little . . . *ingrates*, on the other!

Needless to say, not all Jews are insecure, and Woody Allen notwithstanding, not all Jewish stereotypes are racked by doubt, neurosis, and anxiety. "Jewish American Princesses," for example, are supposed to be just about the opposite of insecure. Many Jewish American mothers and fathers today affirmatively object to making kids feel they have to achieve in order to satisfy their parents. In fact, Jewish insecurity has been lessening on several fronts—Roth's generation may have been the last to feel that their Jewishness made them existential outsiders in the United States—which, as we'll discuss in a later chapter, could portend a decline in Jewish success.

. . .

THE JEWISH CASE may be the most spectacular—or at least the most satirized, and certainly the most psychoanalyzed—but the phenomenon of children feeling they must succeed in order not to disappoint their parents is of course far broader, with special prominence in immigrant families. Children who have seen their mothers and fathers working double shifts as maids or restaurant workers, devoting all their savings to their kids' education, often feel an internal pressure to live up to their parents' dreams and expectations—to make their parents' sacrifices worthwhile.

Florida senator Marco Rubio, speaking at the 2012 Republican National Convention, poignantly (if self-servingly) recalled his Cuban immigrant father "who had worked for many years as a banquet bartender. . . . He stood behind a bar in the back of the room all those years, so one day I could stand behind a podium in the front of a room." Rubio amplifies on this theme in his autobiography. When, as a young man, he thought of how much his parents had sacrificed for him, he would pray at night that he could "make them proud. I pray that You let them live long enough so that they can know that all their hard work and all their sacrifices were not in vain."

In a study of over five thousand immigrants' children, researchers found "again and again" the same anxiety about redeeming parental sacrifice, driving the second generation to excel:

> Perceiving the sacrifices made by their parents, ostensibly on their behalf, not a small amount of guilt tinges the children's sense of obligation toward their parents and spurs their motivation to achieve—a dynamic that in turn can give immigrant parents a degree of psychological leverage over their children.

This anxiety can last a lifetime, even after one's parents have passed away.

In Asian families, the pressure to excel is often intensified. Particularly in Chinese, Japanese, and Korean immigrant families, children are frequently taught that "failing"—for example, by getting a B—would be a disgrace for the whole family. When asked open-endedly in a recent study what makes children do well in school, almost a third of Taiwanese American mothers—compared with zero white American mothers—brought up family honor. Harvard sociologist Vivian Louie observes that in traditional, Confucian-influenced cultures, "ancestors are ever present and 'the individual alive is the manifestation of his whole Continuum of Descent.' When it comes to achievement, then, the accomplishments of the individual are strongly grounded in familial obligation and prestige."

Family honor is only part of the picture. It's important to remember that for Asian immigrant parents (as for so many other immigrant parents), high grades, prizes, brand-name schools, and other visible markers of achievement are a child's best—perhaps only—protection in an uncertain, competitive, potentially hostile world. Hence the stereotype of the Asian parent fixated on rankings. "Harvard #1!" is the Asian parent's constant reminder in this stereotype, which, for many, hits painfully close to home. "Why just an A, not an A plus?" is a query regularly heard in some Asian American households. As one second-generation teenager in a study of Korean American college students put it, "Korean parents are like you have to do this this this to be successful . . . you have to go to medical school or law school and study study study. They think the best colleges are Harvard, Yale, and Princeton. I am not saying white people don't stress education, but Koreans . . . they take it to another level." There are Asian American kids actually named "Princeton Wong" or "Yale Chang"—and not because their parents are alumni.

With an explicitness that would horrify Western parents steeped in self-esteem literature, Asian parents often deliberately put pressure on their children by making pointed comparisons with the wildly successful kids of other family members and friends. All this translates into an insecurity of just the kind we've described: a multilayered anxiety about one's material position, a persistent feeling of not being good enough, where "good enough" is defined in highly conventional fashion, in terms of grades, rankings, wealth, or prestige. From a young South Asian:

> When I was growing up my parents thought I was a bad girl. I had good grades, but it was never good enough. I used to envy my (white American) friends; their parents were so nice to them. Like this one girl made brownies with drugs in it, and her mother only made her write a poem of atonement.

One Asian American who grew up in the Los Angeles projects, the daughter of poor garment workers, went on to graduate from Cal State. Even as she pursued a graduate degree, she felt she was "average" but not "great"; despite her extraordinary upward mobility, her frame of reference remained the kids of her parents' friends "who had gone to Yale." Likewise, a Vietnamese American girl who graduated third in her class with a 4.21 GPA, when asked if her parents were proud, said, "It would have been better if I was first or second."

The same self-dissatisfaction has been found repeatedly in Asian American students, from middle schools to top universities. Asian American students regularly report low self-esteem despite their academic achievements. Indeed, across America, they report the *lowest* self-esteem of any racial group even as they rack up the *highest* grades. In a study of almost four thousand freshmen at twenty-eight selective American colleges, Asians said they were the least satisfied with themselves of any racial group; blacks reported the highest positive attitude toward themselves, followed by Latinos, then whites, then Asians. Since the 1960s, American educators have accepted, and shaped our schools around, the idea that when minority students—indeed, any students—do poorly, the chief culprit is low self-esteem. The facts indicate otherwise.

INSECURITY CAN MOTIVATE. We've talked about some of America's best-known successful groups. We'll close with a more obscure example: Lebanese Americans.

Joseph J. Jacobs, born in Brooklyn in 1916 to poor Lebanese immigrant parents, founded one of the largest engineering and construction companies in the world. In his autobiography, Jacobs describes how he and his fellow Lebanese Americans always felt looked down on by America's WASP elite, which was especially stinging for a people so "intensely proud" of their heritage.

The Lebanese, he writes, are "descendants of the ancient Phoenicians," a Semitic people who, like the modern-day Lebanese, were famous for being commercially successful wherever they went. Both Greeks and Romans singled out the Phoenicians for their superior intelligence, commercial acumen, and master seafaring skills. They appear in the Odyssey as "greedy knaves" with great "black ships," and Cicero says it was they, with their superior "cleverness," who first introduced to Greece "greed and luxury and the unbridled desire for everything." Evidently well rounded, the Phoenicians were also credited with inventing the alphabet, arithmetic, and glass.

But Jacobs, as a Lebanese, was not only a descendant of the multitalented Phoenicians; Jacobs was a Christian Lebanese, a Maronite, as were most Lebanese Americans of his generation. The Maronites, as Jacobs puts it, "like to claim that they descend from the original

Christian disciples." In Lebanon, where Christians until recently dominated the country politically, economically, and socially, the Maronites' superiority complex is well-known. Even today, Muslim Americans from Lebanon sometimes report that their parents "taught us to be ashamed of ourselves, not to be proud. You know, to look down on yourself, and look up at the Christians."

All this gave Jacobs, though a peddler's son, an outsize superiority complex, which, combined with what he perceived as WASP superciliousness, goaded him and other Lebanese Americans to succeed:

> Having no one to speak for us, no family, no references, we worked doubly hard to become accepted, to demonstrate our worth to the establishment, those scions of the *Mayflower*, those third-generation WASPs who looked down on the "Syrian" immigrants and, by their very presence, seemed to taunt us with our lack of status. . . . The rejection we felt was exaggerated but was part of our motivation. . . . [E]nvious of the easy familiarity of our American peers, we were spurred on by dissatisfaction and the need "to show them."

Jacobs's insecurity and drive had another source as well—his mother. Growing up, he recalls, there was no escaping his mother's "driving ambition" for her children: "[W]e should and could be better than anyone else, better even than we ourselves may have wanted to be." The pressure, he writes, was "confusing and traumatic"; his eldest brother rebelled. Jacobs, however, internalized her demands. "Much of my early uncertainty and subsequent aggressive drive were undoubtedly spurred by a need to please my mother"—to satisfy her "demand for success and the admiration of the community."

Nor was Jacobs's mother unique in the Lebanese American community. He recounts how, when his friend and fellow Lebanese entrepreneur Alex Massad—described by *Fortune* as one of America's toughest businessmen—first accepted a position with Mobil Oil, Massad's mother lamented, "[W]hy don't you go into business for yourself? Why don't you open a store?" But Massad stayed on, rising quickly to become one of Mobil's top executives. Finally, on one visit home,

> Alex . . . announced triumphantly, "Mama, I *have* bought a store." Her elderly face brightened at the news. At last! Alex had taken her advice; her son would finally be judged a success in her community. . . . "I bought Montgomery Ward!"
>
> Her smile changed to a disappointed frown. She was unimpressed and said, with despair, "It's not the same thing. I meant your *own* store!"

As Jacobs describes it, he and his second-generation Lebanese American friends "were doubly driven to succeed" to "show our parents" and to show the world.

In Jacobs's account, we find everything we've discussed in this chapter: a group superiority complex combined with acute insecurity generated by both perceived social scorn and unrelenting parental pressure. Over and over in America's most successful groups, these forces converge to produce an intense chip on the shoulder, an "I'll show everyone" mentality—and ultimately, disproportionate group success.

INSECURITY AS A KEY TO SUCCESS—not exactly the lesson taught by America's self-esteem-centered culture or its "just learn to love yourself" popular psychology. But for an individual to be driven, something has to be driving him: some painful spur, some goading lack. Disproportionately successful groups disproportionately feel this insecurity.

Thus the second element of the Triple Package deals another blow to modern American mantras. Insecurity is the enemy—a pathogen targeted for obliteration—in popular and therapeutic psychology, not to mention contemporary parenting. There's an ocean of difference between zealously protecting self-esteem and actively promoting insecurity; between "Just be yourself" and "You're not good enough"; between "You're so amazing—Mommy and Daddy will always be here for you" and "If you don't get straight As, you'll let down the whole family and end up a bum on the streets." Insecurity is not supposed to lead to success, but in America's most successful groups, it seems to do just that.

CHAPTER 5

IMPULSE CONTROL

IF YOU ASK SUCCESSFUL PEOPLE why they're successful, it's striking how often they'll bring up episodes of failure. "I've missed more than 9,000 shots in my career," Michael Jordan once observed. "I've lost almost 300 games. 26 times I've been entrusted to take the game winning shot and missed. I've failed over, and over and over again in my life. And that is why I succeed."

How people respond to failure is a critical dividing line between those who make it and those who don't. Success requires more than motivation, more even than a deep urge to rise. Willpower and perseverance in the face of adversity are equally important.

Led by social and developmental psychologists such as Roy Baumeister, Carol Dweck, and Angela Duckworth, a large and growing body of research has demonstrated that the capacity to resist temptation—including especially the temptation to quit when a task is arduous, daunting, or beyond one's immediate abilities—is critical to achievement. This capacity to resist temptation is exactly what we mean by impulse control, and the remarkable finding is that greater impulse control in early childhood translates into much better outcomes across a wide variety of domains.

This finding was first made—stumbled on, actually—by Stanford psychologist Walter Mischel in his famous "marshmallow test" of the late 1960s. Trying to determine how children learn to resist temptation, Mischel began putting treats in front of three- to five-year-olds. The children were told that they could either eat their chosen treat (often a marshmallow) or, if they waited a few minutes, get another one too. Children who held out for fifteen minutes received a second marshmallow.

A majority ate up; only a minority held out. The great surprise, however, came years later. Although it wasn't part of his original plan, Mischel followed up on the roughly 650 subject children when they were in high school. It turned out that the children who had held out were doing much better academically, with fewer social problems, than those who hadn't.

Now confirmed by numerous studies, the correlation Mischel discovered between impulse control and success is nothing short of jaw-dropping. Kids who "passed" their marshmallow test, waiting the full fifteen minutes, ended up with SAT scores 210 points higher than those who ate up in the first thirty seconds. For college grades, impulse control has proved to be a better predictor than SAT scores—better even than IQ.

In the most comprehensive study to date, researchers in New Zealand tracked over a thousand individuals from birth to age thirty-two. Controlling for socioeconomic status, intelligence, and other factors, the study found that individuals with low impulse control as children were significantly more likely to develop problems with drugs, alcohol, and obesity; to work in low-paying jobs; to have a sexually transmitted disease; and to end up in prison. Those with high impulse control were healthier, more affluent, and more likely to have a stable marriage, raising children in a two-parent household.

There's been another finding too, of equal if not greater importance. Willpower and perseverance can be strengthened. That's where culture comes in. Cultivating impulse control in children—indeed in anyone, at any age—is a powerful lever of success.

But not by itself: impulse control by itself has nothing to do with academic achievement or moneymaking. The Pennsylvania Amish have as much impulse control as any group, denying themselves electricity and every modern convenience. But they aren't academically or economically overachieving, because their culture points them away from those goals.

In other words, impulse control in isolation is mere asceticism. As always, it's the fusion of all three Triple Package elements that creates an engine of economic success.

On October 8, 2012, spectators in Lower Manhattan, as well as Internet viewers around the world, could watch a man complete his seventy-second straight hour of standing atop a narrow, twenty-foot-tall pillar. He also happened to be attached to wires jolting his body with a million volts of electricity. He never fell.

The media were unimpressed. After all, this was the same man who had remained encased in a block of ice for over sixty hours in Times Square (it was a month before he regained the use of his legs), the same man who had fasted for forty-four days inside a plexiglass coffin suspended above the Thames (to the point where his body began consuming its own organs). The world had grown tired of David Blaine's feats. "[F]or once," yawned an English newspaper, "it would be fun to see Blaine do something properly exciting—something which doesn't consist of doing nothing for hours on end."

Readers of Kafka will remember a similar fate befalling his "hunger artist," who, once celebrated, ends up ignored and forgotten, starving himself to death in a cage on the outskirts of a circus. What

Blaine and Kafka's protagonist share is a belief in the virtuosity of endurance—an idea familiar in Western thought at least since the Stoics espoused it twenty-five hundred years ago. The Stoic sage demonstrated his moral perfection by mastering his passions—restraining his impulses—even in the face of extreme hardship. David Blaine is Stoicism on steroids.

The significance of Blaine, Kafka's hunger artist, and Stoicism for our purposes is that they all exemplify a certain fusion of superiority and impulse control—a belief in the superiority *of* impulse control, an achievement of superiority *through* impulse control—that can generate a self-fulfilling cycle of greater and greater endurance. In the extreme case, people with superiority complexes built up around impulse control end up on top of poles enduring electric shocks. But in milder forms, this dynamic is another potent Triple Package specialty. Many of America's most successful groups build impulse control into their superiority complex, priding themselves on their capacity to endure adversity and working hard to instill this capacity in their children.

This is certainly true of Chinese immigrants—and two-thirds of today's Chinese Americans are immigrants. In fact, Chinese Americans not only tend to value impulse control. They have all three elements of the Triple Package, in spades.

IN ASIA, EVERYONE KNOWS about China's massive superiority complex. China's very name—*Zhongguo*, often uncolorfully translated as Middle Kingdom—connotes in Chinese "center of the world" or "center of civilization." Its emperor was the "Son of Heaven."

If all societies are to a certain extent ethnocentric, special circumstances combined to make China an extreme case. Surrounded by natural barriers, China for millennia experienced minimal contact with the great civilizations of Europe, India, and the Middle East. Instead, China's neighbors were scattered, nomadic, and tribal. For most of its history, China was the largest unified population in a vast region, by far the most urbanized, the most literate, and the most technologically, politically, and culturally advanced. As the historian John K. Fairbank put it, "Since ancient China began as a culture island, it quite naturally considered itself superior to the less cultured peoples roundabout, whom it gradually absorbed and assimilated."

Like the Greeks, the Han Chinese viewed all foreign peoples as "barbarians." Those who resisted Chinese culture were "raw" (*sheng*); those willing to accept it were more generously referred to as "cooked" (*shu*). Cambodia, Central Asia, Indonesia, Japan, Korea, and Vietnam were all China's "vassal states" whose rulers were required to pay "tribute" to China and technically expected to prostrate themselves before the emperor. Because Chinese superiority was understood to be moral and cultural, it could be maintained even in the face of military weakness. In 1260, the Mongols conquered China, but Chinese historians—taking a page from the Persians—later recast this defeat as a victory. The Mongols, they observed, were ultimately "sinicized," adopting Chinese names, rituals, customs, and values.

And when the Ming Dynasty (1368–1644) overthrew the Mongols, China rose to heights unprecedented in world history. Its population of over 100 million was at least double that of all the European states put together. While Europe was creeping out of the Middle Ages, China had paper money, printing presses, gunpowder, and the world's largest iron industry. At its height, the Ming navy consisted of more than 3,500 vessels, including leviathans that could accommodate perhaps 1,000 passengers and carry 400 times the cargo of their largest European counterparts. By contrast, the "royal fleet" Henry V assembled to conquer France consisted of four fishing boats each able to ferry one hundred men across the channel at a time;

Christopher Columbus's ship, the *Niña*, was about the size of a Chinese treasure ship's rudder.

When the Ming completed their new capital, the Forbidden City, 26,000 guests enjoyed a ten-course banquet served on fine porcelain; at the wedding feast of Henry V and Catherine of Valois, 600 guests ate salted cod on "plates" of stale bread. In this period and well after, the Chinese belief that other kingdoms and peoples were subservient to the Son of Heaven extended far beyond Asia. As late as 1818, even when the powerful empires of Europe were on the verge of taking over China, Qing dynasty imperial records refer to England, Holland, and Portugal as countries of "barbarians that send tribute" to China.

Today, Chinese kids—in America as in the rest of the world—are typically raised on a diet of stories about how Chinese civilization is the oldest and most magnificent in world history, how China was advanced far earlier than the West, how countries like Japan derived everything from China (including sushi, origami, and kanji writing characters), how the Chinese language is the subtlest, most complex, and most sophisticated—and ditto Chinese cuisine. One Chinese American writer remembers that his father

> used to take every opportunity to show how advanced the Chinese civilization had been in ancient times. He always sneered at the American history books that I studied for class: "You kids have it so easy here in America! You only have to study 200 years of American history! We had to study over 5,000 years of Chinese history in school! The great Chinese civilization has been around far, far longer than America!" Whenever we went out to eat Italian food, he would tell me about how the Chinese had first invented pizza and spaghetti, not the Italians, and how the Chinese had invented just about everything else thousands of years before the Europeans did.

Andrea Jung, former CEO of Avon Products, learned a similar cate-chism as a child: "[I]n our house, everything important in life came from China, was invented in China and owed all to the Chinese." This exaltation of Chinese history, Jung says, gave her an important "advan-tage" growing up. "For me, my Chinese heritage has been a wonderful compass, a fortuitous gift, and an enormous source of strength."

But the long era of Chinese stagnation and decline, which histo-rians like Paul Kennedy date to the fifteenth century, and especially the period of European domination, have created deep insecurities alongside this superiority. The unequal treaties imposed on China by the Western powers, with their "treaty ports" and "foreign conces-sions," brought shame and humiliation. China was opiated, carved up, and dished out, with the British, French, Germans, Americans, Rus-sians, and Portuguese each getting a piece. Allegedly, there were signs saying NO DOGS AND CHINESE ALLOWED in China itself. And any resentment China may harbor toward the West pales in comparison to its attitude toward Japan, which invaded and occupied China in the 1930s, massacring an estimated 300,000 and raping tens of thou-sands of women. "Never will China be humiliated again," declared Mao Zedong in 1949, and even in present-day booming, ever-more-confident China, this collective chip on the shoulder—this need to "show" both West and East—remains a powerful, palpable force.

Thus when, after 1965, Chinese immigrants in the hundreds of thousands began arriving in America, the mentality many brought with them was infused not only with superiority, but with insecurity as well. Once here, that insecurity only increased. Like so many new-comers to America, Chinese immigrants often experience "the twin burdens of foreignness and marginalization"; they're the ones with the funny accents and wrong clothes, who don't know where any-thing is or how things work. Some worry about how they're going to earn a livelihood. One Chinese American recalls his father insisting

that he excel at school so "you won't have to work like me in the restaurant, sweating." Many find it rankling that Americans look down
on them. As a Taiwanese American woman put it, "we don't speak
English that clearly, okay. . . Actually, most Chinese people who
came here to study, came for higher education, they are the elite of
our society. But people here, they don't know. They talk to you like
dumb-dumb."

Then there's the racial dimension discussed in the last chapter.
Anti-Chinese hostility has a long history in the United States, going
back to the Chinese Exclusion Act of 1882. A 2012 Pew study found
that only 24 percent of Chinese Americans felt that discrimination
was "not a problem"; 21 percent said they had personally experienced
discrimination. There are signs, moreover, that China's rising global
power, together with resentment at Chinese Americans' strong performance in schools—they "work too hard," it's sometimes charged—
may start to rekindle anti-Chinese animus in the United States. This
in turn could intensify Chinese American insecurity.

In classic Triple Package fashion, the combination of superiority
and insecurity in Chinese Americans often produces a "how dare
they look down on me?" mentality and an iron will to prove oneself.
A Chinese American woman who arrived in the U.S. as a girl in the
1990s still remembers the sting she felt when a teacher told her
mother that "she'll never learn English." Now a professional, she says,
"I was so driven to prove her wrong that I taught myself English by
reading books, eventually receiving a perfect score on the SAT Verbal
section."

At the height of his NBA fame, Jeremy Lin described the same
need to prove himself:

> I know a lot of people say I'm deceptively athletic and deceptively
> quick, and I'm not sure what's "deceptive." But it could be the fact

that I'm Asian-American. But I think that's fine. It's something that I embrace, and it gives me a chip on my shoulder.

Lin went on: "I'm going to have to prove myself more so again and again and again, and some people may not believe it."

In a recent study of Chinese communities in New York City, similar themes appeared over and over. "Being Chinese will never be a plus, it will always be something against you," one Chinese American respondent remembers her physicist father telling her. "My father wanted us to be smarter, you have to be smarter than other people, than the majority, because unless you have 110, they're not going to take you." Chinese American novelist Anchee Min writes of telling her high-school-age daughter: "It's better that you're taught the truth. If America honored race-blind competition, the nation's elite colleges would be filled with the hard-working Asians. Have you heard of the American saying 'You don't stand a Chinaman's chance'? Chinaman, that's who you are." A Chinese immigrant living in Queens warned his son: "You have to work harder. . . . they're going to look down on you. . . you have to be much better than whites."

How do Chinese immigrant parents expect their kids to do "better than whites"? Through impulse control.

THERE'S A CHINESE TERM, *chi ku,* that a billion Chinese people know, that every Chinese immigrant in America knows, and that probably all their children are deeply familiar with too. Translated literally, *chi ku* means "eating bitterness" and refers to the capacity to endure hardship, which, along with perseverance and diligence, is a cardinal Confucian "learning virtue." For a thousand years, these virtues—which include discipline, self-control, resisting the temptation to complain, wallow, or give up—have been fundamental elements of child rearing and education in China and Confucian-influenced societies.

A cultural chasm separates "learning should be fun" from the idea that learning anything well requires hardship, exhaustion, even pain. Visit just about any primary school in China, Taiwan, or Singapore, and rather than children running around exploring and being rewarded for spontaneity and originality, you'll find students sitting upright, drilling, memorizing, and reciting excruciatingly long passages. Calligraphy, part of the basic school curriculum, is all about patience, mastery, and exactitude. After school and on weekends, it's rare for even very young children to "hang out with friends." Far more typical are hours of additional study and tutoring, followed by more hours of highly regimented music or sports practice. "You need to make more effort, not be so lazy," was a typical comment by a Chinese mother to her child (recorded in a large 2008 comparative study of parenting)—and considered not harsh, but supportive.

That's in Asia. What about the Chinese living in the United States? In theory, the roughly 66 percent of Chinese Americans who are first-generation immigrants could be opposed to traditional Chinese parenting. After all, many were themselves raised by exceedingly harsh parents, and today a growing number say that it's important for parents to express love and to give their kids more freedom. (Even in mainland China itself, some parents today have swung to the other extreme, coddling their only children, raising concerns about a generation of pampered, spoiled-brat "little emperors.") But study after study—and there's been a prodigious amount of research—confirms that overall, Chinese immigrants parent far more strictly than non-Asian Americans, making discipline, high expectations, perseverance, and self-control part of their children's daily lives.

The Asian American child pianist or violinist is so common as to be a cliché; what puzzles many is not only how Asian American parents produce such accomplished musicians, but *why*. What's the point of the grueling hours of practice? In one study, the Asian immigrant parents (primarily Chinese and Korean) of Juilliard Pre-College students identified discipline, focus, hard work, and self-sacrifice as typically "Asian" qualities, but not in the sense of being innate. On the contrary, these were qualities that had to be "learned" and "repeatedly practiced," qualities the parents "hoped to instill in their children *through* classical music training," starting at a very early age. "If they learn an instrument, they don't have any big problems when they are learning science, mathematics," said one parent. "Because if you have this kind of discipline, the concentration to overcome obstacles, do something over and over again, then you will know, "'Oh! If I put in every kind of effort, then I will win.'"

With numerous exceptions, of course, Chinese immigrant parents—whether affluent or poor, well educated or not—tend to force on their children rules and regimens very different from those prevailing in white American households. Here are just a few typical findings from empirical studies. Chinese American preschoolers and kindergartners engage in a "focused activity" at home about an hour a day, compared to less than *six minutes* per day for white American children the same age. Chinese American children watch about one-third less television than white Americans. Asian kids are more likely to attribute success or failure at school to how hard a student works; by contrast, white American kids are more likely to attribute it to innate talent, luck, or teacher "favoritism."

Chinese immigrant mothers tend to make their children do extra work at home, while the majority of white American mothers are more interested in building their children's social skills and self-esteem. Zappos founder Tony Hsieh recalls that his Taiwanese immigrant parents made him start preparing for the SAT in sixth grade. He was supposed to practice the piano, violin, trumpet, and French horn a half hour each per day; on weekends and "[d]uring the summer, it was an hour per instrument per day." He eventually figured out how to play a tape-recording of himself practicing to fool his parents while he read magazines instead.*

Studies also confirm that Chinese immigrant parents tend to be much stricter than Western parents when it comes to their children's social behavior, often imposing restrictions on "hanging out" with friends after school, talking on the phone, and, as their kids get older, dating and going to late-night parties. In a study of both suburban and working-class Chinese immigrants, sociologist Vivian Louie found that "[t]hings that other children took for granted, like sleepovers, were matters of contention" in Chinese immigrant households. As one of her respondents recalled: "for the longest time, I wasn't allowed to sleep over at someone's house. God forbid, like, I get corrupted there [laughs]. But that was when I was young. As I got older, they just really had like, academic expectations." Another respondent said that it made her feel better to know that her Chinese American friends had to struggle with similar rules. "I have a friend,

---

* One recent study was widely reported in the popular media as having shown that strict parenting is "uncommon" among Chinese Americans and that "supportive" parenting is the norm. The actual study, however, is not inconsistent with the well-established finding that Chinese immigrants tend to impose much more discipline and higher expectations on their children than most American parents. The study's methodology involved no inquiry into specific child rearing practices such as what the children did after school, what grades were considered acceptable, or how many hours of study or practice were required; nor was there any comparison between Chinese parents and any other parents. Rather, Chinese American parents were asked to rate themselves according to selected phrases such as whether they "act caring," "listen carefully" to their children, "act supportive," "insult or swear," "change the subject whenever the . . . child has something to say," or "punish . . . with no or little explanation." Based on answers to these questions (by parents and children), the researchers classified a majority (or near-majority) of the parents as "supportive" relative to the other Chinese parents in the survey. But whether these parents were strict by American standards—imposing greater discipline or requiring more intense work habits—is a question the study did not address. For sources and more discussion, see the endnote.

she's older than me, she can't go anywhere, and she's still hiding her boyfriend from her mother. She can't hang out, you know, stuff like that, till twelve, one."

Because of the unpleasantness of imposed impulse control, inter-generational conflict is a frequent theme in Chinese immigrant households. Louie's respondents often referred to arguing with their parents, concealing from them, or having them "flip out." Such parenting can cause frustration, alienation, rebellion, and in extreme cases, as we'll discuss in the next chapter, deeply damaging effects. Nevertheless, the consistent finding is that the children of Chinese immigrants develop habits of discipline and persistence exceeding those of their non-Asian peers—habits they themselves generally see as giving them an "educational advantage." Although Chinese immigrant parents may rarely say so aloud, the willpower they instill in their kids is wrapped up with a sense of the superiority of the Chinese work ethic and ability to endure hardship. As a Juilliard parent put it, the reason high school orchestras throughout the United States are disproportionately Asian is that:

> American kids, they do not have the discipline. They have parties, they have play dates, they have soccer games, outside activities, lots of activities outside. But for music you have to sacrifice. American families cannot do it. . . . So there is orchestra group and band group. Band you don't have to practice a lot, but it's easy to play. . . . You see more American kids in band group. But Asian kids more statistically, you see in orchestra, because they have discipline so well.

Parents of kids who make it into Juilliard may be extreme even in Asian culture, but these views are hardly atypical. In a nationwide Pew survey, over 62 percent of Asian Americans agreed that "most

American parents do not put enough pressure on their children to do well in school." In fact, Chinese (and other Asian) immigrant parents sometimes perceive American parents as lacking in the very work-ethic and diligence they are trying to instill in their children. As a respondent in one study put it, "American parents, if they too hard have to sacrifice, they don't do it. They don't like it, they're kind of children themselves a little bit. . . . So how can I say, we always count first our children. Whatever is good for them, we like to provide for them as much as we can."

Thus cultivating impulse control in their children is woven into the Chinese superiority complex, and the result can be—often to the distress of the children—a full-court, high-pressure Triple Package press that becomes an ingrained part of everyday family life. One young Taiwanese American professional's recollections are typical:

> There wasn't much praise when my brother and I returned home from school with top marks; that was the expectation. After all, we were the only Chinese (or Asian) students in our classes. . . . We hadn't yet been compared against our real competition. . . . There was a lot of "good job, but don't forget that hundreds of Chinese kids younger than you in Taipei, Beijing, and Palo Alto are doing better and working harder"-style commentary. Even at the age of 8, there was a sense that we were already behind. Conversations at the dinner table . . . read like status updates of outstanding Asian kids our family knew. So-and-so's son just got into Stanford; the daughter of an old roommate just graduated first in her class. . . . Embedded within these statements was an assumption that Chinese students (Chinese-American and Chinese nationals) were more diligent, and more likely to succeed in school and beyond.

All three Triple Package forces are on vivid display here. The idea that the children belong to a culturally superior group and are therefore more capable is simply assumed, even as the children are told that they are "falling behind." The only successes that really count are conventional forms of prestigious achievement, like getting into Stanford, graduating first in one's class. In this simultaneous condition of superiority and insecurity, the only way to get ahead—to live up to expectations, to prove yourself—is by working even harder, never slacking off.

MUCH OF WHAT'S JUST been said about impulse control in Chinese American parenting could also be said of other successful immigrant groups. With exceptions as always, first-generation Korean and Indian Americans also tend to impose demanding academic expectations on their children. United States Attorney Preet Bharara, who has called his Indian immigrant father "a tiger dad," was expected to get perfect scores on all his tests. Many of these parents teach their children the alphabet earlier and drill extra math at home from an early age (often because they view the math taught in U.S. primary schools as woefully deficient). As one South Asian undergraduate told us:

> I remember when I was learning decimals in middle school and I just couldn't get the hang of it. My dad stayed up with me all night after coming back exhausted from a 12-hour work day and I remember this so clearly: he would give me 2 seconds to answer the question. I remember being assigned hundreds of problems to do and getting so tired and annoyed but today, math is a breeze for me and I totally attribute that to him.

Typically, these immigrant parents expect their children to come right home from school to do "productive" activities and impose strict rules on socializing. "My parents were obsessed with our not being tainted by loose American values," reported one young Indian American woman. "It was literally like they believed I should do nothing but study for twenty-two years but somehow be married by the age of twenty-five to a nice Indian man." A Nigerian American parent said she always told her children: "We are your parents, we are here to support you, and your job is to study. We are not asking you to pay us money. Your job is to go to school, listen to the teachers, come home, and do your homework." Another Nigerian American mother said, "[G]oing to the mall was forbidden in my house unless you were going to buy something. You don't go to the mall to start walking around as some people do in this country."

As in Chinese American households, the conflict these stringent rules generate can be intense, especially when children reach their teenage years and begin to see how different their lives are from their friends'. As two young Korean Americans explained:

> At school you talk to your white friends . . . [and with them] you go to sleepovers, paint your nails, eat pizza, but then you come home . . . it is like a different world . . . Most of us have first-generation parents. We know what goes on in a Korean house . . . parents' pressure . . . "study study study, marry a Korean, don't talk back."

Why so many immigrants tend to raise their children more strictly is not hard to understand. Immigrants often come from more traditional societies; their economic position is relatively precarious; they're outsiders, frequently more wary of what they see as an excessively permissive American culture; and they view academic achievement as their children's best (or perhaps only) path to a more secure future.

For all these reasons, immigrant parents of all different backgrounds have been found to "place greater expectations on children than do native-born parents" and probably raise their children with relatively greater impulse control than other Americans.

This is certainly true of Asian immigrants, as evidenced by rates of substance and alcohol abuse. Asian American teenagers—and Asian Americans on the whole—have dramatically lower rates of drug use and heavy or binge drinking than any other racial group in the United States. Asian American girls also have by far the lowest rates of teenage childbirth of any racial group (around 11 births per thousand Asian Americans in 2010, as compared with around 56 for Hispanics, 52 for blacks, and 24 for whites). Because giving birth for teenage girls, and being convicted of a drug crime for teenage boys, are so highly correlated with adverse economic outcomes later, Asian Americans' impulse control in these domains contributes to their disproportionate success.

ONE OF THE MOST REMARKABLE THINGS about impulse control is that it transfers over from one domain of life to another.

Impulse control is like stamina. If you ran five miles every few days for several months, you'd build up stamina, which would allow you not only to run farther, but to perform all sorts of unrelated physical tasks better than you could before. As numerous studies have now proved, it's the same with impulse control. If people are made to do almost any impulse-controlling task—even as simple as getting themselves to sit up straight—on a regular basis for even a few weeks, their overall willpower increases. Suddenly they're stronger in all kinds of unrelated activities that also require concentration, perseverance, or temptation resistance.

Which is why stricter child rearing can make such a difference, even when the forms of impulse control emphasized in a particular culture have nothing to do with getting ahead in school. Consider the Mormons.

MOST PEOPLE KNOW THAT Mormons don't drink. As interpreted by the LDS Church, the Word of Wisdom—the Mormon "health code" traced to an 1833 revelation—requires Latter-day Saints to abstain not only from alcohol, but also tobacco, coffee, tea, and of course drugs. Gambling, pornography, adultery, and abortion are considered evil. Premarital and extramarital sex is absolutely forbidden, "a sin exceeded in seriousness only by murder and 'denying the Holy Ghost.'"

Discipline and self-control are inculcated in Mormon children when they are very young. Mormon families spend three hours at church every Sunday, with kids beginning Sunday School at age three. At fourteen, Mormon teenagers are encouraged to attend "seminary," and many get up at five a.m. every day to attend an hour-long scripture study class before regular school starts. These efforts to raise "clean-living" children appear to be effective. According to the four-year National Study of Youth and Religion, Mormon teenagers are less likely to have sexual intercourse, consume alcohol, smoke pot, or watch X-rated films than teenagers of any other faith.

For most male and many female Mormons, however, a central life-defining experience is the mission—a two-year stint in an assigned location that is basically impulse-control boot camp. The experience begins at a Missionary Training Center; the flagship center is in Provo, Utah. There, young men and women spend up to two strenuous months learning languages, studying scripture, and drilling missionary techniques with "military efficiency." ("What's the difference between the MTC and prison?" runs a popular joke. "You can call home from prison.")

Once on location, which can be anywhere from New Jersey to Ghana, missionaries dressed in suit and tie (or neat skirts) work ten to fourteen hours a day, six days a week, cleaning house on the seventh. They receive no financial compensation; they must give up cars, movies, and romance. Most live in relative poverty and experience constant rebuffs, rejections, or insults from the strangers they attempt to convert. "The thing a mission does is teach you persistency," says Gary Crittenden, a former CFO of American Express.

Another extraordinary feature of the Mormon mission is the "companionship" requirement. Latter-day Saints are required to do their missions in pairs. Companions spend twenty-four hours a day together, eating, working, studying, praying, sleeping in the same room—not even allowed to take a walk or go to a store alone. For some, companionship turns into lasting friendship. For others, it is itself an exercise in impulse control. As one missionary put it, "The idea of having a companion, at first, is pretty traumatic. You grow up in a society that stresses individuality and privacy for nineteen years and then . . . that all changes. You are constantly with someone else and this someone else is not of your choosing. In some cases, he is definitely not of your choosing." In almost every case, companionship works for one or both missionaries as a way to monitor and encourage compliance with the mission's arduous regimen.

The strictures of Mormon youth prove too much for some, but there can be no doubt that they translate into a great many sedulous Mormon adults. Mormon businessmen are well known for *not* playing golf, *not* having a beer at the firm barbecue, and *not* going out for even a one-martini lunch. They also refrain from many nonforbidden social activities, reserving their "free" time for church, community, and customs like Family Home Evening, the Monday nights most Mormon parents set aside—religiously—to spend with their children.

Thus Mormons present a fascinating instantiation of the Triple Package. Here we have a group who believe they are literally God's children—"gods in embryo"—placed on earth to lead the world to salvation. They know their way of life is superior to the "sea of moral decay" they see around them. Yet at the same time, in the United States, they remain marginalized, regarded with suspicion, persistently viewed as cultish, deviant, or "creepy" (as Mitt Romney's sons were repeatedly described). Several Protestant denominations have officially stated that they do "not regard the Mormon church as a Christian church." In a 2006 South Carolina poll, 44 percent said they believed that Mormons still practice polygamy; in a nationwide poll, 53 percent said they had "some reservations" about or were "very uncomfortable" with Mormon candidates.

Among many Latter-day Saints, these suspicions give rise to deeply conflicted feelings. On one hand, Mormons may long to refute their doubters, to fit in, to be seen as "normal" Americans. On the other, they want to preserve their identity as God's "peculiar people," with their own distinctive beliefs, values, and practices. In the "cognitive dissonance" that follows, Mormons sometimes present themselves with a clean-cut, all-American image so exaggerated that many Americans actually find it peculiar—which of course Mormons aren't sure they mind.

This inner conflict has been instinct in Mormonism ever since Joseph Smith ran for president even while Mormons were being threatened with extermination. (Orrin Hatch, the veteran Mormon senator from Utah, wears a Jewish mezuzah—a piece of parchment inscribed with biblical verses—around his neck, explaining, "I wear [it] just to remind me, just to make sure that there is never another holocaust anywhere. You see, the Mormon church is the only church in the history of this country that had an extermination order out against it.") Throughout its history, as the sociologist and practicing

Mormon Armand Mauss describes it, the LDS Church has vacillated between assimilation and retrenchment, between respectability and uniqueness.

Entire dissertations could be written (and undoubtedly are being written) on the inner Mormon turmoil revealed by a story in which a chaste, saintly, alcohol- and tobacco-abstaining young woman is seduced by a supernatural being who is simultaneously a blood-sucking vampire and a radiant angel—which is of course the plot of the mega-smash *Twilight* books, whose author is an observant Latter-day Saint. Such is the nature of Mormon insecurity: half in, half out; half normal, half strange; wanting acceptance, wanting to convert others; indeed, wanting to be accepted by the very people they want to convert.

In classic Triple Package fashion, the result among many Latter-day Saints has been an intensely disciplined drive to prove themselves through business success and other badges of esteem. "As somebody who grew up in Utah . . . I have always felt like there was a little bit of a chip on the shoulder," explains Dave Checketts, formerly the CEO of Madison Square Garden. "We feel like we're really good citizens, good people, and misunderstood." "A big part of my drive," says Checketts, "is this sense of needing to prove myself." David Neeleman, founder and CEO of JetBlue, feels that "the Mormon Church is one of the most misunderstood organizations on the planet" and says of his own intense discipline, "It's all about doing better than everyone else."

Just as it was for the early Calvinists, whose work ethic Max Weber studied, success in business is for Mormons not only a way of proving the superiority of their values and way of life. It's also a proof of their divine favor. Harold Bloom was hardly exaggerating when he called Mormonism a "kind of Puritan anachronism," perhaps "the most work-addicted culture in religious history."

. . .

ASCETICISM HAS NEVER LOOMED large in Judaism, but impulse control is without doubt foundational to the Jewish religion and to the traditional Jewish way of life.

The Ten Commandments could have been called the Ten Big Impulses to Control, and they were but a fraction of the hundreds of injunctions—613 to be precise, according to Jewish tradition—given to Moses as everlasting law to bind the Jewish people. Early rabbinic Judaism posited self-control as one of the highest ethical values. "Who is strong?" asks the Talmud. "One who subdues his passions. As it says: 'One who is slow to anger is better than a hero and one who has control over his will is better than one who conquers a city.'" According to Philo, the Jewish philosopher of the Hellenistic era, the Stoic virtue of *enkrateia*—self-control or self-mastery—was at the heart of the entire body of Mosaic law, and in this respect the Jews were a model for all other nations: Moses "exhorted [the Jews] to show [*enkrateia*] in all the affairs of life, in controlling the tongue and the belly and the organs below the belly."

Over the long centuries, rabbinic interpretation and application multiplied the rules of Jewish law, until by the sixteenth century one abbreviated code—the *Shulchan Aruch*—contained thousands of regulations governing every aspect of life, from waking until sleep. Exclusion and anti-Semitism may have intensified Jews' habits of self-control in societies where their position was precarious. Immanuel Kant offered an account of Jewish abstemiousness along these lines in the late eighteenth century. "Jews," wrote Kant, "normally do not get drunk, or at least they carefully avoid all appearance of it." The explanation, he said, lay in the Jews' "civil status," which was "weak" and insecure. As a result, Jews needed to be circumspect, "for which sobriety is required."

It's true that American Jews of the nineteenth century, who were largely German, often favored the assimilationist Reform movement, which repudiated most of the age-old strictures of Jewish law. Circumcision, declared the leading Reform rabbi in America in 1885, was a "remnant of savage African life," the bar mitzvah an obsolete ritual. At an 1883 banquet celebrating the inaugural class of rabbis graduating from Hebrew Union College—the first Jewish seminary in the United States and a crowning achievement of the Reform movement—the menu included littleneck clams, "Salade de shrimps," and frog legs.

But the millions of Eastern European and Russian Jews pouring into New York beginning around 1890 brought with them an orthodox Judaism still committed to all the old restrictions. As described in a 1914 book about Jewish life, the strictly observant Jews in America's new Jewish enclaves, like the orthodox Jews of Eastern Europe, still "regulate[d] every day in their lives, from the cradle to the grave, by the minute and comprehensive laws of the mediaeval codex, the *Shulchan Aruch*." This was the Judaism of the Pale and *shtetl*, where ghettoized Jews lived "bound and shackled"—as one (Jewish) observer of their poor villages put it—by impulse-constraining rules they imposed on themselves:

> [The Jews] tend even to forsake whatever pleasures Jewish law allows them. They are constantly placing new yokes upon themselves. They hide their natural impulses. They renounce the darker elements in their nature. They have ears only for the reading of the law, eyes only for scrutinizing sacred texts, voices only for crying "Hear, O Israel."

Even the Jewish Sabbath—which many Jews cherish as a family-centered day of rest—was a highly disciplined regimen of self-restraint. Observance included hours of services and Torah study. If a family had a horse and wagon, they could not use it (because riding is forbidden on the Sabbath). Under finely nuanced rabbinic reasoning, recreation too could be forbidden. If, for example, Jewish children were lucky enough to have a river or lake nearby, they could not swim in it, because swimming might entail use of a raft, and a raft might need repair, and repairing the raft would be work—forbidden on the Sabbath.

In America, the core Sabbath ban against remunerative work became itself an acute form of impulse control—where the impulse being controlled was the urge to make extra money. In *What Makes Sammy Run?*, the popular 1941 rags-to-riches novel that became a long-running Broadway musical, a Lower East Side Jewish father is horrified when his son returns home on a Saturday with money in his hands:

> "I hadda chance to make a dollar," Sammy said.
>
> "Sammy!" his father bellowed. "Touching money on the Sabbath! God should strike you dead!"
>
> The old man snatched the money and flung it down the stairs . . . .

On top of the innumerable religious restrictions, typical Jewish American parenting in the first half of the twentieth century was a match for any Asian immigrant parenting today. Surprisingly early on, Jewish newspapers in New York began to record objections (from Jews themselves) to the pressures and after-school discipline being imposed on children—objections that sound eerily familiar to some of us. In 1911 a Jewish doctor wrote in the *Forward*:

We push our children too much. After school they study music, go to Talmud Torah. Why sacrifice them on the altar of our ambition? Must we get *all* the medals and scholarships? Doctors will tell you about students with shattered nerves, brain fever. Most of them wear glasses. Three to five hours of studying a day, six months a year, are better than five to twelve hours a day for ten months a year.

Music practice—especially piano practice—played a significant role in these children's lives, as revealed by another critical essay in the *Forward*:

[A] piano in the front room is preferable to a boarder. It gives spiritual pleasure to exhausted workers. But in most cases the piano is not for pleasure but to make martyrs of little children, and make them mentally ill. A little girl comes home, does her homework, and then is forced to practice under the supervision of her well-meaning father. He is never pleased with her progress, and feels he is paying fifty cents a lesson for nothing. The session ends with his yelling and her crying. These children have not a single free minute for themselves. They have no time to play.

Thus the two million Eastern European Jews who immigrated to America in the early 1900s brought with them habits of heightened discipline, religious prohibitions, and hard work that they not only practiced themselves but passed down to their children. Later waves of Jewish immigration—for example, during and after the Second World War—would reinfuse Jewish American culture with the same ethos. The anxiety Jews felt as a result of anti-Semitism in the United States for much of the twentieth century probably strengthened their impulse control, as Kant thought it did in Germany. In the face of such prejudice, the disproportionate success twentieth-century American Jews eventually achieved in professions and careers requiring long years of study or practice—medicine, science, law, music—is testimony to their habits of discipline and perseverance.

Today, it's no longer clear that impulse control is a defining element of Jewish American culture. Orthodoxy has waned, as has anti-Semitism. Contemporary Jewish parents in the United States are much more ambivalent about "pushing" their children (at least openly), especially at upper income levels. The transformation in some circles of bar and bat mitzvahs into $250,000 coming-out parties for thirteen-year-olds is not evidence of a culture of restraint or temptation resistance. (At one early twenty-first-century event, with an estimated seven-figure price tag, the catsuit-clad bat mitzvah girl descended into her party at Cipriani on Wall Street hanging by a wire from the ceiling, and was then "serenaded by Jon Bon Jovi for 45 minutes.") Whether Jewish culture in the United States today has lost its impulse control—and therefore the Triple Package—is a subject to which we'll return.

AMERICA IS THE GREAT wrecker of impulse control. Chinese upbringing traditions, with their emphasis on drilling, obedience, and discipline, survived two thousand years of dynastic cycles, communist upheaval, even the Cultural Revolution with its anti-Confucian, anti-intellectual thrust. But it seems the one thing they can't survive is—America. After a single generation in the United States, traditional strict parenting in Chinese American households softens; parental expectations drop, and a sharp fall-off in academic performance occurs between the second and third generations.

It's hardly news that modern America today is not big on strictness in childhood or impulse control in general, at least as compared to traditional societies. We'll return to this point, too, later on, but the short of it is that American culture today celebrates a powerful live-in-the-moment message.

The not-so-secret truth, however, is that successful people typically don't live that way. On the contrary, the successful are often the ones profiting from the people who do live that way. Executives at America's junk-food corporations are notorious for assiduously avoiding their own products. At some level, American media obviously recognizes and extols the value of impulse control. Interviews with sports heroes often stress their work ethic and resilience. Two of the most popular TV series in recent history are *The Wire* (subject of a Harvard college course) and *Breaking Bad* (winner of multiple Emmys), the former situated in inner-city Baltimore, the latter in Albuquerque. Both are about drugs and addiction, and a common theme is that the kingpins who make the millions—the successful ones—don't "use." Both shows stress the discipline, self-restraint, and perseverance exercised by the drug lords and their best employees, in stark contrast to their dissolute, hand-to-mouth customers.

Nevertheless, the reality of American culture is that it runs acutely counter to the traditional high-discipline parenting and other elements of impulse control characteristic of successful groups. Which once again underlines and explains the fact that America's most successful groups tend to be cultural outsiders. Success in America today comes more often to groups who resist today's dominant American culture.

Thus, impulse control is yet one more point on which America's successful groups aren't listening to the piper. Here's what America likes to tell Americans: Everyone is equal; feel good about yourself;

live in the moment. Meanwhile, America's successful groups tell their members something different: You are capable of great things because of the group to which you belong; but you, individually, are not good enough; so you need to control yourself, resist temptation, and prove yourself.

CHAPTER 6

THE UNDERSIDE OF THE TRIPLE PACKAGE

Having laid out its three elements and their combustible interaction, having seen the Triple Package in operation in America's most successful groups, we turn now to the critical question of whether the Triple Package is a blessing or curse. Is it something one should aspire to re-create in one's own life and family, or something to be avoided?

The Triple Package always comes at a price. It carries with it certain characteristic pathologies, which this chapter will explore in detail. In assessing whether this price is worth paying, there's a simple answer we could give, which wouldn't be false. Namely: the Triple Package works by making people very good at attaining conventional success, so everything depends on how much you think conventional success is worth. For some—poor immigrants, for example, or inner-city parents struggling to give their children opportunities they never had—it can surely be worth a great deal. Indeed, debating the value of conventional success is a luxury many can't afford.

But there's a more complicated answer, too, expressing a deeper truth. The Triple Package is worth aspiring to *precisely in order to break out of it.* Its strictures can take people to a place where they can write their own scripts—conventional or unconventional, achieving success however they define it—provided they're able to cast aside the constraints of the Triple Package once they get to that place. In other words, the Triple Package is like Wittgenstein's paradoxical ladder: you have to throw it away after you've climbed it.

We'll come back to this more complicated side of the story at the end of this chapter. First we turn to the most glaring Triple Package pathologies.

TRIPLE PACKAGE CULTURES take a view of childhood that's out of fashion in youth-obsessed America, a view that conflicts with deeply held contemporary American ideals of what childhood is for and what it should be like.

The celebration of youthfulness may go all the way back to the beginnings of American history, part of the contrast between the "New" and "Old" Worlds. ("The youth of America is their oldest tradition," quipped Oscar Wilde. "It has been going on now for three hundred years.") But since the 1960s, and accelerating with the digital revolution, America's obsession with being or seeming young, along with its disrespect of being or seeming old, has intensified. America is a youth culture, where youth means freedom, creativity, pleasure, happiness—and age the opposite of all these things.

Idealizing childhood goes along with this reverence for youth. Childhood is imagined as the stage in life before the constraints, denials, inhibitions, self-doubts, and discontents of adulthood kick in. Childhood was or should have been the happiest time of our lives. In short, childhood should be fun. Learning should be fun. Good parents try to smooth the way for their children, to make their lives as painless, carefree, and unencumbered as possible.

Triple Package groups take a very different view, at least when they are on the rise. Triple Package parents see childhood as a time of investment, training, preparation for the future. In this view, childhood comes encumbered. Often, children feel a sense of indebtedness to their parents. "I studied nonstop in school," the attorney daughter of Ethiopian immigrants recalls, "because I was well aware that my privilege came at tremendous cost to my parents. I was burdened to excel so that my parents' hardworking past wouldn't be in vain." For many immigrants' children, as one study found, "happiness has to take a back seat." Success means "hard work and monetary gains, rather than . . . an emotional sense of fulfillment." Upward mobility is not just a dream for these children. It's a duty.

In Chinese (and other East Asian) American families, the idea of an encumbered childhood is deeply rooted in the Confucian philosophical and moral tradition. In this tradition, as sociologist and Confucian expert Jin Li explains, every child is to "willingly" and "gladly" show his parents filial piety in exchange for the parents' "unconditional love" and "total commitment." All human morality begins with these obligations, which "endure for life." In the Confucian worldview, there is no idealized carefree childhood. The highest purpose of life is moral self improvement through learning, and children are expected as a moral matter to cultivate the core learning virtues—diligence, enduring hardship, striving for self-perfection—from a very early age.

There is a harmonious logic here, but if the elements of parent-induced insecurity and impulse control in the Confucian approach to childhood are taken to extremes, the result can be intense pain. When it goes wrong, or sometimes even when it goes right, the East Asian version of the Triple Package can make life feel like a prison—a prison of expectations that can never be met. Asked why his mother had never been to one of his runway shows, fashion designer Phillip Lim explained, "You know, I think it comes back to that Asian cultural aspect where you're never good enough. You know what I mean? It's like an 'A' is not good enough, you have to be an 'A+' and I've never felt ready to receive her."

In the Confucian outlook, there is literally no achievement a child can attain that could not be improved on. Celebrity designer Vera Wang—who's made wedding dresses for Uma Thurman, Ivanka Trump, Mariah Carey, Alicia Keys, Kim Kardashian, and Chelsea Clinton—still remembers letting her mother down when she didn't make the U.S. Olympic team as a figure skater. She also speaks painfully of her father's refusal to accept her career choice. Up to the day he died, notwithstanding her fame, she never got "a vote of approval or a 'hurrah for you' or any of that" from him. "If I were to say at any point that I feel really confident or really in control, that would be a mistake. Because I don't. I always see where I didn't do things the right way."

In addition to the pressure, Triple Package children may come to feel that their parents merely instrumentalize them. "I feel like I'm just an investment good for my parents," one young Chinese American put it. "I have to build on the achievements of the first generation—do even better than them—to bring my parents honor, to lift the family from lower-middle to middle to upper class. One always carries the burden of the family's expectations and fortunes." Sometimes the feeling is of being valued only for "bragging rights." As another Chinese American said, "My grandparents are actually proud of me, but they're only proud of me because they can boast about the name of my school." And of course it's much worse if you're the one they *can't* boast about. "If you don't achieve that, it's like, 'Okay, well, then, I'm garbage.'" In the most aggravated cases, genuine trauma or even worse can ensue. Here is novelist Amy Tan describing her childhood:

> [My parents] just didn't understand. They didn't know who I really
> was. They didn't know how much the smallest amount of recognition would have meant to me and how the smallest amount of criticism could undo me. . . .
>
> I remember once one of my playmates from around the corner died, probably of leukemia. My mother took me to this funeral and took me up to see Rachel. And I saw Rachel's hands clasped over her chest, and her face was bloodless, and her hands were flat, and I was scared, because this was the little girl I used to play with. My mother leaned over to me and she said, "This is what happens when you don't listen to your mother."
>
> Talk about pressure. Here was a little girl who didn't listen to her mother. According to my mother, she should have washed her fruit and she didn't. . . . Pesticides might have led to leukemia and killed this little girl. . . . [F]ear was the way to control children for their own good. That's what I grew up with. . . .
>
> I grew up thinking that I would never, ever please my parents. That is a difficult thing to grow up with. . . . It's a horrible feeling, especially when you experience what you think is your first failure and you think your life is over. No more chances. . . .
>
> I reached a point where I had infuriated my mother so much we nearly killed each other. Literally. And I was sick to my stomach, literally. I had dry heaves, and the pain was so enormous that at one point . . . I thought I was going to die.

Empirical research sheds some light on how frequently Asian American Triple Package parenting leads to psychological problems, and how severe these problems can be. But the picture is surprisingly complicated. In 1995, Asian American adolescent girls were found to have "the highest rates of depressive symptoms of all racial groups," and several recent studies confirm that Chinese American students report higher levels of stress and anxiety than white Americans. One study of high-achieving ninth graders in a highly selective East Coast high school found that Chinese American students reported "higher levels of anxiety and depressive symptoms," and that these problems were associated with higher levels of family conflict.

On the other hand, as noted earlier, rates of alcohol abuse and substance dependency—frequently associated with emotional problems—are much lower among Asian American youth. Moreover, a 2010 nationwide psychiatric survey of 6,870 white Americans and 1,628 Asian Americans found that Asian Americans had significantly *lower* rates of social anxiety disorder, generalized anxiety disorder, panic disorder, and post-traumatic stress disorder. Similarly, a 1990s study found that Chinese Americans had substantially lower rates of major depressive episodes than the national average. In part these findings may reflect reluctance to acknowledge psychological symptoms (mental illnesses are still considered shameful by many Asians), but Asian American suicide rates are also lower than the national average. Despite widespread media reports to the contrary, the Asian suicide rate in the U.S. is less than half the white rate.*

---

* For 2000–2010, the overall Asian American suicide rate was 5.6 per 100,000 as compared with 12.3 per 100,000 white Americans. Asian Americans commit suicide at lower rates in every age bracket, but breaking down the data by gender reveals important vulnerabilities. The Asian American suicide rate is lower than the white rate for both men and women, but the gap is considerably narrower for women than for men (8.3 suicides per 100,000 for Asian men, as compared to 20.3 for white men; 3.2 per 100,000 for Asian women, as compared to 5.0 for white women). Among women 15 to 24 years of age, the Asian American suicide rate is very close to the white rate (3.2 per 100,000 for Asians; 3.5 for whites), and among women aged 70 and over, the Asian

Nevertheless, there is undoubtedly a psychological underside to the Asian American Triple Package. As mentioned earlier, Asian American students report the lowest self-esteem of all racial groups even while outperforming others academically. This divergence strikes many as puzzling; as one family-studies professor puts it, "If you're doing well, you should be feeling good." But a core insight of the Triple Package is precisely that *not* feeling good about yourself— or not feeling good enough—is part of what drives success.

"FIVE PERCENT OF JEWS are mildly depressed," runs a Jewish joke. "The rest are basket cases." You might think that the Triple Package pathologies of Jewish Americans would be just like those of Asian Americans, given certain obvious similarities. For example, the commandment to honor your parents is just as central to Judaism as it is to East Asian culture. Guilt, including the feeling that you're never doing enough to satisfy your parents, is another common trope. Parents getting "bragging rights" from their children's success is so widespread in Jewish culture that there's a word for it: *naches*.

But in fact the characteristically Jewish pathologies differ in revealing ways. Perhaps because Jews have a long tradition of questioning authority, a tradition with roots in the Talmud and the Old Testament itself (the Book of Job, for example), a Confucian-style reverence for one's parents, and deference to their authority, is not the norm. In a play by Jules Feiffer, a young Jewish man describes his debt to his parents: "I grew up to have my father's looks—my father's

---

American rate (6.8) is higher than the white rate (4.1). Overall, however, and in virtually every age/gender bracket, the Asian rate is lower than the white rate (and the national average). For sources and more detail, see the endnote.

speech patterns—my father's posture—my father's walk—my father's opinions—and my mother's contempt for my father."

The Jewish propensity to challenge authority has produced extraordinary, sometimes revolutionary new ideas in philosophy, social theory, and science, but it has figured in Jewish family unhappiness as well. At least since Freud began diagnosing them, Jewish neuroses have stereotypically featured children with—to put it mildly—attitudes less than reverent toward their parents. Children who defy parental authority are common figures in Jewish American culture, often celebrated, but often also causing painful strife and irreparable family rupture; *Fiddler on the Roof* is just one of many examples.

In more extreme cases, Jewish Triple Package striving seems to give rise to a destructive competitiveness between parents and children. Sons may want to outdo their fathers (as Freud saw it), or fathers may be jealous of their more successful sons, a theme that pops up with regularity in Jewish culture—as in Joseph Heller's *Good as Gold* or the 2011 prizewinning Israeli film *Footnote*, about father and son Talmudic scholars—but rarely if ever in East Asian culture.

Particularly for early-twentieth-century Jewish immigrants to America, the acute scorn and discrimination they faced, even as it spurred drive in classic Triple Package fashion, exacerbated conflict between fathers and sons. Sons in many families came to see their tradition-bound fathers, who barely scratched out a living as peddlers or shop clerks, as disappointments, embarrassments, or hindrances to their own success. "In my time," writes Lionel Trilling, "we all were trying to find a release from our fathers." Conversely, if these fathers felt like failures themselves, their stinging resentment was occasionally directed against their own sons, triggered when the son threatened to become more successful than the father had ever been. (Hollywood's first "talking" feature, *The Jazz Singer*, played on this theme.) Either way, for the "bulk of Jewish immigrants," Daniel Bell has written, the "anxiety" felt about making it in America "was translated into the struggle between fathers and sons."

In twenty-first century America, these intense immigrant generational struggles have a dated, Lower East Side feeling for many Jews. The Oedipal conflict that so dominated twentieth-century Jewish American life and culture just doesn't seem as central any longer (which may explain the waning of Freudianism). Extremely few American Jews under the age of thirty today are in a position to be embarrassed by having had a peddler for a father. The opposite problem is more likely: having hyper-successful parents who rose from very little—a frustratingly hard act to follow.

Which is not to say that all the old Jewish Triple Package insecurities and pathologies are gone. On the contrary, the fear of persecution, which has pushed Jews to success throughout the world for centuries, has long had its own corrosive effects on the Jewish psyche. This fear shaped the lives of many Jewish immigrants to the United States.

Consider, for example, Meyer Guggenheim, the German Jewish immigrant who literally rose from rags to riches, first peddling stove polish on his way to founding one of the great fortunes of the Gilded Age. Guggenheim, as described by his biographer John Davis, was "a caricature of the nineteenth-century Jew. He was a small, reticent, suspicious loner" "single-mindedly devoted to making money." He trusted only his family members and was "always on the alert for an ulterior purpose on the part of both friend and foe." Davis continues:

> It was a cardinal point of Meyer Guggenheim's creed—conditioned by centuries of oppression and tyranny in Germany and

Switzerland—that safety and happiness in this world lay only in money. . . . [O]nly money could protect you from being devoured or swept away.

Many contemporary American Jews, notwithstanding the safety and affluence they may enjoy, continue to suffer from a similar anxiety. This is particularly true of Jews with family connections to the Holocaust. "I have a bit of a phobia," George Soros once said. "Why do you think I made so much money? I may not feel menaced now but there is a feeling in me that if . . . I were in the position that my father was in in 1944, that I would not actually survive." A Jewish American attorney recalls how Germany's attempt to exterminate the Jews hung like a cloud over his family's table:

My grandparents were Holocaust survivors who literally met each other on the way to Auschwitz. Their entire families were wiped out (with the exception of one uncle). It was odd for anyone in my parents' generation to have grandparents. Food was not to be left on our plates. If we intermarried, we were finishing Hitler's work. The Holocaust was omnipresent.

The precarious nature of the Jewish people was often emphasized during my youth. Would America just spit us out if it became convenient? Would Israel be wiped off the map with one push of a button? Did that police officer give my father a ticket just because he was wearing a *yarmulke*?

The fiercely protective, defensive attitude so many Jews hold toward Israel is probably due in large part to Israel's near-mythic status (among Jews) as a bulwark against worldwide anti-Semitism—the one country on earth that will never turn on its Jews, the one place where Jews can really be at home. With Triple Package indefatigability, Jews have gained in wealth and power, but the fear underlying it all should not be minimized. Radio host Dennis Prager says:

> Jews are probably the most insecure group in the world. This may come as a surprise to most non-Jews since Jews are widely regarded as particularly powerful. But Jews' power and Jews' insecurity are not mutually contradictory. In fact, Jews' power derives in large measure from their insecurity. The stronger the Jews' influence, Jews believe, the less likely they are to be hurt again.

If insecurity is a spur to Jewish success, it comes at a psychological price.

So FAR WE'VE EXAMINED pathologies associated with extreme insecurity and (in the case of some Asian Americans) impulse control. The pain these pathologies inflict is typically suffered by the individuals afflicted with them, like the child made miserable by internalizing too much parental pressure. By contrast, the harms associated with superiority complexes tend to be inflicted on others, which makes this element of the Triple Package potentially the most nefarious. Group supremacy claims have been an unrivaled source of oppression, war, and genocide throughout history.

To be sure, a group superiority complex somehow feels less ugly when it serves as an armor against majority prejudices and hostility. America's disproportionately successful groups are all ethnic or religious minorities; it's easier to get away with a superiority complex when your group is a minority in a society that harbors lingering hostility or suspicion toward you. America's Triple Package groups are fighting fire with fire, so to speak. But the dark underside of their superiority feelings shouldn't be forgotten.

Being "deeply proud of Chinese culture" can easily shade into "We'll disown you if you marry someone non-Chinese." Anti-Zionists such as Noam Chomsky point out that early-twentieth-century Jewish settlers in Palestine viewed the Arab population with contempt, calling them "half-savage peoples," "wild men," "dishonest," and "cowardly." Persian superiority implies the inferiority of all other Middle Easterners. The insistence by many Cubans that they are "not Hispanic" is loaded with unspoken premises. West Indian and African immigrants are often accused of looking down on American blacks.

Superiority complexes don't have to be invidious. They don't have to espouse notions of innate group differences. But the unfortunate reality is that many members of immigrant groups in the United States have embraced all-too-familiar racist attitudes. "[T]he move into mainstream America," Toni Morrison writes, often "means buying into the notion of American blacks as the real aliens. Whatever the ethnicity or nationality of the immigrant, his nemesis is understood to be African American."

The Mormon superiority complex is no more invidious than any other—in some ways much less, because Mormonism has opened its doors to individuals of all races and made millions of converts all over the world—but it highlights a feature worthy of special attention. A group's superiority complex isn't always distributed equally among all group members. It can raise one class of people over another *within* the group.

Officially, Mormonism holds that women and men are spiritual equals, but as in Catholicism, women are excluded from the priesthood and therefore from the higher offices of Church leadership. No woman has ever served in the Quorum of the Twelve Apostles, a primary governing body of the LDS Church, or in the First Presidency, the highest Church body of all. The Church has not been reticent about its expectations for women. In 1987, Church president Ezra Taft Benson preached, "In the beginning, Adam—not Eve— was instructed to earn the bread by the sweat of his brow. Contrary to conventional wisdom, a mother's calling is in the home, not the marketplace."

In the 1990s, several women professors who criticized Church positions on women's issues were forced out of Brigham Young University. A few Mormon women have been excommunicated because of their feminist views. In 1995, the Church issued a "Proclamation to the World," regarded as near-scripture by most Mormons, stating that "[m]arriage between man and woman is essential to [God's] eternal plan," and that "[b]y divine design, fathers are to preside over their families" and "are responsible to provide the necessities of life," while "[m]others are primarily responsible for the nurture of their children."

Joanna Brooks, author of *The Book of Mormon Girl*, writes movingly of her position as a faithful Mormon whose belief that women should have the same opportunities and aspirations as men put her at odds with her Church:

> For years, I cried every time I set foot in a Mormon ward house.* . . .
> Crying that the Church had punished women like me, people like
> me, leaving us exiled among our own. . . . How badly I wanted to
> belong as I had when I was a young Mormon girl, to be simply a
> working part in the great Mormon plan of salvation, a smiling exemplar of our sparkling difference. But instead I found myself a
> headstrong Mormon woman staking out her spiritual survival . . . .

* The terms "ward house," "meetinghouse," and "church" are often used interchangeably to refer to "the basic building type used by the LDS Church for holding general worship services" and other activities.

Mormonism by no means forbids professional success to women—there are many prominent Mormon women in academia and the arts—and recently the Church took steps to create more opportunities for young women in missionary work. In 2012 the Church announced that women could begin serving missions at age nineteen (previously it was twenty-one), and a new leadership position was created for women missionaries ("sister training leaders"). Nevertheless, Mormon patriarchy still exists, adding an extra layer to the superiority complex of Mormon men. According to the theory of the Triple Package, this should help Mormon men succeed, but it can potentially come at some expense to Mormon women.

If sex inequality in Mormonism seems backward to modern Americans, it should be remembered that inequality and even intolerance characterize virtually all of America's successful groups. The ugly corollary of a superiority complex is all too often a propensity toward bigotry, exclusivity, insularity, or parochialism—an intolerance of other groups and other ways of life.

In fact, such intolerance in many cases is a condition of the group's very existence. At the most concrete level, ethnic or religious groups can maintain their identity only by condemning marriage "outside the tribe." For just this reason, many American Jews, who as a group out-marry at very high rates, worry that Jews will out-marry themselves into oblivion. East Asian and Indian cultures traditionally had strong taboos against marrying outside one's group, although in the United States such traditions are hard to maintain. (For Chinese who still remembered the Rape of Nanjing, marrying a Japanese was anathema; today, especially in California, inter-Asian marriages are becoming more and more common.) In their values, beliefs, self-conception, and marriages, these groups survive only by being intolerant, by denying the complete equality of all mankind; and the groups with the strongest superiority complexes have the best chance of surviving.

. . .

EVEN WHEN THE TRIPLE PACKAGE works relatively unproblem-
atically as an engine of success, without causing obvious harms or
neuroses, it can still be imprisoning—because of the way it defines
success.

Triple Package cultures tend to channel people into conventional,
materialistic careers. This is a direct result of the insecurity that
drives them. The "chip on the shoulder," the need to show the world
or prove yourself—these typical Triple Package anxieties tend to
make people crave obvious tokens of success such as top grades, merit
badges, high salaries, luxury cars, and "respectable" careers.

Triple Package cultures often seem like they're in a defensive
crouch, more concerned with avoiding failure than promoting meaningful, fulfilling, or path-breaking success. The floor that Triple
Package cultures impose on acceptable achievement—a selective college, respectable graduate school, good professional job—may inad-
vertently create a ceiling as well. If you're aiming to do well by
conventional standards, there's a reasonably clear path; a kind of box-
checking approach might be all you need. Get good grades: check. Be
a diligent student: check. Be polite to powerful people who can help
you: check. The path isn't easy, but it's clear. If, however, you want to
be an innovator or artist, if you're looking for meaning rather than
money—if you want to make a difference in the world—there's no
predetermined path. On the contrary, getting there may require
flouting convention, taking risks, and infuriating authority figures—
not the sort of values embraced by most Triple Package cultures.

To be crystal clear, there's nothing wrong with a high salary or a
luxury car. "Respectable" professions are often the most socially valuable. To state the obvious, doctors save lives; even lawyers occasionally help people. A Nobel Prize is certainly a merit badge, but that

doesn't make it degrading to win one. (Conversely, there's a lot of conventionality in what is considered an "unconventional" career.)

The danger, rather, is judging your own worth solely by external measures—allowing your life to be defined not by values, interests, or aspirations of your own, but by others' expectations, or more precisely by the fear of failing to satisfy those expectations. Unfortunately, promoting this fear and those external measures of self-worth is another Triple Package specialty. "When I was younger, I thought achievement had to do with gaining approval from other people—my parents, my teachers, then higher-ups," says Amy Tan. "That was what achievement was. . . . People would give you the feedback and tell you if you had done the achievement." Children raised this way may well, as they grow up, make important life decisions based primarily on parental and social expectations. "[T]hey won't have the guts," one young Indian American woman commented, referring to her peers' unwillingness to reject career paths selected by their parents. "[A] lot of them will always have other interests in other things, [but] they won't be willing [to pursue those things]. A lot of them will be like, I might as well just do this, make money and all, because I want to make my parents happy." The result, observes an Indian American sociologist, "is often self-denial, guilt, and frustration." Another result can be a sense of not having lived your own life—of having spent your whole life striving for goals you didn't even want.

Which is why the best thing about the Triple Package may be that it can empower people to break out of it.

BORN TWO WORLD WARS and a hemisphere apart, the novelist Saul Bellow and the filmmaker Ang Lee had something in common: a father contemptuous of his son's chosen occupation. Bellow's father wanted him to be a businessman. "You write, and then you erase," he said disdainfully. "You call that a profession? *Was meinst du* 'a writer'?" Lee's father wanted him to be a professor. When Lee won an Academy Award, his father told him there was still time. "You're only 49," he said. "Get a degree and teach in universities, and be respectable."

Bellow and Lee shared something else too. They both grew up in extremely demanding, high-expectation, Triple Package households. Lee's father, the principal of a prestigious high school in Taiwan, imposed on Lee at a young age the strictures of Confucian discipline; "the only purpose in life," Lee's younger brother would later recall, was to succeed academically. "I didn't know what I wanted from life," Lee has said, "but I knew I had to please my father." According to biographer James Atlas, Bellow's mother was a "devout believer in the gospel of self-improvement"; her husband's failure "only intensified her aspirations for her youngest son." Like many Jewish immigrant parents of that era, she forced on Bellow an exacting regimen of classical music practice:

> She was always nagging him to practice—a typical scenario on the Northwest Side of Chicago circa 1927. . . . Once a week, Bellow got on the trolley with his violin and made the long trip down to the Fine Arts Building on South Michigan Avenue, where he was subjected to the harsh tutelage of Grisha Borushek, a stout, gloomy refugee from Odessa who trained his pupils "by whipping them on the buttocks with his bow."

Both men rebelled against their family's expectations. Bellow, escaping from what he would later describe as the "suffocating orthodoxy" of his childhood, got himself fired from the family's coal business and began writing stories and novels. Even when published, these stories barely paid his rent, forcing him to ask his brothers for money. The derision this provoked in his family would last a long time. After he won the National Book Award in 1954 (for *The Adventures of Augie March*), his older brother commented, "[M]y name won't go down in the *Encyclopaedia Britannica*, but I have money, and he doesn't."

Much to the embarrassment of his father, Lee enrolled in a three-year arts program in Taiwan after twice taking, and twice failing, the national university entrance exam. "I had a lot of guilt," says Lee, "that I didn't follow his path." He then came to America and studied film at New York University, where he worked on the crew of a movie made by fellow student Spike Lee. After graduation, he "struggled through six years of agonizing, hopeless uncertainty"; he sent one of his screenplays to more than thirty production companies, only to be rejected by all of them. After his in-laws offered his wife money so that he could start a restaurant—she refused—he very nearly gave up filmmaking.

Bellow and Lee are almost textbook Triple Package types. A friend of Bellow's from the 1940s remembered him as "carrying a chip on his shoulder." Bellow's son Greg has written about his father's "insecurities" and describes him "throwing down the gauntlet to the American literary establishment personified by Ernest Hemingway." And Bellow was famously "disciplined." Lee too has spoken of his family's "insecurity"; as the "first son," he recalls, he "felt everything rested on my shoulders." Lee's need to prove himself powered him through years of failure and rejection that would surely have derailed most.

Both men would of course go on to groundbreaking achievement—groundbreaking in both content and in relation to their own ethnic identity. Bellow was "the first Jewish-American novelist to stand at the center of American literature," as the influential critic Leslie Fiedler put it in 1957—and that was before his Pulitzer and Nobel prizes. Other successful American novelists had been Jewish

(Norman Mailer's 1948 Hemingwayesque *The Naked and the Dead* became a sixty-two-week *New York Times* bestseller), but Bellow was the first to break through to a large readership "without departing from an American Jewish idiom"—opening the door for a whole generation of Roths and Hellers who would follow. Lee was the first Chinese to win the Acadamy Award for best director even once, much less twice, and his films have spanned a head-spinning diversity of genre and subject matter that lesser men would never have dared.

In short, both Bellow and Lee kicked away the ladder—but not before having climbed it. The Triple Package had been instilled into them as children, producing its strange combination of confidence and insecurity, its relentless drive, its discipline—and the unquenchable need to prove themselves. In this sense, they never left the Triple Package entirely behind. On the contrary, they attained their heights of achievement only because of the qualities it imparted to them. Yet both men, because they had these qualities, were able to defy the constraints on the kind of life, the kind of success, they were expected to pursue. The Triple Package empowered them to break free from its confines—and write their own, original scripts.

America is a central player in this story too. For both men, it was America that released their creativity and galvanized their self-invention. Both Bellow and Lee were immigrants; both would eventually become naturalized U.S. citizens. Yet both, importantly, found themselves doubly outsiders—first in their own families, then in the United States.

"I never belonged to my own family," Bellow once recalled; "I was always the one apart." His Jewish protagonists may proclaim their Americanness ("I'm an American, Chicago-born" are the first words of *Augie March*), but Bellow himself described Jews in the United States as "*métèques*"—outsiders or "resident aliens." Lee has made similar remarks: "I'm a drifter and an outsider. There's not one single environment I can totally belong to," he has said. "Every movie I make, that's my hideout, the place I don't quite understand but feel most at home."

In America, individuals raised in Triple Package cultures occupy a strange position, pulled simultaneously by the demands of their home culture and by the allure of American freedom. Poised on the edge of both worlds, they belong fully to neither. This cultural edge has its own price, exacting its own psychological toll, but it may be one of the most liberating and creatively productive places a person can inhabit.

A SIGN OF THIS LIBERATING POTENTIAL is the intense opposition to "model minority" stereotypes that has become widespread among young Asian Americans. A growing number of Asian college students in the United States major in the humanities and want nothing to do with math or classical music. Many are finding their role models in breakout Asian American stars who have gone against type, like John Cho and Kal Penn (a.k.a. "Harold and Kumar"), Mindy Kaling (of "The Office" and "The Mindy Project"), YouTube sensation Freddie Wong, stand-up comic Aziz Ansari, basketball player Jeremy Lin, and celebrity fashion designer Jason Wu (who designed Michelle Obama's inaugural ball gown).

Still others are looking at public-interest careers, which is also unconventional from a Triple Package point of view. Immigrant parents may be dismayed by the prospect of their children taking relatively low-paying jobs with the government or nonprofit organizations, which they view as neither prestigious nor financially secure. Here too America—because of its strong encouragement to public service, especially for young people—is a large part of the story. Moreover, immigrant communities may come to take pride in their members who forgo riches and work instead for a greater good. When the United States Attorney for the Southern District of New York, Preet Bharara, the first Asian American to become a U.S. Attorney, gained fame for taking on high-level Wall Street corruption, like the billion-dollar insider trading scandal at the Galleon Group hedge fund, he was celebrated in the Indian American community (and in India as well).

But the very same Galleon insider trading scandal also brought consternation and shame to the Indian American community. One of the defendants U.S. Attorney Bharara brought down in that case—former McKinsey head Rajat Gupta—was himself Indian American. Which brings us to a final Triple Package pathology: the pathology of drive itself.

Triple Package striving is by nature insatiable; it has no built-in limit. Whatever arena you're competing in, there's almost always someone higher than you are, and even if you managed to make it to the top, the fear of losing that spot would keep pushing you on. As a result, the Triple Package can make some individuals prone to over-weening ambition and greed. At the extreme, the longing to rise can become desperate or monomaniacal, oblivious to law, ethics, or the harm caused to others. Gupta and Bernie Madoff are just two of hundreds of names exemplifying this Triple Package disease.

Often people find a mystery in these cases. Many have wondered why Gupta, whose estimated net worth was already over $80 million, who attended a state dinner at the White House, and who was re-vered in his community, would engage in illegal information-trading for benefits that even now remain unclear. "The Court can say with-out exaggeration," pronounced the judge at Gupta's sentencing hear-ing, "that it has never encountered a defendant whose prior history suggests such an extraordinary devotion, not only to humanity writ large, but also to other individual human beings in their time of need." But perhaps for some, the urge for more and more recognition—the ceaseless chip on the shoulder, the need to show the world by rising higher and higher—becomes irresistible.

Even in much less virulent forms, Triple Package drivenness by definition makes it difficult to live a nondriven life. A simple, decent existence—with no scrambling to climb any ladders, without caring whether anyone thinks you're successful enough—may be the most admirable life of all. But it is rarely available to people afflicted with the Triple Package.

CHAPTER 7

IQ, INSTITUTIONS, AND UPWARD MOBILITY

"Horatio Alger Is Dead": so proclaimed a *Daily Beast* headline in 2012. Obituaries reporting the demise of upward mobility in America are ubiquitous; they have become the new mythology of American upward mobility. And while the new myth, that mobility in America no longer exists, contains important kernels of truth, it is just as misleading as the old myth, that America is a classless society.

The central statistic cited when American upward mobility is pronounced dead—found in a 2008 Pew study—is that 42 percent of people raised in the lowest economic quintile remain there as adults. The first point to observe about this figure should be obvious. If 42 percent stay in the lowest income bracket, 58 percent don't. In other words, a *majority* of Americans born to the poorest families in the United States escape such poverty. (A different Pew study, using a larger data set, found that two-thirds of Americans born in the lowest quintile rose to a higher quintile as adults.) Rising remains the rule in America, not the exception.

The 42 percent figure has also been cited repeatedly to show that

America compares badly on the upward mobility front with other developed countries. In Denmark, one study found, only 25 percent of men born in the poorest quintile remain there as adults. What's more, according to the same study, only 8 percent of American men rose from the bottom fifth to the top fifth, as compared, say, with 12 percent in Britain. Thus America isn't a country of exceptional rags-to-riches opportunity; it's a country where rising from poverty to wealth is exceptionally difficult and rare.

These statistics can be found everywhere today, online and in print. What's infrequently mentioned is that these studies largely *exclude immigrants*. (As the Pew Foundation explains, when researchers examine intergenerational mobility in a given country, "[i]mmigrant families are not included in the surveys for the simple reason" that immigrants' "parents were born in another country.") In America's case, this omission is quite substantial; there are more than 40 million immigrants in the United States. The overwhelming majority arrived after 1965, but they *and their children* are almost entirely excluded by the upward-mobility studies conducted thus far. Indeed, the same Pew study repeatedly cited for the death of upward mobility in the United States expressly cautions that its findings do not apply to "immigrant families," for whom "the American dream is alive and well."

America offers exceptional upward mobility to an exceptional number of immigrants and their children (which is why so many people want to come here). This is true across the board—the children of poor Hispanic, African, and Asian immigrants generally experience strong upward mobility on a variety of measures—although it's especially pronounced in certain Asian groups. This phenomenon demands attention and substantially complicates the claim that upward mobility is dead in the United States.

Shrinking upward mobility in America is a serious problem, and being born to rich parents is of course a huge advantage. If you're born with the proverbial silver spoon, you can grow up to be wealthy without hard work, discipline, drive, or any other Triple Package quality. (The Triple Package is not a prerequisite to wealth; someone wins the lottery every week.) But to the extent that a group passes on its wealth that way, it's likely headed for decline.

At the other end of the spectrum, for many nonimmigrant groups in America, poverty *is* deeply entrenched. For example, Owsley County, Kentucky, in the heart of Appalachia, called a "pauper county" in the 1890s, remained in 2010 one of the poorest counties—according to some, *the* poorest county—in the continental United States.

This chapter is about upward mobility and the Triple Package. Much of this book has been about America's most successful groups and what they have in common. But what about America's poorest groups? This chapter looks at some of America's persistently low-income groups. It shows not only that these groups lack the Triple Package, but that immigrant upward mobility (as well as the recent dramatic rise of Mormons) is a quintessential Triple Package phenomenon.

This chapter is also about other causes of success and nonsuccess. The Triple Package is a set of cultural forces, but culture doesn't exist in a vacuum. Below we'll explore how the Triple Package stands in relation to other theories and determinants of differential group performance—including IQ, immigrant selection, institutions, and class rigidity—reinforcing some of them, working independently of others, and explaining important phenomena the other theories can't account for.

By statute, Stuyvesant High School in Lower Manhattan accepts students based solely on standardized test scores. Although tuition is free, Stuyvesant is one of the best high schools in the country. Like Phillips Exeter Academy in New Hampshire (tuition: $35,750), Stuyvesant reportedly sends upwards of 25 percent of its graduates to Ivy League or other top-tier universities. In 2013, the school's new admittees consisted of 9 black children, 24 Latinos, 177 whites, and 620 Asians. At Bronx Science, another of New York City's superelite public high schools, 64 percent of the students are Asian American.

It bears emphasizing that many of the Asian kids getting into these schools are from poor immigrant families. In 2012, more than three hundred students got into Stuyvesant and New York City's other selective public high schools from three Brooklyn zip codes covering neighborhoods like Sunset Park, with heavily immigrant Chinese American populations and average incomes low enough to make schoolchildren eligible for a free lunch. Higher-income neighborhoods like Manhattan's Upper West Side supplied most of the other students.

The lower working-class background of these Chinese students is significant because Asian success in the United States is frequently explained as a simple by-product of the head start that Asian immigrants bring with them. Since 1965, America has granted a large percentage of entry visas on the basis of technical skills and educational attainment. Asian immigrants are much more likely to have gained entry this way than are, say, Mexican immigrants, who more often come in under the "family reunification" program, as relatives of U.S. citizens. This gives some Asian immigrants a big leg up economically.

For example, many Indian immigrants graduated from that country's incredibly competitive Institutes of Technology, arriving in America with highly marketable technical skills. Some Chinese immigrants were top scorers on China's nationwide university exam—

the cream of the intellectual crop. It's hardly surprising when members of this cohort go on to get PhDs and become professionals or academics in the United States. Thus immigrant selection criteria have contributed substantially to Indian and Chinese American success.

But not all Indian and Chinese immigrants fall into this category. The Chinese parents in Sunset Park sending their children to Stuyvesant don't tend to have PhDs; they're more likely to be restaurant or factory workers. (In fact, a majority of Chinese immigrants do not obtain entry visas under the skills or education criteria.) The same phenomenon has been documented across the country. In Los Angeles, as elsewhere in the United States, the Chinese immigrant community is "bimodal": one segment came from highly educated backgrounds, while another arrived with only an elementary school education. Yet the children of both groups are enjoying exceptional academic success, outperforming their peers. Starting their lives in working-class households, these children are on their way to professional, high-skill jobs.

Thus immigrant selection criteria can't fully explain Chinese success in the United States. It can't explain the upward mobility of second-generation Chinese Americans whose parents came to this country with no educational or economic advantages. As a result, some have claimed that IQ must be the answer. But the data don't support this notion either. In a comprehensive review of fifty years of studies, IQ expert James Flynn found that higher IQ could not explain disproportionate Chinese American success. According to Flynn, notwithstanding their economic and academic outperformance of white Americans, Chinese Americans' mean IQ is no higher than that of white Americans.

Rather, Chinese Americans are getting more bang for their intelligence buck: Chinese Americans with an IQ of (say) 103 get significantly better grades in school, scores on tests, and ultimately higher-paying jobs than do white Americans with an IQ of 103. This is true even with respect to performance in math, which many consider the most IQ-correlated; thus Chinese Americans have been found to score significantly higher on calculus, analytic geometry, and SAT math tests than white Americans *of the same IQ level.* Which is to say: IQ is not the differentiating factor. As one developmental psychologist quips, "If Asian students were truly genetically superior to other students, they would not be spending twice as much time on homework each week as their peers in order to outperform them."

The real explanation of how Chinese American families in Brooklyn are sending their children to Stuyvesant and Bronx Science isn't complicated. They're working harder at it. Visit Sunset Park's Chinatown on a Saturday afternoon, and you'll find that the kids aren't enjoying a day off from lessons. They're studying math and English, often attending tutoring sessions offered by test preparation companies. Tutoring takes place not only on weekends, but during the summer. "I would say, in this area, probably 95 out of 100 Chinese students attend one of these programs," said the associate director of one such company.

Asian studying habits and their disproportionate admission to New York's elite public schools have provoked considerable backlash. In 2012, the NAACP Legal Defense Fund filed a federal complaint against the city, objecting to the vast underrepresentation of blacks and Hispanics, and claiming that admitting students solely on the basis of test scores was racially discriminatory. Some call Asian success unfair because other groups can't afford the test-prep programs. Still others object to the whole practice of sending kids to after-school and weekend tutoring, saying that Asians study "excessively" and that the regimen is too hard on children.

These objections are all understandable, but it's difficult to fault

Asian families for what they're doing. Many Chinese parents who pay for tutoring—which can cost $2,000 and up for a single student—are poor, like the couple who worked every day to send their three kids to tutoring, at a cost of $5,000 a year despite their combined income of only $26,000. Moreover, New York City offers free tutoring programs for poor families; roughly 43 percent of those taking advantage of these sessions are Asian (although Asian Americans make up only 14 percent of the city's public school students). As to the charge that they study "excessively," Chinese parents often say their kids are being blamed for having a good work ethic and good values; they disagree with the notion that they're working their kids too hard. One Bensonhurst parent of Chinese heritage said her daughter's tutoring sessions were nothing compared to what she had seen growing up: her father and brothers starving to death in Cambodia; her mother's struggles as a garment worker in America when they emigrated. "This is the easy part," she said.

Other Asian Americans in New York City express similar attitudes. "Most of our parents don't believe in 'gifted,'" said a Bangladeshi American boy of fifteen, a student at Bronx Science who had spent summers as well as weekends at a storefront "cram school," improving his math and reading. His immigrant parents, a taxi driver and a drugstore cashier, had instilled in him the familiar message of how much they had endured, both in their native country and in America. "You try to make up for their hardships," he said. "It's all about hard work."

In other words, Asian American success at New York's elite public schools is a classic example of Triple Package success. Both parents and children, with the usual combination of superiority, insecurity, and impulse control, are driving themselves to make it. It's also a classic example of old-fashioned American Dream upward mobility—

of rising from relative poverty to relative affluence, and not through luck or connections, but through hard work. Upward mobility is alive and kicking in the United States; it's just that not all groups are equally able to take advantage of it.

ONE PLACE WHERE MANY are not upwardly mobile is Appalachia, particularly Central Appalachia—a region that includes about half of Kentucky and Tennessee.

It's hard to write or talk about Appalachia even if you're from there. When in 2009 Diane Sawyer, who was born in southern Kentucky, did a *20/20* segment on poverty in Appalachia, she received not only national praise and a Peabody Award, but also an onslaught of furious criticism from the region she was trying to help. While some appreciated the documentary, far more reviews read like this: "[G]et out, stay out of people's lives." Or this: "[Y]our elite group is so condescending." (For some, once you leave Appalachia, you're no longer Appalachian.) Or: "Why don't you plan a trip to meet some of us with Master's Degrees, Professional Degrees, or advanced certifications. There are plenty of us here who are health conscious, self-sufficient, non-welfare drawing, non–Mountain Dew guzzling, non-addicted, intelligent, educated, and still have our natural teeth who would be willing to speak with you and show you the REST OF APPALACHIA!!!!!!!!!"

If you ask Appalachians where they're from, they don't tend to use the term Appalachia. They're more likely to name a state or part of a state, like West Virginia or Perry County, Kentucky. And they often describe deeply contradictory impressions. On the one hand, the aching beauty of mountain streams and rolling hills; on the other, the ugly reality of dying towns and derelict trailers, of mountains decapitated for strip mining. On the one hand, resentment at the negative stereotypes of rednecks, hicks, and white trash; on the other, a tendency to embrace those terms defiantly. Historian Anthony Harkins, of Western Kentucky University, has observed that "southern mountain folk" have both condemned the word *hillbilly* "as a vicious slur and embraced it in defense of their value system and in celebration of their cultural heritage."

There are plenty of myths about Appalachia—the region does not, for example, seem to have a disproportionate number of crystal meth addicts—but there are also facts. Rates of cancer, heart disease, and diabetes are higher in Appalachia than throughout the rest of the country. In Central Appalachia, fewer than 12 percent of adults have college degrees (compared to 27.5 percent nationwide). Although gains in many parts of the region have been significant, much of rural Appalachia—and the region is 42 percent rural, as compared with 20 percent for the rest of the country—remains mired in a poverty unusual by American standards. Owsley County's status as one of the nation's poorest is not unique; the same is true of four of its neighboring Kentucky counties. In fact, of America's one hundred lowest median-income counties, roughly a quarter are in Appalachia, making the region as a whole probably the poorest in the United States.

To the extent there's such a thing as "Appalachian culture," it is not a Triple Package culture. Insecurity is certainly rife throughout the region, in terms of both economic worries and a perception of being looked down on by the rest of the country. But while many Appalachians take fierce pride in their region and heritage, it's safe to say that most do not have a deeply internalized sense of superiority vis-à-vis the rest of America.

Like African Americans, Appalachians have been relentlessly subjected to stereotypes of inferiority—genetic, moral, cultural, intellectual. (It's far more socially acceptable today to insult and look down on "white trash" than the poor of any other racial group.) These stereotypes are well-known among Appalachians. They've been fighting against them for years, often with power and insight, as in bad boy Jim Goad's *Redneck Manifesto*:

> Our magazines and sitcoms and blockbuster films are crammed with slams of hicks and hayseeds and hillbillies and crackers and trailer scum. . . . Gradually, we come to believe that working-class whites are two-dimensional cartoons—rifle-totin', booger-eatin', beer-bellied . . . homo-hatin', pig-fuckin', daughter-gropin' slugs. . . .
>
> America's hate affair with white trash is, ultimately, self-hatred. Guilt projection. A convenient way for America to demonize itself, or, rather, to exorcise the demon and place it somewhere outside of itself.

One common response to all the redneck jokes, as Kentuckian writer Ann Shelby says, is "just to go along. For every briar hopper joke they tell you, tell them two. 'Why did they build a bridge across the Ohio River?' 'So the hillbillies could swim over in the shade.' Yuck it up. Pretend not to be all that bright. Manufacture and exaggerate your hillbilly credentials. I have a friend who practices this method and recommends it highly. 'You're pulling their leg all the time,' he says. 'And they don't even know it.'" Another response is "anger. . . . It can come across as mere testiness or oversensitivity, but it is the product of a thousand insults, small and large. In this huge and diverse region, it is, perhaps more than anything else, what we have in common."

In Cincinnati, which has a large Appalachian population, educators have established community schools to "counter the 'dumb hillbilly' stereotype." As the coordinator of one school puts it, "What we have consciously tried to do, and have had some success in doing, is to help people feel and know it is OK to be who they are." Needless to say, none of this reflects a deep-seated superiority complex.

What about impulse control? The frontier tradition of mountain toughness and indomitable spirit has not disappeared, but something different seems to have settled into rural Appalachian culture. Many observers, perhaps contributing to the stereotypes just mentioned, have described a "defeatism," "dejection," or "surrender" prevalent in poor Appalachian communities. In his memoir *At Home in the Heart of Appalachia*, West Virginia writer John O'Brien described "Appalachian fatalism" as a "profound sense that you are fundamentally inferior and that life is absurd and hopeless."

Religious attitudes may play a role here. "We're a religious bunch, Appalachians," writes Denise Giardina, who grew up in West Virginia's Black Wolf coal camp. "And so a lot of people in the mountains have adopted this attitude of: 'Ain't no use worrying . . . because Jesus is going to come anyway!' . . . . People pray and get saved and wait for the Lord to fix everything." Sociologist Kai Erikson says that "the mountaineer" is apt to feel "helpless before the God who reigns over Appalachia, helpless before the crotchety ways of nature, and helpless before the crafty maneuvering of those who come to exploit him and his land." Others have argued —notwithstanding the fact that many Appalachians are working two and three jobs just to make ends meet—that the once-strong mountain work ethic has been undermined by government welfare.

Whatever the source, the fact is that Appalachia shows disturbing signs of impulse control's opposite. Obesity is common, and while many factors contribute to obesity, cultural attitudes seem to play a significant role. Author J. D. Vance, whose family is from Breathitt County, Kentucky, writes in his compelling memoir:

As kids, we never learned about portion control, or to otherwise connect what we put into our bodies with any actual consequences. We all made fun of certain types of exercise—"running is for dorks; they don't even use a ball!"—but celebrated our young kids for finishing four heaping plates of biscuits and gravy at the local breakfast buffet. It was like excess was a cultural value.

Substance abuse is also widespread. Oxycontin is the narcotic of choice; Appalachia's abuse rates of prescription opioid painkillers are among the highest in the country. So-called "pill mills" can be found throughout Appalachia, giving rise to the new sobriquet "pillbillies." In one Ohio county in 2010, nearly 1 in 10 newborns tested positive for drugs. The region has some of the highest teen pregnancy rates in the country, perhaps because young women are expected to be mothers rather than breadwinners. In Appalachian Kentucky, the birth-rate among fifteen- to nineteen-year-old women in 2009 was almost four times that of New Hampshire (and about 50 percent higher than the nationwide figure).

In short, Appalachian culture is not a Triple Package culture—but that's not the cause of Appalachian poverty. This point is so important we're going to repeat it: *The absence of the Triple Package did not cause poverty in Appalachia.*

Geography, history, tectonic shifts in the economy, and a version of the "resource curse" (with its associated corruption, exploitation of cheap labor, and highly skewed distribution of resource-extraction wealth) were the chief causes of Appalachian poverty. Traditional industry declined. Salt and timber reserves were extracted without a rechanneling of wealth into the region. Coal mining became increasingly mechanized. Mountaintop removal, while enormously profitable for the coal companies, has not only left behind environmental wastelands; it has destroyed jobs. In West Virginia alone, mining jobs have dropped from 120,000 to 15,000. According to one study, the returns on a high school education in Appalachia are falling. As the study's author concludes, Appalachia is a "double jeopardy" region of low skills and low returns to skill—a recipe for economic stagnation. At the same time, catastrophic industrial accidents in Appalachia, like the Buffalo Creek flood of 1972 or the Martin County sludge spill of 2000, do not provoke the same outcry as similar disasters elsewhere in the country (the Martin County spill was thirty times larger than that of the *Exxon Valdez*, but it was the latter that became worldwide news), and the victims too often go uncompensated.

In these circumstances, dejection and fatalism are natural, predictable consequences. So is the grabbing of immediate gratification where it can be found.

By now almost everyone knows about the marshmallow test, but one of the most important recent findings has not received enough attention. Many interpret the marshmallow test as a measure of character, identifying the kids with the strongest self-control. But in 2012, researchers reran the test with a new wrinkle: before the children received their first marshmallow, some of the kids had an encounter showing the adults to be unreliable. The adult administering the test would tell the kids, for example, that if they waited a few minutes they would get some "exciting art supplies" to play with, but then wouldn't follow through. After this kind of encounter, virtually all the kids "failed" the marshmallow test; that is, they gobbled up the first marshmallow instead of holding out for two. (Other kids had the same encounter with the adult, except that the adult *did* follow through; most of these kids passed the test.) But the kids who "failed" weren't displaying a lack of self-control. They were responding rationally to the fact that the adults who promised them a second marshmallow had proved themselves untrustworthy.

If people don't trust the system, if they think society is lying when it tells them that discipline and hard work will be rewarded—if they don't think that people like them can really make it—they have no incentive to engage in impulse control, sacrificing present satisfactions in hopes of future gain. This is as true in America's inner cities as in rural Appalachia. Studies have shown that childhood poverty and abuse make people more likely to choose immediate gratification over greater delayed rewards. Just as American society can grind down a group's superiority narrative, so too can it grind away their impulse control. Thus the same conditions that cause poverty can also suck the Triple Package out of a culture.

But once that happens, the situation worsens, and poverty becomes more entrenched. Girls who become pregnant as teenagers end up, statistically, with far worse economic outcomes. So do kids who don't get a higher degree. So do people with drug addictions. The Appalachian Regional Commission recently reported "growing concern that substance abuse is eroding the economic and social fabric of the Appalachian region." The same is true of many groups all over America, whether white or black, rural or urban. They did not originally fall into poverty because they lacked the Triple Package, but now that they do lack it, their problems are intensified, and the hole they're in is even harder to climb out of. Under these conditions, it takes much more grit, more resilience—perhaps a more exceptional individual—to escape the cycle.

A few groups in America are not upwardly mobile because they don't want to be. The roughly two hundred thousand Amish living mainly in Pennsylvania, Ohio, and Indiana are a prime example. The theory of the Triple Package fits very well with such groups; they clearly don't have it. They also throw a fascinating light on the relationship between the Triple Package and Christianity.

Impulse control is not a problem for the Amish, even if the different orders vary to some degree in austerity. The Renno Amish drive black-topped horse buggies; the Byler Amish, yellow-topped; the rakish Indiana Swiss Amish go topless. The Beachy Amish actually drive cars. At the other extreme, the Swartzentruber Amish live with no modern heating, no indoor plumbing or continuous hot water, no refrigerators, no tractors, and almost no contact with the outside world.

But most Amish children, starting "when they are in diapers," are taught "complete obedience," discipline, and unconditional compliance with the strictures of their faith—not just plain clothes and no electricity (obviously no TV or video games), but also no exploring or cavorting around, no questioning authority. Fairy tales, science fiction or fantasy, and stories with talking animals are frowned upon. At age three or four, Amish children begin working around the house or farm. They also have to sit through hours of adult-oriented church services, conducted in "high" German. The difference between Amish and "English" children is striking. As one local orchard owner observed, Amish children who come to pick fruit "usually keep to the task till the parents are satisfied. . . . English children, on the other hand, may enthusiastically pick cherries for five to ten minutes till they get bored and end up in cherry fights, tree climbing, or sulking."

In other words, the Amish have the third element of the Triple Package in spades. Do they have a superiority complex?

The answer the Amish would give is surely no. Their creed is one of extreme humility; pridefulness is a sin. A favorite biblical passage is "Do not be haughty, but associate with the lowly; never be conceited." Unlike Mormons, the Amish view man as infinitely below God, and they are taught to abhor any effort by one individual to rise above others. Indeed, the reason they oppose education beyond eighth grade is that they believe "*high* school and *higher* education produce

Hochmut, or high-mindedness and pride . . . the antithesis of simplicity and humility."

Nevertheless, claims of humility can themselves be upside-down superiority stories (the most humble people on earth). Nietzsche—a man with a superiority complex if ever there was one—believed that all of Christianity was founded on a reverse superiority claim of this sort, exalting the poor and pitiable in order to destroy aristocratic civilization, ultimately replacing it with modern democracy and the equality of all mankind. In the Amish case, the truth may be somewhere in between.

According to leading Amish scholar John Hostetler, the Amish "are not highly ethnocentric." They accept other people "without attempting to judge or convert them." They don't even believe they are saved (they hope so, but believing it would be prideful). On the other hand, the Amish implicitly view the rest of America as morally "contaminat[ed]." Comments like "Letting children go unsupervised to watch the trash that comes over the TV every day has to be the greatest form of child abuse" indicate that Amish parents view their way of raising children as better. At some level, consciously or not, most Amish presumably take pride in how unprideful they are and believe that living a simple, frugal, industrious life, seeking no superiority over anyone else—makes them superior, or at least is a superior way to live.

But whether or not the Amish have a superiority complex, there's one thing they definitely don't have: Triple Package insecurity.

Insecurity in Triple Package cultures is a discontented, anxious uncertainty about your worth or place in society, a feeling that what you've done or what you have is in some fundamental way not good enough. This insecurity is stingingly personal, goading you to constantly compare yourself with others, to crave recognition and respect. This is what creates the hunger to rise, to "show the world."

It's barely an exaggeration to say that the entire thrust of Amish culture is to *not* instill Triple Package insecurity. The Amish teach their members to be indifferent to their place in American society. They reject the idea that their worth is measured by American material values. They aim not to "show the world," but "to be separate from the world." They have successfully turned inward, shunning the rest of the country's judgment, values, and criteria of success.

More than this, the Amish are taught precisely to feel that the few, simple things they have *are* good enough. They rigorously try to suppress the kind of thinking that leads individuals to strive against one another, to compare themselves against their fellows, to compete over who has accomplished more.

For example, when Amish boys began playing and winning in local softball leagues, Lancaster bishops opposed it. "What if the motive is no longer relaxation and diversion, but a spirit of competition?" one Amish newspaper asked. "What if these teams compete in tournaments and win the state championship in their class, and then go on to the nationals?" Ball games are not per se sinful, but they are to be condemned "when batting and catching a ball becomes more than play—when it becomes serious competition."

Of course the Amish have their worries about making ends meet, about their crops, about their kids—but the overwhelming impression described by visitors is one of peacefulness and acceptance. The Amish don't seem to have any chip on the shoulder. They aren't stung by "English" perceptions of them. They aren't worried about proving themselves in America's eyes at all.

Thus the Amish demonstrate one way a group can avoid the Triple Package: by using religion to stay outside the system altogether.

America's most successful groups, as we've said, are all outsiders, but they are outsiders *who want in*—or, at a minimum, who want success as America defines success. Groups like the Amish, turning their backs on America, find security within their group and within their faith, suppressing the insecurity that makes others crave outward signs of success like money, prestige, and power.

THE AMISH ALSO HIGHLIGHT how foreign the Triple Package is to the millennia-old strand of Christian teaching that repudiates all striving for worldly success and wealth. ("The love of money is the root of all evil"; "the last shall be first"; "it is easier for a camel to go through the eye of a needle than for a rich man to enter the kingdom of God.") Christianity excelled at teaching the poor to seek security in faith, love, and salvation, not in worldly possessions.

A theological revolution was required to bring the Triple Package into Christianity, and—at least according to Max Weber's famous account of early Protestantism—precisely such a revolution took place at the time of the Reformation. When it did, those who followed the new Christian faith rose to economic dominance all over the world, including America.

Weber's classic study of the Puritans and other early Calvinists, *The Protestant Ethic and the Spirit of Capitalism*, was arguably the first systematic sociological inquiry into disproportionate group success. That book famously began by documenting, as Weber put it, a "remarkable" fact: that Protestants dominated Catholics economically throughout Europe and the United States. Weber explained this phenomenon by uncovering in early Protestantism an unusual set of cultural traits that, while of course not expressed in the same terms, essentially track what we're calling the Triple Package.

According to Weber, Protestant sects such as the Puritans, heavily influenced by Calvinist doctrine, viewed themselves as a chosen people, a tiny minority elected by God for salvation while the rest of humanity foundered in darkness and sin. But they also had a whopping case of insecurity. Calvinist doctrine taught that mankind could not know who was saved and who wasn't. God had already predestined your fate, and nothing you could do—no good deeds, no confession of sins, no magical sacraments—could change His plans. "The question, *Am I one of the elect?* must sooner or later have arisen for every believer and have forced all other interests into the background. *And how can I be sure . . . ?*"

To make matters worse, doubt about one's own salvation was evidence of imperfect faith and therefore a sign of damnation. This condition of personal, theological, and epistemological insecurity led to a "doctrine of proof" through worldly success. The purpose of material wealth was not luxury. Rather, the early Protestants believed that men's occupations were or should be their "calling"—the work God intended them to do on earth. Thus a person could prove his faith and election through relentless hard work, thrift, self-discipline, and ultimately profit. "Above all," said Weber, a Protestant businessman "could measure his worth not only before men but also before God by success in his occupation."

In other words, Protestant success in America was a version of Triple Package success. The reason the Puritans toiled and saved (impulse control) was that they simultaneously believed they were special (superiority) and literally had something to prove (insecurity) at every moment of their lives. They had to show everyone—Catholics, other Protestant sects, their congregation, themselves—that they were indeed God's chosen, and the way they would do so was through the accumulation of wealth.

. . .

Part of what led Weber to a cultural explanation of Protestant economic success was that no other explanation worked. Inherited wealth, Weber observed, might explain some of the disparity between Protestants and Catholics, but not all. Nor could political or national differences account for Protestant success. It wasn't, for example, that all Protestants were English, and the English were richer and more industrialized than everyone else. No, whether you started with Huguenots in France or Puritans in America, and whether you looked at countries with Protestant minorities or majorities, you found the same puzzling pattern of Protestant economic dominance—requiring an explanation within Protestant culture itself.

It is a striking feature of Mormon success in America that it too defies all the leading explanations of group economic performance. Obviously Mormon accomplishment isn't a by-product of immigrant selection bias, because most Mormons aren't immigrants. And what is the proponent of IQ explanations to say—that a mass genetic mutation in Utah made Mormons smarter than other Americans? Scholars like Jared Diamond have shown how geography can shape people's economic fate, but it's hard to think what a geographical explanation of Mormon success would even look like.

Mormon success defies biological, geographical, and demographic explanations because it's Triple Package success. In fact, the Mormons' Triple Package parallels the early Protestants'. Mormons have inherited the Puritans' sense of chosenness, their immense self-discipline, and their faith not only that they have been placed on earth with a calling, but that, as the author and historian James Carroll puts it, achieving worldly success is a "sign of divine favor."

Even the timing of Mormons' rise to prominence conforms with the theory of the Triple Package. It was only in the twentieth century that the Triple Package consolidated in Mormon culture, laying the groundwork for their present disproportionate achievement.

Today, strict self-discipline is a fixed star of Mormon culture, manifest in their abstemiousness, their grueling two-year missions, and their sexual conservatism relative to the permissiveness more typical in America. But none of these elements of Mormonism was clearly established until the twentieth century. The Word of Wisdom, which is the textual source of Mormonism's injunctions against alcohol, tobacco, coffee, and drugs, was viewed as merely good advice until 1921, when the Church made it a code that Mormons had to follow in order to participate in temple rituals. The Mormons' "Family Home Evening"—a night set aside each week for family hymn singing, prayer, and religious reading—also began in the early twentieth century, as did a new emphasis on encouraging Mormon youth to participate in church activities.

Missionary work increased stunningly as the century wore on. From 1830 to 1900, the Church estimates that roughly 13,000 Mormons went on mission; from 1900 to 1950, 50,000; from 1950 to 1990, 400,000; since 1990, well over 650,000. And as for sexual permissiveness, until the Church banned polygamy in 1904, the Mormons were the libertines.

The Mormons' renunciation of polygamy also brought to a close what might be called their separatist "Utah phase," a period of isolation in which Mormons sought refuge in the remote Salt Lake Valley from a country that spurned and attacked them. As they sought to put the polygamy issue behind them, Mormons turned outward, engaging with America, hanging U.S. flags in the Salt Lake City tabernacle, forming alliances with the Boy Scouts. In Triple Package terms, nineteenth-century Mormons were like the Amish, finding security in their own faith, values, and insularity, heedless of what the rest of the country thought of them. But in the twentieth century,

Triple Package insecurity became much more predominant. As this outward turn came to define twentieth-century LDS culture, Mormons became eager to prove themselves to America, to show they could succeed in America—indeed that they could lead America, that they were the most American of Americans.

It's impossible, of course, to prove that the arrival or strengthening of these Triple Package elements in Mormon culture drove their success, but there's some interesting indirect support. A century ago, some Mormon communities refused to go along with the renunciation of polygamy and splintered off from the main church. One such group, the Fundamentalist Church of Jesus Christ of Latter-Day Saints (FLDS), retains a significant number of followers even today (perhaps 10,000), located mainly in Colorado City, Arizona. The FLDS broke away before the main church made the Word of Wisdom binding and in other ways began moving Mormon culture toward increased self-discipline and self-denial. As a result, the FLDS don't refrain from alcohol, cigarettes, or coffee. They retained, moreover, the inward focus of earlier Mormon history, living their own way regardless of what the rest of America thinks of them. And as it happens, Colorado City is one of poorest cities in the United States.

THE CHILDREN OF HOLOCAUST SURVIVORS—so-called second-generation survivors—have been studied extensively over the last several decades. To no one's surprise, symptoms of post-traumatic stress disorder have been found in this group, both in their behavior and in higher levels of cortisol, a stress hormone. (It has even been suggested, on the basis of new discoveries in epigenetics, that the parents' trauma may have affected their children's genetic inheritance, leaving so-called genetic scars on the next generation.) The surprise, rather, came in a very different finding: that second-generation survivors outperform other groups economically, not only non-Jewish groups, but Jews as well, including Jews who immigrated at about the same time.

This unexpected finding prompted sociologists to take a closer look at the causes of success in Holocaust-survivor families, and what they found was—the Triple Package.

One of the most powerful and poignant impressions conveyed in interviews with the children of survivors is the extent to which they felt they had to hold themselves in check, to suppress their own feelings, needs, or rebelliousness, out of care for their parents:

> I remember as a child always worrying. I felt I had to take care of my parents. I think [my mother] looked up to me as being a mature person and someone she could count on even though she didn't want to count on me. I felt children around me had much more freedom, a carefree attitude. . . . I was more burdened in feeling responsible for my parents. I was very mature. I was never a child, able to play and have fun.

In his book on Holocaust survivors' children, psychologist Aaron Hass, himself a member of the "second generation," writes:

> 'For this I survived the Nazis? For this I survived the camps?' This was my parents' frequent anguished refrain—if I talked back to them or if I came home later than I said I would. . . .
>
> I, my needs, seemed to slip away in the face of their horrible past. Given my understanding of what they had experienced, how could I cause them more grief? How could I electrically prod an already exposed, frayed nerve? . . . Their reservoir of pain was already straining the limits of the fragile structure enclosing it. My parents, I believed, were always on the edge. . . .

Acts of rebellion were not a part of my childhood or adolescence. . . . After having learned more details about what actually happened to Jews in Europe, it has been even more difficult to act contrarily, to criticize, to say no.

In agonizing fashion, these statements describe a heightened need for impulse control, communicated to children at a very young age. Researchers have even found that as young children, second-generation survivors often seem to have been "inhibited from making noise."

As to insecurity, the anxiety of survivors' children is in some ways too obvious to require description. In addition to worrying about their parents' fragility, they were often taught that the world might turn on them, that they needed to work harder at school and in life precisely because of life's precariousness. Helen Epstein, also a second-generation survivor who has conducted extensive interviews with other survivors' children, quotes a twenty-nine-year-old Canadian man:

My parents always said that a person can lose everything, but what's inside his head stays there. I had to acquire an education, they said, because our enemies could take everything away from us but that. Little by little, as I became aware of who the enemies were, I began to understand.

In her moving memoir, *The Watchmaker's Daughter*, Sonia Taitz (again, a second-generation survivor) writes:

When college ended, I had fears of going back to Washington Heights, my mother's onion-potato kitchen and my own set of pans. I didn't want to be sent to the wrong line—the death line, then heaped in a pile. I wanted to strive, to win.

One of Hass's respondents, a television scriptwriter, remembers:

> My mother raised us to be self-sufficient, prepared, and success-oriented because she was afraid . . . we wouldn't be able to take care of ourselves in case something like the Holocaust happened again. We had to do well to protect ourselves, more than to make ourselves happy.

On top of this insecurity, another was layered: survivors' children frequently describe a pressure to do well in life in order to redeem their parents' hardships, to make good their sacrifices. One study summarizes this attitude as follows: "They felt it was up to them to bring joy, pride, and pleasure into their parents' lives. This obligation to make parents happier through their own lives was described by all participants. . . . The adult children described feelings of constant pressure to fulfill goals that parents had set for them."

If this burden was in some ways typical of second-generation immigrants, it was especially acute in the case of survivors' children. This was so not only because of the intensity of their parents' suffering. A "need to resurrect their lost families" ran deep among many survivors. Their children were somehow expected to replace the family members whom the Germans (or Poles or Ukrainians or others) had murdered. Children were often named after the murdered, parents would impress on them a "strong need to make up for the loss of their deceased family members by telling them such things as 'you are all we have left,' while pointing out similarities to those relatives who perished." To quote just two second-generation survivors:

> I was not David Greber, but my father's brothers Romek and Moishe and Adamek, and his father David; my brother wasn't Harvey, but

Herschel, my mother's beloved brother, or Aharon, her father; my sisters were named for our grandmothers and aunts Sarah and Leah and Bella and Molly, loved ones our parents last saw when they . . . were being separated for transportation to camps from which they never emerged. Representing six million dead is a grave responsibility, and a terrible burden for a child to carry.

I wondered what I could do in my life that would even register. . . . At times, my life seemed to be not my own. Hundreds of people lived through me, lives that had been cut short in the war. . . . My parents, too, were living through me. They saw in my life the years they had lost in the war and the years they had lost in emigrating to America. My life was not just another life, I thought often when I was a child, it was an assignation.

And yet despite these impossible-to-meet expectations, despite the need to suppress their own impulses, survivors' children repeatedly describe their peculiar heritage in terms not of victimization, but of pride. A corollary of the much more famous "survivor's guilt," it turns out, is survivor's superiority. As another survivor's child told Hass:

I make a differentiation between Jewish survivors and Jewish non-survivors. I always felt almost proud that my parents were survivors. Perhaps I thought they were better Jews because of all they had sacrificed and been through. I always felt I was better than other Jews. I felt proud, almost as if *I* was there.

Epstein reports a similar comment from one of her interviewees:

I may have said *my parents were in concentration camps* calmly, smoothly, but in my ears the sentence rang like a declaration of loyalty. It put me squarely on the side of "those people," far away from the complacent, untouched Americans—Jews or Gentiles—who seemed to be so quick to make assumptions about things they did not understand. I answered their pity or embarrassment or confusion with pride.

Taitz expressed a sentiment along the same lines:

My father and mother were both concentration camp survivors. Not victims—survivors, people who had looked death in the face and rebutted it. . . . I never thought of them as weak, but as God-like warriors themselves, however wounded.

Thus, in the children of Holocaust survivors, we find the complete Triple Package, if in a strange and painful guise. Indeed, reading descriptions of this cohort, you'll find almost express Triple Package terminology: the "sense of superiority" that many survivors' children have; the "lingering insecurities" that drive them "toward educational and occupational success." Out of this combined superiority and insecurity came an almost compulsive need to work, to persevere, to achieve: "Permeated by an intense drive to build and achieve, the home atmosphere of fighter survivors was filled with compulsive activity. Parents forbade any behavior that might signify victimization, weakness, or self-pity. Pride was fiercely held as a virtue, relaxation and pleasure were deemed superfluous."

IT'S FITTING TO CLOSE THIS CHAPTER with the possibility of group decline. In 2012, Ron Unz, former publisher of *The American Conservative*, assembled striking data that, he argued, demonstrated a recent "dramatic collapse in Jewish academic achievement." Using admittedly imprecise surname-based analysis, Unz estimated that Jews dropped from more than 40 percent of U.S. Math Olympiad top scorers in the 1970s to 2.5 percent since 2000. From 1950 to 1990, Jewish high schoolers made up roughly 20 percent of the finalists in the prestigious, nationwide Intel Science Talent Search; since 2010, only 7 percent. In the Physics Olympiad, Jews accounted for 25 percent of the top scorers as recently as the 1990s; in the last decade, just 5 percent.

If Jewish academic performance is declining, whether or not as dramatically as Unz says (many have criticized his methodology and claimed his data exaggerate), this drop-off would throw a wrench in a belief held by lot of people: that Jews do better academically for the simple reason that they have higher IQs. By contrast, a decline in Jewish performance would be wholly consistent with the theory of the Triple Package.

The Triple Package and claims of higher Jewish IQ are by no means mutually exclusive. If it were true that Jews had a higher mean IQ—which has not been established*—that would feed directly into the Jewish superiority complex, reinforcing their Triple Package. But higher IQ couldn't by itself explain Jewish success. To repeat: study

---

* Widely cited but not universally accepted IQ studies have shown higher scores in Ashkenazi Jews (although not Sephardic Jews), with a mean ranging from 108 to 115 (as opposed to a U.S. mean of 100). To explain higher Ashkenazi IQ, the theory most in the news today holds that centuries of anti-Semitic restrictions in medieval Europe "naturally selected" for Jewish intelligence. Barred from ordinary occupations, Jews could survive only if they were good at commerce, money-lending, or other jobs "that people with an IQ below 100 essentially cannot do." Wealthier (and thus more intelligent) Jews had more offspring, while some less intelligent Jews converted to survive. Ironically, in one version of this theory, the same genes that produce enhanced Ashkenazi intelligence are responsible for Tay-Sachs and other "Jewish diseases." At present, these claims remain speculative and hotly debated. "[I]t's bad genetics and bad epidemiology," said a prominent geneticist of the leading paper. A historian was more circumspect: "I'd actually call the study bullshit." Other commentators have been more receptive. For sources, see the endnote.

after study has proved that IQ is not a complete predictor of success. IQ without motivation lies fallow. Drive predicts accomplishment better than IQ, and the Triple Package generates drive.

But drive, unlike IQ, is something a group can lose in a single generation. And if the rise and fall of past successful groups in America is any guide, Jews are long overdue for a fall.

Immigrant success in the United States is almost always a two-generation affair. For most immigrant groups, including Asians, Africans, Hispanics, and Afro-Caribbeans, decline sets in at the third generation. For example, "first- and second-generation Asian students outperform whites, whereas there is no performance difference between the third-generation Asian and white students." All this is exactly what the theory of the Triple Package would predict: assimilation and success weaken the insecurities and other cultural forces that drove the first and second generation to rise.

Today, the majority of Jews in the United States are third-, fourth-, or fifth-generation Americans. Most are the children of lawyers, doctors, bankers, or other white-collar workers, relatively secure both economically and in their identity as Americans. Bellow and Roth, who wrote "Jewish American" literature, with predominantly Jewish characters speaking in Jewish voices, have given way to Matthew Weiner, who created *Mad Men*. No longer the outsiders their immigrant forebears were, today's affluent American Jews should be ready to follow the seemingly inevitable Triple Package trajectory into decline—not unlike the mid-twentieth-century WASP establishment that *Mad Men* depicts.

But it's possible that because of the Jews' unique history, their Triple Package is less dependent on immigrant status and more durable even in the face of wealth. In no other group has the coupling of a superiority complex and insecurity been so core to its historical identity. In America today, Jews may be part of the economic and cultural elite, but the memory yet lingers of Jews just as affluent and assimilated in Weimar Germany. And while Jews may no longer fear persecution in the United States, Israel—with which American Jews often feel an almost ethnic identification—remains surrounded by 220 million people many of whom deny its right to exist and some of whom openly call for its annihilation.

Perhaps the best evidence that Jews are still insecure is the consistently apoplectic reaction to any suggestion of Jewish decline. By contrast to WASPs, who seem to accept their plight almost cheerfully—time to trade in the sailboat for a canoe!—Jews don't seem ready to throw in the towel yet. There are no books with titles like *Cheerful Money: The Last Days of Jewish Splendor.* On the contrary, Jews are insecure about losing their insecurity and as anxious as ever about their population, their future, their religion, their identity.

From this point of view, looking at math scores or science competitions for evidence of Jewish decline may be misplaced. Concentrating on science and math would have been perfectly logical for American Jews throughout much of the twentieth century, when they had less English proficiency and faced greater barriers to success in fields like law, politics, or publishing (just as East Asian immigrants focus on science and engineering today). It bears remembering that Jews in America top the charts not only in terms of standard metrics like household income or net worth or even Nobel Prizes. To a degree shocking given their minuscule population percentage, Jews are also among America's preeminent poets and jurists, directors and journalists, comics and comic book artists, opinion leaders of both the left and right.

In other words, Jews in the United States are further along the Triple Package trajectory than America's other disproportionately successful groups; Jews have confronted the "problems" of success and assimilation in a way that America's other Triple Package groups have not. In a sense, the fundamental difference between American Jews and Asian Americans is that the Jews are three generations ahead of them. As mentioned at the outset of this book, there are several fates that can befall a successful Triple Package group in America. One is disappearance through assimilation and out-marriage. Another is decline. But the more intriguing possibility is that a group might find a way to turn its Triple Package capacities and energies in new directions, maintaining its identity but achieving previously uncharted (as well as conventional) forms of success. All these possibilities are open to American Jews today; all might be realized, to one extent or another.

The same issues are already facing Mormons, a growing number of whom—especially in the younger generation—are bridling at the culture of conformity that has become part of the Latter-day Saint image. Chinese American business leaders are increasingly pointing to an "Asian" reluctance "to speak up, stand out and make waves" as a factor "limiting their own upward opportunities as corporate officers." Indian Americans are already looking beyond corporate success; "no longer confined to walking the corridors of corporate America," they are achieving renown as public intellectuals, judges, or authors. Whether breaking out of type in this fashion will allow these Triple Package groups to avoid the common trajectory of decline, in which disproportionate success fades by the third generation, remains to be seen.

To return to the Jews, the Achilles' heel in their Triple Package may turn out to be impulse control. If superiority and insecurity are almost hardwired into Jewish culture, impulse control doesn't seem to have quite the same purchase. As we've seen, self-control and discipline have long been associated with Jewish culture, but Jews in America are much less observant than they used to be, and Jewish parents today are often described as "permissive." Jews in the U.S. are no longer hyphenated Americans. They're just Americans; as such they are drawn to American attitudes toward impulse control. And these attitudes do not favor the systematic sacrifice of present satisfaction for future returns, which is the hallmark of every Triple Package culture.

The United States has a Triple Package trajectory of its own. Once a quintessential Triple Package nation, America has in recent decades moved in a very different direction. This trajectory—along with its reversibility—is the subject of the next and final chapter.

CHAPTER 8

AMERICA

WE TURN NOW TO a very different kind of question: whether America itself, as opposed to one of its many groups, has the Triple Package.

At the most basic level, the answer is simple. American culture today is not a Triple Package culture. On the contrary, the Triple Package runs directly counter to major tenets—almost mantras—of contemporary American thinking. That's why Triple Package groups do so well in the United States. Triple Package groups have an edge in America precisely because America as a whole lacks the Triple Package.

But in reality, the story is much more complicated—for two reasons. First, nations aren't groups. A nation as large and diverse as America can't be analyzed the same way that an ethnic or religious group can. America doesn't have only cultural variety. It has culture *wars:* protracted, impassioned debates about the country's basic values.

Second, there is the fact of historical change. As we'll show below, America's relationship to the Triple Package has shifted dramatically in the nearly two and a half centuries since the nation's birth. To cover this subject in any detail would take a book of its own.

Nevertheless, in this chapter we're going to offer a broad-brush portrait of the Triple Package in America. Despite America's diversity and all its political disagreements, there is still an overarching, recognizable American culture—otherwise we wouldn't hear so much about the "Americanization" of other parts of the world—and it's important to see, even if just in rough outline, when and how American culture broke away from the Triple Package. We're only going to book-end this history, focusing on America at the founding and today, and even so we can't possibly do justice to the details.

The United States was born with the Triple Package. All the usual trappings were on display: the drive, the grit, the chip on the shoulder, the longing to rise. These cultural forces helped propel Americans across a continent, turning thirteen ragtag colonies into an industrial and commercial giant, a military juggernaut, and eventually the most powerful country in the world.

But then something changed. After two hundred years, America lost its Triple Package, damaging our economy, our health, our relationship to future generations. This chapter is about how that loss took place. In the last half of the twentieth century, America declared war on both insecurity and impulse control. By 2000, all that remained of the American Triple Package was the superiority complex—which, by itself, leads not to success, but to swagger, complacency, and entitlement.

IT'S FUNNY THAT TODAY everything is bigger in America than in Europe—bigger cars, refrigerators, buildings, bathrooms, and especially beverages—because in Thomas Jefferson's time the idea was that everything American was punier.

"In America . . . all the animals are much smaller than those of the Old Continent," wrote the eminent French naturalist George-

Louis Leclerc, Comte de Buffon. "No American animal can be compared with the elephant, the rhinoceros, the hippopotamus, the dromedary, the camelopard, the buffalo, the lion, the tiger, &c." Having no great animals, reasoned Abbé Guillaume Raynal, another influential Enlightment writer, America could never be a great nation; any species transplanted from Europe to America would likewise grow small and feeble, including Europeans and their offspring. No wonder, concluded Raynal, that America had not "produced one good poet, one able mathematician, one man of genius in a single art or a single science."

Everyone knows about American "exceptionalism": the belief dating back to the Massachusetts Bay Colony that America was destined to be a New Israel, a light unto nations. John Winthrop told his fellow Puritans that their colony was to be a "City upon a Hill." A century and a half later, the first Federalist paper reminded Americans that it had been "reserved to the people of this country" to determine whether a nation based on political freedom could exist on earth. "[W]e Americans," as Melville would put it, "are the peculiar, chosen people—the Israel of our time . . . we bear the ark of the liberties of the world."

What's often forgotten is that side by side with America's superiority complex came a deep insecurity and massive chip on the shoulder—a need to prove itself to a supercilious Europe.

From Paris, Jefferson wrote home exhorting his hunter friends to send him a giant moose to refute Buffon. As president, he sent Lewis and Clark on their expedition in part to collect bones of large mammals, especially a mammoth—and sent one proof after another to the Natural History Museum of France. In his *Notes on Virginia*, Jefferson included a chart demonstrating that America's quadrupeds were bigger than Europe's, and observed that the "tremendous" mammoth (no sign of which had ever been found in Virginia) was by far "the largest of all terrestrial beings," at least "five or six times the cubic volume" of the Old World elephant.

Throughout the nineteenth century, Americans nursed the feeling—probably justified—of being looked down on by the older, more cosmopolitan nations of Europe. Andrew Carnegie, even after becoming "King of Steel" and the wealthiest man on earth, wrote an entire book cataloging America's "ascendancy in every department," including manufacturing, commerce, education, literature, and fecundity. "It is, I think, an indisputable fact," wrote Henry James, that Americans are "the most self-conscious people in the world, and the most addicted to the belief that the other nations of the earth are in a conspiracy to undervalue them."

In addition to this underdog's determination to prove itself, a second, more personal kind of insecurity developed as well. The radical difference between America and Europe, as Tocqueville observed in the 1830s, was that in America there were no "stations": America had no rankings of lord and commoner, no birth-based restrictions on what a man could own or what occupations he could pursue (slavery being the obvious and massive exception). In theory, any man could do anything, rise to any height. Thus each man's place in society became dependent on his own conduct, his worth a reflection of personal economic performance. As the economy matured it became possible, perhaps for the first time in history, for every man to be judged—and to judge himself—a "success" or "failure" depending on how well he did, how much he made.

We might call this the insecurity of capitalism itself—or perhaps even the insecurity of individualism. It was a new form of insecurity, unknown in aristocratic societies where people knew who they were based on the family into which they were born. From this followed the irony so central to Tocqueville's *Democracy in America*. In a society where so many had so much, each was afflicted with a restless desire for more. In a society committed at its core to equality, every man suffered from a "longing to rise."

If Jefferson and his mammoth exemplify early America's simultaneous feelings of superiority and insecurity vis-à-vis Europe, the indefatigable Ben Franklin epitomizes American-style impulse control. Franklin's parents were Puritan, and his practical homilies in *Poor Richard's Almanack* cajoled his countrymen with the values of moderation, self-control, industry, saving for the future, never wasting time, and refusing to give up in the face of adversity:

> Industry, Perseverance & Frugality, make Fortune yield.
> Dost thou love Life, then do not squander Time, for that's the
>   Stuff Life is made of.
> There are no Gains, without Pains.
> No man e'er was glorious, who was not laborious.
> Be at War with your Vices, at Peace with your Neighbors.
> He that can have patience can have what he will.
> To lengthen thy life, lessen thy meals.
> Leisure is Time for doing something useful.
> Diligence is the mother of good luck.

Thus the United States came into the world a quintessential Triple Package nation: with a chosen-people narrative rivaling that of the Old Testament; an acute insecurity simultaneously collective and individual; and a Puritan inheritance of impulse control.

BUT THAT'S ONLY PART of the picture. Right from the beginning, alongside Triple Package impulse control, there has always been another side of America: its vibrancy, its dynamism, its individualism, its rebelliousness. The penchant for defying authority and going your own way is as deeply rooted in American history as Puritanism. After all, the United States was created through an act of rebellion, and Revolutionary America was bursting with antiauthoritarian ferment at every level.

Contemporary observers reported that the Revolution had "loosened the bonds of government everywhere": "children and apprentices" had become "disobedient," "Indians slighted their guardians, and Negroes grew insolent to their masters." As social historian Claude Fischer observes, local elections became more "boisterous," involving more "common folk" and "more crowd attacks on the authorities." More wives filed for divorce. College students grew "rowdier." Young couples no longer believed they needed their parents' permission to marry—or needed to be married in order to have sex. A third of the brides in several New England towns were already pregnant.

Uniting all this ferment was an urge to throw off the yoke of the past. America was "a country of beginnings," said Ralph Waldo Emerson, where men could leave their past behind, as Alexander Hamilton had when he came to a country that didn't know he was a bastard. "With the past," wrote Emerson, "I have nothing to do; nor with the future . . . I live now." Emerson was giving voice to a new way of thinking about how people should live: imprisoned neither by the past nor the future; living rather *in the present*.

The desire to live in the present runs deeply against the grain of impulse control. As a result, the Triple Package in America has always had to fight against—to find a way to rein in—America's live-in-the-present dynamism. To get just a glimpse of how profoundly this conflict has shaped American social and political life, consider the country's two most important founding documents: the Declaration of Independence and the Constitution. Most Americans may not realize it, but there's a deep tension between these two foundational charters of American liberty.

The Declaration of Independence was the consummate expression of America's live-in-the-present rebelliousness. The Declaration threw off the yoke of America's past, insisting on the right of the people to be governed by nothing but their own present will. "[T]he earth belongs to the living," wrote Jefferson, the Declaration's author. Indeed Jefferson went so far as to argue that one generation of Americans could not make law for the next; all laws would automatically expire after nineteen years.

But rebellion, once begun, is hard to contain. Today, with democracy so well established, it's easy to forget that America was an unprecedented experiment, which many expected to fail, and which very nearly did fail. The years immediately following the Revolution were full of lawlessness, ineffectual government, and popular insurrections, such as Shays's yearlong rebellion in Massachusetts. The fledgling United States came perilously close to anarchy.

This is where the Constitution came in. If the Declaration captured America's throw-off-the-past, antiauthoritarian streak, the Constitution was the Triple Package writ large. At its core was impulse control.

The political theory behind the Declaration of Independence was simple. Government gained its just power from one source alone. the "consent of the governed." But if the majority themselves turned tyrant, what then? An inflamed populace might turn into a mob. It might want to persecute heretics or seize the property of the rich. If the people were sovereign, there appeared to be no check against what the founding generation called popular "passions."

It took the Framers over a decade of upheavals and disintegration before they found a solution to this problem—the only solution possible without denying the principle of popular sovereignty itself.

Because the people were sovereign, the people had to agree to restrain their own passions. That was exactly how James Madison understood the Constitution's purpose. The "passions . . . of the public" could not be permitted to govern, wrote Madison in *The Federalist*; but "[a]s the people are the only legitimate fountain of power," they had to agree to "control" and "regulate" themselves. Countless Americans since then have seen it the same way. "Constitutions are chains with which men bind themselves in their sane moments that they may not die by a suicidal hand in the day of their frenzy."

In other words, the Constitution's very purpose was to bring the structure and restraint—the checks and balances—of impulse control to the vibrant energies of democracy. The Constitution established a structure of government and a set of principles that would check majoritarian intemperateness and impulsiveness for generations to come. At the same time, the Constitution strengthened the rule of law— and after all, in a nation, the rule of law *is* impulse control—creating a powerful national government with vastly expanded lawmaking and law-enforcing powers.

Jefferson saw immediately the departure from the principles of government by the present, for the present, in which he believed. The Constitution, he would argue, was itself a form of tyranny—the tyranny of one generation imposing law on the future. But the future was, precisely, the Framers' principal concern.

It's no coincidence that the Constitution didn't mention "the pursuit of happiness," which the Declaration of Independence called an inalienable right. Triple Package cultures do not focus on happiness. The Constitution's proclaimed objects were to forge "a more perfect Union, establish Justice, insure domestic Tranquility, provide for the common defence, promote the general Welfare, and secure the Blessings of Liberty to ourselves and our Posterity." These are future-oriented, objectives, looking to the nation's prosperity, freedom, and success over ensuing generations. They are, in other words, Triple Package objectives; individual happiness is not mentioned.

THUS THE CONSTITUTION cemented the American Triple Package, helping to launch the nation on its extraordinary rise to continental and then global preeminence. But as a Triple Package nation, America had Triple Package pathologies. Characteristically, the Achilles' heel was superiority.

Lofty as its ideals may have been, Revolutionary America had all the moral failings of its time. The American superiority complex of the late eighteenth century, and for a long time afterward, did not accept the idea that all men are created equal. A great many Americans believed in the superiority of one race over the rest; of "civilization" over "savagery"; of men over women; and of Christianity—specifically Protestant Christianity—over all other religions.

As a result, America's Triple Package drive and ambition also contributed to some of the darkest stains on American history: at least two centuries of race slavery and, even after the Civil War, another century of segregation, violence, and systemic discrimination; the decimation, subjugation, and relocation of Native Americans, including the forced and fatal march of several tribes halfway across the continent in the 1830s; the Chinese Exclusion Act of 1882; and the forgotten lynchings of Mexicans and Mexican Americans well into the twentieth century.

BUT FOR THOSE WHOM America's Triple Package benefited, it was an engine of extraordinary potency. The American Triple Package was not the same as the Puritans'. Infused with the American desire to live in the present, it was both more materialist and more rule-breaking.

This gave the American Triple Package a distinctive character. It was as liberating as it was constraining. In some Triple Package cultures, the path prescribed for individuals is rigidly conventional and risk-averse. America's Triple Package was different.

Americans want to make the world anew—in every generation. Long after Jefferson said the "earth belongs to the living," Henry Ford would famously declare, "History is more or less bunk. It's tradition. We don't want tradition. We want to live in the present." One generation of Americans after another has pushed into new frontiers, torn down the old, created new religions, new rights, and new kinds of commerce, forever engaging in what economists, following Schumpeter, call creative destruction. If America was the quintessential Triple Package country, it was also a country where people constantly climbed the Triple Package ladder only to kick it away—drawing on its drive and discipline to defy convention, to break uncharted ground.

More than any other country, America managed to fuse two seemingly opposite impulses: Triple Package deferred gratification and living in the present. It did this in part through the alchemy of the American Dream. The formula was simple. Work hard during the week, and you can play hard on the weekend; work hard for years, and you can have the house and cars and family you dreamt of.

Through this productive fusion, America succeeded in becoming not only the world's hardest-working country, but, just as famously, the land of letting loose and living in the moment—of Coca-Cola, keggers, and California. An American could live the Triple Package five days a week—but declare independence on Friday afternoon. No other society in history has ever had it more both ways.

Then, in the latter part of the twentieth century, something hap-

pened. America turned against both insecurity and impulse control. With those Triple Package elements gone, what remained was
superiority and the desire to live in the present—a formula not for
drive, grit, and innovation, but for instant gratification.

MANY TODAY CLAIM THAT welfare has destroyed Americans'
work ethic. "Even as the welfare state has improved the material
comfort of low-income Americans," write Robert Rector and Jennifer
Marshall of the Heritage Foundation, its "result has been the disintegration of the work ethic." In Triple Package terms, this argument in
essence asserts that welfare eliminated insecurity: freed from the
chastening anxiety of starvation, poor Americans live happily on
their unemployment checks, watching their flat-screen TVs instead
of taking a minimum-wage job at McDonald's.

And as a matter of historical fact, the modern American welfare
state did originate with the goal of eliminating insecurity. "There is
still today a frontier that remains unconquered," said Franklin D.
Roosevelt in a 1938 radio address: "the great, the nation-wide frontier
of insecurity, of human want and fear." In his famous "Second Bill of
Rights" speech in 1944, Roosevelt declared that the Constitution had
"proved inadequate to assure us equality in the pursuit of happiness";
what America needed was "the freedom from insecurity." Not for
nothing was FDR's groundbreaking welfare program called Social
Security.

But for members of the comfortable classes to blame America's
decline on the lower classes is, at best, myopic. Lamenting the disintegration of the work ethic in our inner cities when in some Baltimore, Detroit, and New York neighborhoods, the chances of
graduating high school are under 50 percent, but gang leaders can
earn as much $130,000 a year selling drugs—and die reportedly at

the average age of twenty—is missing pretty much everything that's important about the big picture. So is telling Central Appalachians to stop whining and apply at McDonald's, when changes in the economy essentially wiped out all the sectors that previously provided livelihoods for entire towns.

America did lose its insecurity—and hence the Triple Package—in the late twentieth century, but the real culprit was success itself. Insecurity faded because of *prosperity*, some of it undeserved, but much of it hard-earned and reflecting the best of America—the fact that we have good institutions and great incentives for innovation. All the partisan finger-pointing—It's welfare! It's tax cuts! It started with Reagan! No, it's totally Obama's fault—detracts from the undisputed most important fact: the extraordinary, inexorable rise in American income and standards of living, both at the top and on average, over the twentieth century, culminating after 1980 in one of the greatest wealth explosions in the history of mankind. The 1980s were so loaded and overloaded that the entire decade came to be symbolized by Tom Wolfe's *The Bonfire of the Vanities* and Oliver Stone's *Wall Street*. And the 1990s were astronomically even richer than the decade before.

With the dot-com boom and the growth of venture capitalism, twentysomethings barely out of college became multimillionaires overnight. Between 1995 and 2000, the stock market tripled, and by 2000, Silicon Valley was producing up to sixty-four new millionaires per day. Corporate compensation soared, but few complained, because employment was strong and shareholder profits were soaring too. People with no special skills or insight suddenly believed they were stock-picking geniuses because everything they bought turned to gold. Across the country, family net worth climbed; by 2007, it was over $120,000—a record high. Homes were worth twice as much as people paid for them only a few years earlier.

In some ways, perceptions of how flush Americans were in those boom years outran the reality. In fact, the top 1 percent of U.S. earners reaped the lion's share of the new wealth. But according to the Congressional Budget Office, the vast majority of Americans (all the way down to the eightieth percentile) enjoyed substantial income growth as well, and widely available credit increased ordinary Americans' spending power beyond their actual income—sometimes well beyond it. At the same time, it's easy to forget that today's focus on inequality and skepticism about whether Wall Street riches are good for Main Street were not prevalent in America in the decades before 2008. As economist Robert Shiller observes, "In the boom years of the 1990s, overt resentment by the general public of their own country's corporations was at a historic low. Businessmen were lionized."

Meanwhile, globalization seemed to herald even more money for Americans; it was practically synonymous with the spread of Starbucks, Disney, Coca-Cola, Nike, and the Gap. On top of all this, everyone seemed to revere America. With the collapse of Communism, governments from countries all over the world were suddenly begging U.S. "advisers" (some of whom were still law students at Georgetown) to help them implement U.S.-style constitutions and foreign-investment laws. Francis Fukuyama wrote *The End of History*, positing that Western liberal democracy—with America as Exhibit A—represented the "end point of mankind's ideological evolution" and the "final form of human government." In the nineties, more than thirty-five countries, including Tanzania and Swaziland, started U.S.-style stock exchanges. The world, it seemed, wanted to be American. The United States was the sole superpower left standing—a "hyperpower," dominant economically, militarily, and culturally, master of a suddenly unipolar universe.

All this swelled Americans' superiority complex, but had the opposite effect on their insecurity. For the first time, the United States had no serious rival or enemy. America had nothing more to prove to the world; on the contrary, the world was scrambling to catch up to America.

Meanwhile, there were other forces—more personal, more psychological—attacking insecurity in America as well.

"I CANNOT THINK of a single psychological problem—from anxiety and depression, to underachievement at school or at work . . . to spouse battering or child molestation . . . —that is not traceable to the problem of low self-esteem," wrote psychotherapist, Ayn Rand disciple, and father of the American self-esteem movement Nathaniel Branden. It's an extraordinary fact about Branden's extraordinary claim that it will strike many American readers as not implausible— as at least arguable. Such has been the astonishing triumph of self-esteem thinking in the United States.

Branden's 1969 *The Psychology of Self-Esteem* became a bestseller. By the 1980s, educators had seized on his ideas, broadening them still further. Not only were all *psychological* problems caused by a lack of self-esteem; "virtually every *social* problem can be traced to people's lack of self-love," said the reformers. "[M]any, if not most, of the major problems plaguing society have roots in the low self-esteem of many of the people who make up society."

Schools all over the United States made it their mission to instill self-esteem in their charges. Parents embraced the same goal. "In the 1990s," writes psychologist Carol Dweck, "parents and teachers . . . decided that self-esteem was the most important thing in the world—that if a child had self-esteem, everything else would follow."

Psychologists had talked about self-esteem before. William James saw its importance in an interesting discussion in 1890. But for James, accomplishment remained central to self-esteem. In the new movement, the causality was reversed: self-esteem was declared necessary to accomplishment. That means self-esteem has to *precede* achievement. Kids have to be given self-esteem before they've achieved anything.

In other words, the self-esteem movement severed self-esteem from esteem-worthy conduct. If self-esteem is the cure for doing poorly and acting badly, then people must be taught to feel good about themselves—to like themselves, to accept themselves—no matter what they have or haven't done.

The self-esteem movement made Americans feel much more satisfied with themselves, but it didn't cure their psychological, academic, or social problems. In fact, its claims have proved jarringly false.* But it did go a long way toward destroying the Triple Package in everyone who embraced it—and their children.

The self-esteem movement promotes the exact opposite of Triple Package insecurity. No one should feel that they're not good enough or that they have to "prove themselves" by doing better; self-acceptance

---

* As is now well established, increasing self-esteem does not improve academic performance. Asian American students have the lowest self-esteem of any U.S. racial group, yet do the best. Overall, American students are among the world's leaders in self-esteem; they're also among the lower-scoring. In a controlled experiment, students who received self-esteem-boosting messages did *worse* than other students. In another study, repeatedly praising children for how intelligent they were *lowered* their scores on standardized test questions—and made them lie when asked how many questions they'd gotten right. Moreover, the basic claim that sociopathic behavior is caused by low self-esteem also proved false. Racists and criminals do not "secretly feel bad about themselves," researcher Nicholas Emler found. Serial rapists have "remarkably high levels of self-esteem." Meanwhile, psychologists report that kids raised on a high-self-esteem diet often suffer depression and anxiety as adults, along with higher rates of narcissism. For sources, see the endnote.

is the first rule of a successful life. This has become the message of an explosion of self-help books, of wildly popular self-improvement "seminars," and of much psychotherapy as well (a "primary task of psychotherapy is to help strengthen self-esteem," wrote Branden in 1969). Above all, children must never feel they failed at something or even came in second. Hence the proverbial trophies-for-everyone, the parents who ask teachers not to use red pens because it's "upsetting to kids when they see so much red on the page," the ubiquitous "Great job!" and "You're amazing!"

At the same time, the self-esteem movement erodes impulse control. "People with incredibly positive views of themselves," researchers have found, are more willing to "do stupid or destructive things" and more likely to satisfy their own desires even when the "costs are borne by others." Children brought up in self-esteem-centered schools and families are not taught to endure hardship or to persevere in the face of failure. They're sheltered from disappointment and rejection by devoted, exhausted parents who monitor their every move, desperate to make their kids feel "special." They're encouraged to believe that they have a right to have and do everything they want when they want it—and tend to be frustrated when they can't.

The erosion of impulse control in America extends far beyond child rearing. It has affected every domain of life. The 1960s—the ultimate anti-inhibition, live-in-the-moment decade—probably played a role. But more important, once again, was rising affluence. Comfortable people have no pressing, life-threatening reason to exercise discipline and restraint. It's hard to live like you're in the Depression when you grow up in the suburbs with two SUVs. The result—when vanishing impulse control combines with diminishing insecurity and excess disposable income—is a society increasingly defined by instant gratification.

. . .

WE'RE HARDLY THE FIRST to call attention to immediate gratification as a corrosive force in America today, but the facts remain bracing.

U.S. public debt, which in 1980 was about 30 percent of gross domestic product (roughly the same percentage as in 1790), is today a barely graspable $17 trillion, over 100 percent of GDP. America's infrastructure, the physical embodiment of a nation's capacity to invest in its future, was once the envy of the world; today, it earns a D+ from the American Society of Civil Engineers, including solid Ds for drinking water, hazardous waste, and transit. What's most surprising is that America's infrastructure investment declined (as a percentage of GDP) even during the stupendous prosperity of the 1980s and '90s.

In the 1960s, the insecurity created by the country's rivalry with the Soviet Union drove the United States to stay ahead of the world in basic science and cutting-edge technology. Since then, American research and development spending has dropped by over 50 percent (again, as a percentage of GDP). By 2009, Americans were spending more on potato chips than their government spent on energy research and development.

Even as U.S. public debt skyrocketed, Americans' personal savings rate dropped from about 12 percent in the early 1980s to 2.5 percent in 2009 (compared with a reported 38 percent in China). The drop occurred steadily and was not merely an artifact of the 2008 financial collapse. Gambling has skyrocketed too, increasing an estimated 6,000 percent between 1962 and 2000. Drug habits that were still countercultural in the Sixties have become practically mainstream. Today, according to recent studies, almost one in five high schoolers drink, use drugs, or smoke during the school day.

All these ailments are examples of gratifying present wants at the expense of the future. This trend is by no means limited to the poor. Over half the students at America's private high schools say that drugs are trafficked, kept, or used at their schools. A study in California found drug and alcohol abuse higher among "upscale youth"; adolescents in a suburb where the average family income was over $120,000 reported "higher rates of . . . substance abuse than any other socioeconomic group of young Americans today."

Moreover, self-esteem parenting is much more common among wealthier Americans. As the principal of a Silicon Valley prep school put it, "Avoiding discipline is endemic to affluent parents." Psychologists are observing an explosion of narcissism in America's children, particularly among the better-off. The so-called millennials—with their sense of entitlement, their expectation of being "CEO tomorrow," their belief that the workplace should adjust "around our lives instead of us adjusting our lives around work"—are for the most part not lower-income or minority youth. They're the children of well-off white baby boomers.

As these children grow up and begin entering the workforce, they are not overwhelming employers with their attitude. "Millennials don't always want to work," the consultant Eric Chester told Forbes.com, "and when they do, their terms don't always line up with those of their employers. All too often, the young worker shows up ten minutes late wearing flip-flops, pajama bottoms, and a T-shirt that says, 'My inner child is a nasty bastard.' Then she fidgets through her shift until things slow down enough that she can text her friends or update her Facebook page from her smartphone."

And the worst of it is that the demand for instant gratification has spread into our political and economic institutions.

. . .

ALL DEMOCRACIES RUN the risk that politicians will focus on short-term gains, rather than long-term national interests. Politicians who want to get reelected have to satisfy their constituents now; the future is someone else's problem. In this sense, the real question isn't why the national debt has grown to its currently unthinkable proportions; it's what kept it from doing so earlier. Part of the answer has to do with the surrounding culture and whether that culture promotes norms of restraint and responsibility, discipline and investment. In the 1980s and '90s, American culture was not promoting those norms. The dangers of this development should not be underestimated. For politicians, even war can be a form of instant gratification; the grievous costs are often borne long after those who started it are out of office.

A false sense of security, no impulse control, and immediate gratification: these forces also played a part in the 2008 financial collapse. Instead of living within their means, average Americans— spurred on by teaser rates and easy credit—bought $500,000 houses with money they didn't have and loans they couldn't afford. Instead of following traditional lending practices, banks offered mortgages to almost anyone who walked in the door, collecting fast fees and handing off the risk to someone else. Instead of exercising a disciplining function, credit-rating agencies earned hefty profits by awarding AAA grades to high-risk mortgage-backed securities. Instead of providing long-term sound investments for their clients, Lehman Brothers, Bear Stearns, and other investment banks raked in billions by packaging these subprime securities into extremely complex instruments that masked their true risk, without bearing any of the long-term costs; as one financier would later describe these derivatives,

"They could explode a day later and you are not impacted one single iota."

The collapse of 2008 undoubtedly had multiple causes, including fraud, corruption, ineffective regulation, and old-fashioned greed. But to some extent it was a pure Triple Package implosion. At its heart was a bubble—the housing bubble—and as Shiller and fellow economist Nouriel Roubini have shown, every such "boom and bust" is fueled by a contagious "excessively optimistic" conviction that the boom "will never end" and disastrous risks will never materialize. In other words, rather than being driven by *in*security, people were driven by a false sense of *security*. And because of this false sense of security, people threw impulse control out the window in pursuit of instant wealth. Those who—like Shiller and Roubini—counseled impulse control and insecurity in the years before the crisis were treated like Cassandras.

Historians will long debate why Americans fell so headlong into the savings and loan bubble of the late 1980s, the dot-com bubble of the 1990s, and then, even after the bursting of those two, the mother of all bubbles in the 2000s. Without minimizing other causes, part of the reason is simply that America had lost the Triple Package, and this loss had infected the whole system. At every level of the economy, from borrowers to bankers to hedge-fund billionaires, people didn't feel anywhere near enough insecurity or exercise anywhere near enough self-control. Thus America failed the marshmallow test—and paid the price.

Where do we go from here?

To recover from its instant gratification disorder, America would have to recover its Triple Package. Throughout this book, however, we've stressed the dark underside of the Triple Package in many if its forms. If, therefore, America is to aspire to a reinvigorated Triple Package, it has to be a Triple Package worth aspiring to. To that end, we close by addressing each element of the Triple Package in turn.

First, superiority. This is the one Triple Package element America hasn't lost. Americans still believe in America's exceptionality. They still routinely and publicly call their country "the greatest nation on earth"—a phrase you'll almost never hear Europeans (or anyone else) using. Tony Blair once called the United Kingdom "the greatest nation on earth," and the British press roundly lampooned him for it. By contrast, when Barack Obama said he believed in American exceptionalism "just as I suspect that the Brits believe in British exceptionalism and the Greeks believe in Greek exceptionalism," he was criticized for being insufficiently and inauthentically exceptionalist.

National superiority complexes can be a tremendous source of confidence, cohesion, and willingness to sacrifice. They are also among the most dangerous forces in human history. Nazi Germany had a whopping superiority complex. So did imperial Japan.

The only justifiable national superiority complex is one true to America's constitutional ideals of equality and openness. America remains today the country most open to the talents and dreams of all. That is a superiority worth aspiring to—a superiority that includes rather than excludes, and at its best restrains rather than fosters imperialism.

This kind of superiority complex is not available to ethnic or religious groups, because they can't open themselves equally to people of all backgrounds and beliefs. To do so would mean losing their identity. But again, America is not an ethnic or religious group. It's a vast and diverse nation. In this respect, America has an advantage over its Triple Package groups. America can and must champion principles of equality and inclusion that ethnic and religious minorities can't. But it's precisely this equality and inclusion, on America's part, that will keep drawing Triple Package groups, with their tremendous economic energy, to America's shores.

Second, insecurity. Insecurity is in many respects destructive at the national level. For one thing, it too can make nations dangerous; the belief that "national security" is in danger can easily become a justification for cracking down on civil liberties or making war on the basis of unsubstantiated factual claims. Moreover, whatever it is that made the country insecure—the attacks it suffered, the economic threats it faces—will have exacted their own, sometimes dreadful price. Nevertheless, the Triple Package holds, in diametric opposition to self-esteem thinking, that adversity and self-questioning can be good things.

For most of its history America was an upstart, an underdog. Only toward the end of the Cold War did America emerge unrivaled, on top of the world. China is exploding today in part because it's so insecure. (As China experts Orville Schell and John Delury put it, a deep sense of "humiliation"—of being "stepped on" and outpaced by the West—has "served as a sharp goad urging Chinese to sacrifice" so that their country can recover its former grandeur.) China has the Triple Package in spades, with an outsize superiority complex, a Confucian tradition of impulse control, and above all a determination to prove itself once again to the world. What Americans needed in the 1980s and '90s was *more* insecurity.

Thus the horror of 9/11, the unwon wars that followed, the rise of China, even the financial collapse—all this has had, paradoxically, one beneficial consequence: the return of Triple Package insecurity. But the insecurity Americans need is not one of fear or belligerence. Triple Package insecurity is the hunger to prove oneself, tempering superiority with the feeling that one is not, at least yet, good enough. Historically, the United States has risen to its greatest achievements when Americans have felt the call to prove their country's mettle, morality, and ability to win out over grievous challenges. Americans did just this after Pearl Harbor, when the country had not only suffered an attack on its soil but had barely emerged from the deepest depression in its history. For better or worse, America has that opportunity again today.

Finally, impulse control. Going forward, impulse control may be the most important element of the Triple Package to focus on because it offers a path into the Triple Package for everyone, regardless of background. The Triple Package isn't members-only. It's not the exclusive property of Triple Package groups. The way in—not that it's remotely easy—is through grit: by making the ability to work hard, persevere, and overcome adversity into a source of personal superiority.

This kind of superiority isn't zero-sum; it's not ethnically or religiously exclusive. It doesn't come from being a member of a group at all. It's the pride a person takes in his own strength of will and his own accomplishments. Like a national superiority complex based on equality, this too is a superiority worth aspiring to.

Born in the South Bronx to struggling Puerto Rican parents, Justice Sonia Sotomayor was not raised in a Triple Package group. Nor was she raised in a high-discipline, high-expectations Triple Package family. Her father was an alcoholic, and her mother's "way of coping was to avoid being at home," Sotomayor recalls in her magnificent autobiography, *My Beloved World.* But Sotomayor—who gave herself painful insulin shots for juvenile diabetes starting around age eight—writes that despite "the fragile world of my childhood," she was "blessed" with a "stubborn perseverance" and the belief that she could overcome whatever obstacles life threw at her. She wasn't always a high achiever in school. But in fifth grade, she "did something very unusual for a child" and "decided to approach one of the smartest girls in the class and ask her how to study." Soon her teachers had reseated her in the row "reserved for the top students," and a few years later she would be applying to Princeton—against the advice of her guidance counselor, who recommended "Catholic colleges."

The point of this example is not, "See, it's easy to climb out of poverty in America—Sotomayor did it." On the contrary, Sotomayor's story illustrates just how extraordinary a person has to be to overcome the odds and institutions she had stacked against her.

The difference between Triple Package individuals, like Sotomayor, and Triple Package groups is that members of the latter are pushed by family and culture to work hard and strive, whereas a Triple Package individual may have no resources to draw on other than his or her own. (As was true in Sotomayor's case, a single relative or mentor can make an enormous difference.) Sometimes Triple Package individuals may even be disparaged by members of their own group.

Many have drawn attention to an "oppositional" strand of contemporary black urban culture that disdains studiousness and "getting straight A's" as "acting white." (Interestingly, however, Harvard economist Roland Fryer's important 2006 study found that this phenomenon did not exist at all-black schools.) But what's rarely observed is the strangely parallel disparagement of discipline and academic striving that has emerged among America's affluent classes.

For example, in response to overwhelming evidence that America's math and science skills are plummeting, putting our innovation lead at serious risk, one school of thought has become bizarrely influential: the one that says—hey, don't worry, the solution for America is to do less, not more. In a widely read recent piece on the secret to innovation, the author implored American parents to let their kids "do less," surf the Internet, "drifting from fad to fad, website to website," and just "follow their passions," adding pointedly: "Remember that Bill Gates and Mark Zuckerberg dropped out of college to follow their passions."

This is the Disney version—the instant gratification version—of creativity and learning (as when Pocahontas "looks into her heart" and can suddenly speak English). American kids already spend seven to eight hours daily on entertainment media and 25 percent more time watching television than in school. The core of wisdom in the "do less," "just follow your passions" view is that it's hard to succeed unless you love what you do. But this view conveniently forgets that real achievement and real creativity—whether artistic, professional, or entrepreneurial—requires as much drive and grit as inspiration. Asked about the qualities required to get through medical school, the novelist and physician Khaled Hosseini answered, "Discipline. Patience. Perseverance. A willingness to forgo sleep. . . . Ability to weather crises of faith and self-confidence. Accept exhaustion as a fact of life." Asked about the qualities required to be a novelist, he said, "Ditto."

Gates and Zuckerberg—not to mention Steve Jobs—were among the hardest-working, most driven people their peers knew. Obviously creativity also requires the freedom to question and challenge authority (which is why China has so far trailed us in inventiveness), the space to wonder and free-associate. But the fact remains that you can't invent Google, Facebook, or the iPod unless you've mastered the basics, are willing to put in long hours, and can pick yourself off the floor when life knocks you down the first ten times.

The next string theory may well hit someone as they're strolling on the beach, but you can be sure that person will have known his quantum physics—and banged his head for years against the equations he's about to throw out the window. (Picasso and Mondrian were masters of painterly technique before they invented forms of painting no one had seen before.) The real prescription for groundbreaking innovation and entrepreneurialism is the Triple Package ladder. Jeff Bezos founded Amazon when he was "dead broke" in 1995, committing himself—and his venture capital investors—to plow every penny of return back into the company for a minimum of five years. He kept his word; the rest is history. Thus the birth of one-click shopping depended on a consummate act of impulse control.

None of which is to say that the Triple Package will make you happy. If in fact successful people tend to feel both superior and inadequate, then success to some extent necessarily implies a trade-off with happiness. Feeling like you're not good enough is painful. But a life that doesn't include hard-won accomplishment and triumph over obstacles may not be a satisfying one. There is something deeply fulfilling, even thrilling, in doing almost anything difficult extremely well. There is a joy and pride that come from pushing yourself to another level, or across a new frontier.

A life devoted only to the present—to feeling good in the now—is unlikely to deliver real fulfillment. The present moment by itself is too small, too hollow. We all need a future, something beyond and greater than our own present gratification, at which to aim or to which we feel we've contributed. Happiness, wrote the Austrian psychologist (and Holocaust survivor) Victor Frankl, "cannot be pursued; it must ensue . . . as the unintended side-effect of one's personal dedication to a course greater than oneself." The Triple Package doesn't promise a meaningful life, but it makes such a life possible, because it allows people to seize the reins of time—to live not only in the present but also for the future, to devote their full capacities to changing themselves or the world, in small ways or large.

At the end of the day, the Triple Package is a form of empowerment, which can be used for selfish gain or for others' good alike. People who have it are not guaranteed anything, and they run the risk of real pathologies. But they are in a position to transform their own and others' lives.

.   .   .

To BE SURE, calling for America to recover its Triple Package creates a paradox. For if America's Triple Package rests on a superiority of tolerance, opportunity, and equality, then the success of the *nation's* Triple Package will always be in tension with that of its Triple Package *groups*. To one degree or another, the superiority complex of every Triple Package group is almost by definition intolerant. Hence America's Triple Package will conflict with and tend to undercut the superiority complexes of its Triple Package groups. Ironically, this conflict will ensure that America's Triple Package groups remain cultural outsiders to some extent—at least for the first or second generation —which will in turn assist their Triple Package (by giving them insecurity), boosting their success.

In the long run, however, the American national Triple Package will be too strong for these group chauvinisms. It will consume them all. The real promise of a Triple Package America is the promise of a day when there are no longer any successful groups in the United States—only successful individuals.

ACKNOWLEDGMENTS

Our parents, Leon and Diana Chua and Sy and Florence Rubenfeld, were inspirations for this book.

We are deeply grateful to our friends, family members, and colleagues Bruce Ackerman, Susan Rose-Ackerman, Tony Kronman, Nancy Greenberg, Henry Hansmann, Marina Santilli, Daniel Markovits, Sarah Bilston, Ian Ayres, Jennifer Brown, John Morley, Erin Morley, David Grewal, Jordan Smoller, Alexis Contant, Sylvia Austerer, Michelle Chua, Viktor Rubenfeld, and Katrin Chua, all of whom provided brilliant criticisms of earlier drafts of this book. We also profited enormously from conversations with Elizabeth Alexander, Susan Birke-Fiedler, James Bundy, Adam Cohen, Anne Dailey, Steve Ecker, Paul Fiedler, Jin Li, and Anne Tofflemire. Special thanks to Ademola Adewale-Sadik, Nabiha Syed, Hal Boyd, Damaris Walker, Tom and Keya Dannenbaum, the Rawls family, and the Swett family for their detailed comments on specific sections of the book.

Some of the key ideas in this book were first developed in a 2008 Yale Law seminar called "Law and Prosperity," and thanks are due to the following remarkable students for the insights they provided during this book's formative stages: Ligia Abreu, Yaw Anim, Monica

Bell, Bridge Colby, Jacqueline Esai, Ronan Farrow, Jon Finer, Jim Ligtenberg, Patricia Moon, Nick Pyati, Amelia Rawls, Ben Taibleson, Lina Tetelbaum, Natalia Volosin, Shenyi Wu, and David Zhou. We also owe a great debt to the indomitable Jeffrey Lee, who saved us from embarrassments we can't describe.

We couldn't have written this book without an amazing group of research assistants. In particular, we'd like to thank Halley Epstein, Tian Huang, Rebecca Jacobs, Christine Tsang, Jordana Confino, Stephanie Lee, Ida Araya Brumskine, Bert Ma, Rich Tao, and Meng Jia Yang, each of whom devoted dozens, in some cases hundreds, of hours to this book, as well as Sam Adelsberg, Casey Arnold, Justin Lo, Renagh O'Leary, Vidya Satchit, Wanling Su, Avi Sutton, J. D. Vance, Ryan Watzel, and Eileen Zelek. In addition, the following students provided critical assistance on particular chapters: Jasmeet Ahuja, Barrett Anderson, Matt Andrews, Ariela Anhalt, Josef Ansorge, Nana Akua Antwi-Ansorge, Amar Bakshi, Rachel Bayefsky, Megan Browder, Walker Brumskine, Christine Buzzard, Kathryn Cherry, Usha Chilukuri, Celia Choy, Charlie Dameron, Bicky David, Rachel Dempsey, Alley Edlebi, James Eimers, Aditi Eleswarapu, Arthur Ewenczyk, Adele Faure, David Felton, James Flynn, Yousef Gharbieh, Dana Stern Gibber, Noah Greenfield, Natalie Hausknecht, Stephanie Hays, Daniel Herz-Roiphe, Maya Hodis, Jane Jiang, Lora Johns, Tassity Johnson, Nathan Hake, Diane Kane, Jesse Kaplan, Sam Kleiner, Robert Klipper, Harrison Korn, Philipp Kotlaba, Doug Lieb, Ming-Yee Lin, Dermot Lynch, Sarah Magen, Nick McLean, Jennifer McTiernan, Dahlia Mignouna, Yannick Morgan, Erica Newland, Luke Norris, Aileen Nowlan, Ifeanyi Victor Ojukwu, Jonathan Ross-Harrington, Emily Schofield, Conrad Scott, Reema Shah, Sopen Shah, Lochlan Shelfer, James Shih, Jon Siegel, Matthew Sipe, Alex Taubes, Caitlin Tully, Chas Tyler, Anna Vinnik, John Wei, Luci Yang, Justin Zaremby, and Ben Zweifach.

Michael VanderHeijden and especially the spectacular Sarah Kraus of the Yale Law Library awed us with their energy and resourcefulness; they have our great admiration and gratitude. We'd also like to thank Patricia Spiegelhalter, Karen Williams, and Rosanna Gonsiewski for their assistance and support.

Our love and thanks to Sophia and Louisa Chua-Rubenfeld for their patience, insights, and sometimes scathing editorial critiques.

Finally, we are deeply indebted to our extraordinary agents Tina Bennett and Suzanne Gluck, to Ben Platt, Sarah Hutson, and the entire team at Penguin, and most of all, to our brilliant editor, Ann Godoff, who understood what we were trying to say in this book better than anyone.

NOTES

CHAPTER 1: THE TRIPLE PACKAGE

5 current or recent CFOs or CEOs: Jeff Benedict, *The Mormon Way of Doing Business: How Nine Western Boys Reached the Top of Corporate America* (New York and Boston: Business Plus, 2007), pp. ix–xii; James Crabtree, "The Rise of a New Generation of Mormons," *Financial Times*, July 9, 2010; Richard N. Ostling and Joan K. Ostling, *Mormon America: The Power and the Promise* (New York: HarperOne, 2007), pp. 137–43; "The Mormon Way of Business," *The Economist*, May 5, 2012; "The List: Famous Mormons," *Washington Times*, Oct. 21, 2011; Caroline Winter, "God's MBAs: Why Mormon Missions Produce Leaders," *Business Week*, June 9, 2011; "Mormons in Business," Businessweek.com, June 9, 2011; see also Matthew Bowman, *The Mormon People: The Making of an American Faith* (New York: Random House, 2012), pp. 223–5 (noting Mormon success in politics and business); Claudia L. Bushman, *Contemporary Mormonism: Latter-day Saints in Modern America* (Westport, CT, and London: Praeger, 2006), p. 187 (noting the growing success of Mormons in finance, corporations, law, and government).

5 hard to find a Mormon on Wall Street: Crabtree, "The Rise of a New Generation of Mormons."

5 death of upward mobility: See, e.g., Timothy Noah, "The Mobility Myth," *The New Republic*, Feb. 8, 2012; Josh Sanburn, "The Loss of Upward Mobility in the U.S.," *Time*, Jan. 5, 2012; Rana Foroohar, "What Ever Happened to Upward Mobility?," *Time*, Nov. 14, 2011.

6 very much alive for certain groups, particularly immigrants: See Pew Research Center, *Second-Generation Americans: A Portrait of the Adult Children of Immigrants* (Washington, DC: Pew Research Center, Feb. 7, 2013), p. 7; Julia B. Isaacs, "Economic Mobility of Families Across Generations," in Julia B. Isaacs, Isabel V. Sawhill, and Ron Haskins, *Getting Ahead or Losing Ground: Economic Mobility in America* (Washington, DC: Economic Mobility Project, The Pew Charitable Trusts, 2008), pp. 15, 17–19; Ron Haskins, "Immigration: Wages, Education, and Mobility," in Issacs et al., *Getting Ahead or Losing Ground*, pp. 81, 86–7; Richard Alba and Victor Nee, *Remaking the American Mainstream: Assimilation and Contemporary Immigration* (Cambridge, MA, and London: Harvard University Press, 2003), pp. 240, 242, 244–5; Renee Reichl Luthra and Roger Waldinger, "Intergenerational Mobility," in David Card and Steven Raphael, eds., *Immigration, Poverty, and Socioeconomic Inequality* (New York: Russell Sage Foundation, 2013), pp. 169, 183, 201; see also Rubén G. Rumbaut, "Paradise Shift: Immigration, Mobility, and Inequality in Southern California," Working Paper No. 14 (Austrian Academy of Sciences, October 2008), p. 37 (summarizing a study that showed enormous differences across different immigrant groups but noting that "compared to their parents, all groups show inter-generational educational progress").

6 findings do not apply to "immigrant families": Isaacs et al., "Economic Mobility of Families Across Generations," p. 6. Chapter 7 will address the American upward mobility data and the exclusion of immigrants in more detail.

6 most arriving destitute: See Miguel Gonzalez-Pando, *The Cuban Americans* (Westport, CT, and London: Greenwood Press, 1998), pp. 20–1, 35–7; Guillermo J. Grenier and Lisandro Pérez, *The Legacy of Exile: Cubans in the United States* (Boston: Pearson Education, 2003), p. 48.

6 NO DOGS, NO CUBANS: Interview with José Pico, director and president, JPL Investments

Corp., in Miami, Fla. (conducted by Eileen Zelek on Jan. 6, 2012) (on file with authors); see also Gonzalez-Pando, *The Cuban Americans*, p. 37.

6 dishwashers, janitors, and tomato pickers: María Cristina García, *Havana USA: Cuban Exiles and Cuban Americans in South Florida, 1959–1994* (Berkeley and Los Angeles: University of California Press, 1996), p. 20; Gonzalez-Pando, *The Cuban Americans*, p. 36.

6 helped transform sleepy Miami: Grenier and Pérez, *The Legacy of Exile*, pp. 46–7.

6 household incomes over $50,000 was double that of Anglo-Americans: See Kevin A. Hill and Dario Moreno, "Second-Generation Cubans," *Hispanic Journal of Behavioral Sciences* 18 (1996), pp. 175, 177.

6 4 percent of the U.S. Hispanic population: U.S. Census, American Community Survey, Table S0201: Selected Population Profile (2010 3-year dataset) (population group codes 400-Hispanic; 403–Cuban).

6 five of the top ten wealthiest: See "Magazine Publishes List of Richest U.S. Latinos," HispanicBusiness.com, May 8, 2002, http://www.hispanicbusiness.com/news/newsbyidfront .asp?id 6775.

6 two and a half times more likely: U.S. Census, American Community Survey, Table DP03: Selected Economic Characteristics (2010 5-year dataset)(population group code 403 – Cuban) (4 percent of Cuban American households had annual incomes over $200,000); ibid., Table DP03: Selected Economic Characteristics (2010 5-year dataset)(ethnic group 400 – Hispanic or Latino) (1.6 percent of Hispanic American households had annual incomes over $200,000).

6 two Harvard professors: Sara Rimer and Karen W. Arenson, "Top Colleges Take More Blacks, but Which Ones?," *New York Times*, June 24, 2004.

6 Immigrants from many West Indian and African: Douglas S. Massey, Margarita Mooney, Kimberly C. Torres, and Camille Z. Charles, "Black Immigrants and Black Natives Attending Selective Colleges and Universities in the United States," *American Journal of Education* 113 (2007), pp. 243, 249–50.

6 A mere 0.7 percent of the U.S. black population: U.S. Census, American Community Survey, Table S0201: Selected Population Profile in the United States (2010 3-year dataset) (population group code 004 – Black or African American) (38,463,510); ibid. (population group code 567 – Nigerian) (260,724).

6 at least ten times: Nigerians already made up about 4.6 percent of the black freshmen at selective American universities in 1999, when they represented about .48 percent of the U.S. black population. See Massey et al., "Black Immigrants," pp. 248, 251 (immigrants accounted for 27 percent of black freshmen at selective universities, and Nigerians made up 17 percent of immigrant black freshmen). As we indicate in chapter 2, Nigerian overrepresentation at top American schools appears to be greater—perhaps significantly greater—today.

7 Nigerian Americans are already markedly overrepresented: Nigerians appear to be over-represented at America's top law firms by a factor of at least seven, compared to their percentage of the U.S. black population as a whole. Study commissioned by authors, July/August, 2013. For more detail, see chapter 2. See also Patricia Ngozi Anekwe, *Characteristics and Challenges of High Achieving Second-Generation Nigerian Youths in the United States* (Boca Raton, FL: Universal Publishers, 2008), p. 129 (quoting an interviewee who said that Nigerians "dominate" investment banking).

7 had to be examined to be worth living: Edith Hamilton and Huntington Cairns, eds., *The Collected Dialogues of Plato Including the Letters* (Princeton, NJ: Princeton University Press, 1989), p. 23 (*Apology*, 38a).

7 success "in its vulgar sense": Oliver Wendell Holmes, *Ralph Waldo Emerson*, in *The Works of Oliver Wendell Holmes*, vol. 11 (Boston and New York: Houghton Mifflin & Co, 1892), p. 201.

7 Indian Americans have the highest income: U.S. Census, American Community Survey,

Table S0201: Selected Population Profile in the United States (2010 3-year dataset) (population group code 013 – Asian Indian) (estimating median Indian household income of $90,525 as compared to $51,222 for U.S. population overall).

7 Chinese, Iranian, and Lebanese Americans: U.S. Census, American Community Survey, Table S0201: Selected Population Profile in the United States (2010 3-year dataset) (population group codes 016 – Chinese; 540 – Iranian; 509 – Lebanese) (estimating median household income of approximately $68,000 for Iranian Americans and $67,000 for Chinese and Lebanese Americans).

7 overrepresented at Ivy League schools: Arthur Sakamoto, Kimberly A. Goyette, and Chang Hwan Kim, "Socioeconomic Attainments of Asian Americans," *Annual Review of Sociology* 35 (2009), pp. 255, 256; Ron Unz, "The Myth of American Meritocracy," *The American Conservative*, Nov. 28, 2012.

7 "new Jews," and . . . tacit quotas: See Unz, "The Myth of American Meritocracy."

7 even the children of poor and poorly educated East Asian immigrants: See, e.g., Laurence Steinberg, *Beyond the Classroom: Why School Reform Has Failed and What Parents Need to Do* (New York: Simon & Schuster, 1996), pp. 83, 88; Min Zhou, *Contemporary Chinese America: Immigration, Ethnicity, and Community Transformation* (Philadelphia: Temple University Press, 2009), pp. 224–5; Min Zhou and Jennifer Lee, "Frames of Achievement and Opportunity Horizons," in Card and Raphael, *Immigration, Poverty, and Socioeconomic Inequality*, pp. 206, 210–2, 215–6; Rumbaut, "Paradise Shift," p. 12; see also Margaret A. Gibson, "The School Performance of Immigrant Minorities: A Comparative View," *Anthropology and Education Quarterly* 18, no. 4 (Dec. 1987), pp. 262–75 (focusing on children of lower-income Punjabi Sikhs); Alba and Nee, *Remaking the American Mainstream*, p. 240 (noting that the children of often illiterate Hmong immigrants "nevertheless achieve high grades").

7 Nobel Prizes: Raphael Patai, *The Jewish Mind* (Detroit: Wayne State University Press, 1996), pp. 342, 547–8 (documenting that between 1901 and 1994 Jews, while "less than half a percent of mankind," won 35 percent of the Nobel Prizes for Economics, 27 percent for Physiology and Medicine, 22 percent for Physics, and 20 percent of Nobel Prizes overall).

7 Pulitzer Prizes: Steven L. Pease, *The Golden Age of Jewish Achievement* (Sonoma, CA: Deucalion, 2009), p. viii, table 1 (Jews have won 51 percent of the Pulitzer Prizes for nonfiction, 13 percent for fiction).

7 Tony Awards: Stewart F. Lane, *Jews on Broadway: An Historical Survey of Performers, Playwrights, Composers, Lyricists and Producers* (Jefferson, NC, and London: McFarland & Company, 2011), p. 190 (69 percent of Tony-winning composers have been Jewish).

7 hedge-fund billions: Nathan Vardi, "The 40 Highest-Earning Hedge Fund Managers," *Forbes*, March 1, 2012, http://www.forbes.com/lists/2012/hedge-fund-managers-12_land.html.

7 among Jewish respondents, it was $443,000: Lisa A. Keister, *Faith and Money: How Religion Contributes to Wealth and Poverty* (Cambridge, UK: Cambridge University Press, 2011), p. 86 and Table 4.1.

8 just 1.7 percent of the adult population: Pew Forum on Religion & Public Life, *U.S. Religious Landscape Survey—Religious Affiliation: Diverse and Dynamic* (Washington, DC: Pew Research Center, 2008) p. 8.

8 twenty of *Forbes*'s top: Jacob Berkman, "At Least 139 of the Forbes 400 are Jewish," October 5, 2009, http://www.jta.org/2009/10/05/fundermentalist/at-least-139-of-the-forbes-400-are-jewish.

8 Protestants still dominated: Max Weber, *The Protestant Ethic and the Spirit of Capitalism* (London and New York: Routledge, 1992), p. 35.

8 Today, American Protestants: Lisa A. Keister, *Getting Rich: America's New Rich and How They Got That Way* (New York: Cambridge University Press, 2005), pp. 164–7, 172; see also Pew Forum on Religion & Public Life, *U.S. Religious Landscape Survey*, p. 60 (13 percent of Evangelical Protestants make $100,000 a year or more, as compared to 18 percent of the total population, 15 percent of Protestants in America generally, 43 percent of Hindu Americans, and 46 percent of Jewish Americans).

8 religious, as in the case of Mormons: Bushman, *Contemporary Mormonism*, p. 3; Terryl L. Givens, *People of Paradox: A History of Mormon Culture* (Oxford and New York: Oxford University Press, 2007), pp. xiii, xvi; Jan Shipps, *Mormonism: The Story of a New Religious Tradition* (Urbana and Chicago: University of Illinois Press, 1985), p. 125.

8 magnificence of your people's history: See, e.g., Kenneth M. Pollack, *The Persian Puzzle: The Conflict Between Iran and America* (New York: Random House, 2004), p. 3; Robert Graham, *Iran: The Illusion of Power* (London: Croom Helm, 1978), pp. 190–2; John K. Fairbank, "China's Foreign Policy in Historical Perspective," *Foreign Affairs* 47, no. 3 (April 1969), pp. 449, 456; Q. Edward Wang, "History, Space, and Ethnicity: The Chinese Worldview," *Journal of World History* 10, no. 2 (Fall 1999), pp. 285, 287–8.

9 "priestly" Brahman caste: Ramesh Bairy T.S., *Being Brahmin, Being Modern: Exploring the Lives of Caste Today* (London, New York, and New Delhi: Routledge, 2010), pp. 87, 280–1; Louis Dumont, *Homo Hierarchicus: The Caste System and Its Implications* (Chicago and London: University of Chicago Press, 1980), pp. 79–80; see also Narendra Jadhav, *Untouchables: My Family's Triumphant Escape from India's Caste System* (Berkeley and Los Angeles: University of California Press, 2007), pp. 1, 4.

9 entrepreneurial Igbo: Chinua Achebe, *There Was a Country: A Personal History of Biafra* (New York: Penguin Press, 2012), pp. 74–6; Donald L. Horowitz, *Ethnic Groups in Conflict* (Berkeley, Los Angeles, and London: University of California Press, 1985), pp. 27–8, 154–5, 164–6, 243–9; Amy Chua, *World on Fire: How Exporting Free Market Democracy Breeds Ethnic Hatred and Global Instability* (New York: Doubleday, 2003), pp. 108–9.

9 Jews are the "chosen": See generally Avi Beker, *The Chosen: The History of an Idea, and the Anatomy of an Obsession* (New York: Palgrave MacMillan, 2008); David Novak, *The Election of Israel: The Idea of the Chosen People* (Cambridge, UK: Cambridge University Press, 1995); Daniel H. Frank, ed., *A People Apart: Chosenness and Ritual in Jewish Philosophical Thought* (Albany: State University of New York Press, 1993); Arnold M. Eisen, *The Chosen People in America: A Study in Jewish Religious Ideology* (Bloomington: Indiana University Press, 1983).

9 a moral people, a people of law: See, e.g., Louis Dembitz Brandeis, "The Jewish Problem: How to Solve It" (speech delivered in June 1915), reprinted in Steve Israel and Seth Forman, eds., *Great Jewish Speeches Throughout History* (Northvale, NJ, and London: Jason Aronson Inc., 1994), pp. 69, 74; Patai, *The Jewish Mind*, pp. 8–9, 324, 339; Nathan Glazer, *American Judaism* (2d ed.) (Chicago and London: The University of Chicago Press, 1972), pp. 136–7.

9 To be an immigrant: See, e.g., Nancy Foner, *From Ellis Island to JFK: New York's Two Great Waves of Immigration* (New Haven, CT, and London: Yale University Press, 2000), pp. 72, 90; Eleanor J. Murphy, "Transnational Ties and Mental Health," in Ramaswami Mahalingam, ed., *Cultural Psychology of Immigrants* (Mahwah, NJ: Lawrence Erlbaum Associates Publishers, 2006), pp. 79, 81; Vivian S. Louie, *Compelled to Excel: Immigration, Education, and Opportunity Among Chinese Americans* (Stanford, CA: Stanford University Press, 2004), pp. 123–5.

9–10 diagnostically recognized symptom: American Psychiatric Association, *Diagnostic and Statistical Manual of Mental Disorders* (4th ed., Text Revision) (Arlington, VA: American Psychiatric Assn., 2000), p. 720.

10 Freud speculated: Sigmund Freud, "'Civilized' Sexual Morality and Modern Nervous Illness," in James Strachey, ed. and trans., *The Standard Edition of the Complete Psychological Works of Sigmund Freud* (London: Hogarth Press, 1986), vol. 9, p. 186 ("Generally speaking, our civilization is built up on the suppression of instincts").

11 youth culture: See Jed Rubenfeld, *Freedom and Time: A Theory of Constitutional Self-Government* (New Haven, CT, and London: Yale University Press, 2001), p. 34; Jon Savage,

Teenage: The Creation of Youth Culture (New York: Viking, 2007); see also Patricia Cohen, *In Our Prime: The Invention of Middle Age* (New York: Scribner, 2012), p. 168.

12 large marble pig: Amy Chua, *Day of Empire: How Hyperpowers Rise to Global Dominance—and Why They Fall* (New York: Doubleday, 2007), p. 38; Anthony R. Birley, *Hadrian: The Restless Emperor* (London and New York: Routledge, 1997), p. 276.

12 filthy and degenerate: Irving Howe, *World of Our Fathers* (New York and London: Harcourt Brace Jovanovich, 1976), pp. 51 ("human garbage"), 230 ("slovenly in dress, loud in manners, and vulgar in discourse").

12 "chip on the shoulder": Stephen Birmingham, *"The Rest of Us": The Rise of America's Eastern European Jews* (Boston and Toronto: Little, Brown and Company, 1984), p. 17; see also Hannah Arendt, *The Origins of Totalitarianism* (New York: Harcourt, Brace & World, 1966), p. 82 (describing "Jewishness" for Proust in *Remembrance of Things Past* as "at once a physical stain and a mysterious personal privilege").

12 New York intellectuals: See Richard M. Cook, *Alfred Kazin: A Biography* (New Haven, CT, and London: Yale University Press, 2007), pp. 32–3 (describing what Kazin called the "humiliation" he suffered at the hands of those with "a long-bred talent for sociability," which he never outlived and which made him "bitter, bitter"); Peter Manso, *Mailer: His Life and Times* (New York: Simon & Schuster, 1985), pp. 340–1 (quoting Rhoda Lazare Wolf as saying that Norman Mailer wanted "to conquer the world. He doesn't want to be the boy from Crown Street"); Ross Wetzsteon, *Republic of Dreams: Greenwich Village: The American Bohemia, 1910–1960* (New York: Simon & Schuster, 2002), p. 491 (describing how Delmore Schwartz "flipped . . . from grandiose self-esteem to histrionic self-loathing"); Greg Bellow, *Saul Bellow's Heart: A Son's Memoir* (New York: Bloomsbury, 2013), p. 54 (describing his father, Saul Bellow, as "an unknown Jewish novelist" who—whatever his "personal insecurities"—threw "down the gauntlet to the American literary establishment personified by Ernest Hemingway"); James Atlas, *Bellow* (New York: Random House, 2002), pp. 10, 23, 32, 50; Florence Rubenfeld, *Clement Greenberg—A Life* (New York: Scribner, 1997), p. 83 ("Although [Greenberg] appeared confident, his arrogance concealed a self-doubt that had to be hidden at all costs, especially from himself"); Wetzsteon, *Republic of Dreams*, p. 533 (describing Clement Greenberg as "arrogant and insecure"); Norman Podhoretz, *Making It* (New York: Random House, 1967), p. 5 ("my 'noblest' ambitions were tied to the vulgar desire to rise above the class into which I was born . . . [and] to an astonishing extent . . . were shaped and defined by the standards and values and tastes of the class into which I did not know I wanted to move") and p. 38 (describing the "disdain" he encountered at Columbia from "those whose wealth was inherited"); see also Alan M. Wald, *The New York Intellectuals: The Rise and Decline of the Anti-Stalinist Left from the 1930s to the 1980s* (Chapel Hill and London: University of North Carolina Press, 1987), p. 34 (noting that Lionel Trilling "was acutely sensitive to the way in which society attempted to exclude Jews," that "he felt a special antagonism toward . . . genteel German Jews who were proud of their high degree of acculturation"); see generally Alan Bloom, *Prodigal Sons: The New York Intellectuals & Their World* (Oxford and New York: Oxford University Press, 1986), p. 308 ("Essential" to the "success of the New York Intellectuals was their desire to achieve something in the society which had excluded their parents"); and pp. 17, 21–23, 27, 155.

12 Nietzsche taught: See Friedrich Nietzsche, *On the Genealogy of Morals and Ecce Homo* (Walter Kaufmann, ed. and trans.) (New York: Vintage Books, 1969), pp. 54, 73, 124 (describing *"ressentiment"* as "inexhaustible and insatiable"), 126–8; see also Robert C. Solomon, "Nietzsche and the Emotions," in Jacob Golomb, Weaver Santaniello, and Ronald Lehrer, eds., *Nietzsche and Depth Psychology* (Albany: State University of New York Press, 1999), pp. 127, 142.

12 moving up the economic ladder: Foner, *From Ellis Island to JFK*, p. 91.

12 steep fall in status: See Susan Eva Eckstein, *The Immigrant Divide: How Cuban Americans Changed the US and Their Homeland* (New Haven, CT, and London: Routledge, 2009), p. 83;

David Rieff, *The Exile: Cuba in the Heart of Miami* (New York: Simon & Schuster, 1993), pp. 48, 81; Gonzalez-Pando, *The Cuban Americans*, pp. 34–6; Tara Bahrampour, "Persia on the Pacific," *The New Yorker*, November 10, 2003; see also Foner, *From Ellis Island to JFK*, p. 91.

13 high academic expectations: See, e.g., Jin Li, *Cultural Foundations of Learning: East and West* (New York: Cambridge University Press, 2012), p. 71; Rebecca Y. Kim, *God's New Whiz Kids? Korean American Evangelicals on Campus* (New York and London: New York University Press, 2006), p. 79; Zhou, *Contemporary Chinese America*, pp. 194–5; see also Richard R. Pearce, "Effects of Cultural and Social Structural Factors on the Achievement of White and Chinese American Students at School Transition Points," *American Educational Research Journal* 43, no. 1 (2006), pp. 75, 94–5; Wenfan Yen and Qiuyun Lin, "Parent Involvement and Mathematics Achievement: Contrast Across Racial and Ethnic Groups," *The Journal of Educational Research* 99, no. 2 (2005), pp. 116, 120–1; Bandana Purkayastha, *Negotiating Ethnicity: Second-Generation South Asian Americans Traverse a Transnational World* (New Brunswick, NJ, and London: Rutgers University Press, 2005), pp. 91, 93; Clara C. Park, "Educational and Occupational Aspirations of Asian American Students," in Clara C. Park, A. Lin Goodwin and Stacey J. Lee, eds., *Asian American Identities, Families, and Schooling* (Greenwich, CT: Information Age Publishing, 2003), pp. 135, 149–50.

13 Comparisons to cousin X: See Li, *Cultural Foundations of Learning*, p. 207 (noting that Chinese parents seek out "good students in the community" and "refer their own children to them often, in clear comparative terms, urging their children to emulate these models"); Jin Li, Susan D. Holloway, Janine Bempechat, and Elaine Loh, "Building and Using a Social Network: Nurture for Low-Income Chinese American Adolescents' Learning," in Hirokazu Yoshikawa and Niobe Way, eds., *Beyond the Family: Contexts of Immigrant Children's Development*, no. 121 (2008), pp. 9, 18 (in a study of low-income Chinese American adolescents, 77 percent of the children said their parents frequently compared them to higher-achieving relatives or peers); see also Lee and Lee Zhou, "Frames of Achievement and Opportunity Horizons," p. 216.

13 lower- and higher-income: Louie, *Compelled to Excel*, pp. 97–8; Lee and Zhou, "Frames of Achievement and Opportunity Horizous," p. 216; Li et al., "Building and Using a Social Network," pp. 15, 18–20; Kyle Spencer, "For Asians, School Tests Are Vital Steppingstones," *New York Times*, Oct. 26, 2012.

13 "In Chinese families": Ruth K. Chao, "Chinese and European American Mothers' Beliefs about the Role of Parenting in Children's School Success," *Journal of Cross-Cultural Psychology* 27, no. 4 (1996), pp. 403, 412; see also Louie, *Compelled to Excel*, p. 48 (noting that in Confucian-influenced cultures, "the accomplishments of the individual are strongly grounded in familial obligations and prestige"); Peter H. Huang, "Tiger Cub Strikes Back: Memoirs of an Ex-Child Prodigy About Legal Education and Parenting," *British Journal of American Legal Studies* 1, no. 2 (2012), pp. 21–3 ("My mother was not amused . . . She told me that I had not only embarrassed myself, but also her, my entire immediate family, all Chinese people, all Asian people, all humans, and in fact all carbon-based life forms").

13 "It was not for myself alone": Alfred Kazin, *A Walker in the City* (New York: Harcourt, Brace and Company, 1951), p. 21.

13 "If there were Bs": Cook, *Alfred Kazin*, p. 13; see also Kazin, *A Walker in the City*, p. 21 (recalling from his Brooklyn childhood that "anything less than absolute perfection in school always suggested to my mind that I might . . . be kept back in the working class forever").

14 Mormon teenagers are less likely to drink: See Bushman, *Contemporary Mormonism*, p. 47 (citing the four-year National Study of Youth and Religion).

14 two hundred thousand people in poverty: U.S. Census, American Community Survey, Table S0201: Selected Population Profile in the United States (2010 3-year dataset) (population group code 013 – Asian Indian) (showing a poverty rate of 8.2 percent and a population of over 2.7 million).

14 "Let me summarize my feelings": Wesley Yang, "Paper Tigers," *New York Magazine*, May 8, 2011.

15 "[H]e will never become World Champion": Natalia Pogonina, "The Art of Defense," Chess.com, March 22, 2011, http://www.chess.com/article/view/art-of-defense.

15 psychological armor: See Min Zhou, "The Ethnic System of Supplementary Education: Nonprofit and For-Profit Institutions in Los Angeles' Chinese Immigrant Community," in Marybeth Shinn and Hirokazu Yoshikara, eds., *Toward Positive Youth Development: Transforming Schools and Community Programs* (New York: Oxford University Press, 2008), p. 232; John U. Ogbu and Herbert D. Simons, "Voluntary and Involuntary Minorities: A Cultural Ecological Theory of School Performance with Some Implications for Education," *Anthropology and Education Quarterly* 29, no. 2 (1998), pp. 155–88.

15 "Yes, I am a Jew": Beker, *The Chosen*, p. 85 (quoting Benjamin Disraeli).

16 resilience, stamina, or grit: See Roy F. Baumeister and John Tierney, *Willpower: Rediscovering the Greatest Human Strength* (New York: Penguin Press, 2011); Angela L. Duckworth, "The Significance of Self-Control," *Proceedings of the National Academy of Sciences* 108, no. 7 (2011), pp. 2639–40; Angela L. Duckworth, Christopher Peterson, Michael D. Matthews, and Dennis R. Kelly, "Grit: Perseverance and Passion for Long-Term Goals," *Journal of Personality and Social Psychology* 92, no. 6 (2007), pp. 1087–1101; see also Bob Sullivan and Hugh Thompson, *The Plateau Effect: Getting from Stuck to Success* (New York: Dutton, 2013), pp. 68–74.

16 They tend to believe: See, e.g., Louie, *Compelled to Excel*, pp. 47–8; Grace Wang, "Interlopers in the Realm of High Culture: 'Music Moms' and the Performance of Asian and Asian American Identities," *American Quarterly* 61, no. 4 (December 2009), pp. 892–4, 896; Purkayastha, *Negotiating Ethnicity*, p. 89; Maryam Daha, "Contextual Factors Contributing to Ethnic Identity Development of Second-Generation Iranian American Adolescents," *Journal of Adolescent Research* 26, no. 5 (2011), pp. 543, 560–1.

16 When on mission: See Bushman, *Contemporary Mormonism*, pp. 63–4; Keith Parry, "The Mormon Missionary Companionship," in Marie Cornwall, Tim B. Heaton, and Lawrence A. Young, eds., *Contemporary Mormonism: Social Science Perspectives* (Urbana and Chicago: University of Illinois Press, 1994), pp. 182, 183–85; Benedict, *The Mormon Way of Doing Business*, pp. 1–21.

16 binge-drinking culture: In a recent national study, 37 percent of American college students reported heavy drinking. Lloyd D. Johnston, Patrick M. O'Malley, Jerald G. Bachman, and John E. Schulenberg, *Monitoring the Future: National Survey Results on Drug Use, 1975–2012*, vol. 2 (Ann Arbor: The University of Michigan Institute for Social Research, 2012), p. 27.

17 institutions rather than culture: Daron Acemoglu and James A. Robinson, *Why Nations Fail: The Origins of Power, Prosperity, and Poverty* (New York: Crown, 2012).

17 geography ultimately explains: See Jared Diamond, *Guns, Germs and Steel: The Fates of Human Societies* (New York and London: W. W. Norton and Company, 1999); Jared Diamond, "What Makes Countries Rich or Poor?" *New York Review of Books*, June 7, 2012.

18 Triple Package cultures tend to focus: See, e.g., Bloom, *Prodigal Sons*, p. 17 (noting that among early-twentieth-century Jewish immigrant families "material worth became the tangible sign of American success"); Purkayastha, *Negotiating Ethnicity*, p. 91 (South Asian Americans "often emphasize achievement in terms of a narrow range of white-collar careers in medicine, technology, law, and the sciences"); Mitra K. Shavarini, *Educating Immigrants: Experiences of Second-Generation Iranians* (New York: LFB Scholarly 2004), p. 148 (stressing the importance to Iranian Americans of "symbolic status" as well economic status); Daha, "Contextual Factors Contributing to Ethnic Identity Development of Second-Generation Iranian American Adolescents," pp. 560–1; Lisa Sun-Hee Park, "Ensuring Upward Mobility: Obligations of Children of Immigrant Entrepreneurs," in Benson Tong, ed., *Asian American Children: A Historical Handbook and Guide* (Westport, CT: Greenwood Press, 2004), p. 126.

18 "teach us how to make a living": James Truslow Adams, "To 'Be' or To 'Do': A Note on American Education," *Forum* LXXXI, no. 6 (June 1929), pp. 321, 325.

18 nothing-is-ever-good-enough mentality: See. e.g., Zhou, *Contemporary Chinese America*, p. 195.

19 America's persistently low-income groups became poor: See, e.g., Thomas J. Sugrue, *The Origins of the Urban Crisis: Race and Inequality in Postwar Detroit* (Princeton, NJ: Princeton University Press, 1996), pp. 5–14; Dwight B. Billings and Kathleen M. Blee, *The Road to Poverty: The Making of Wealth and Hardship in Appalachia* (Cambridge, UK: Cambridge University Press, 2000), pp. 8–24.

19 little reason to engage in impulse control: Celeste Kidd, Holly Palmeri, and Richard N. Aslin, "Rational Snacking: Young Children's Decision-Making on the Marshmallow Task Is Moderated by Beliefs About Environmental Reliability," *Cognition* 126 (2013), pp. 109–14.

20 American Huguenot community: See Jon Butler, *The Huguenots in America: A Refugee People in New World Society* (Cambridge, MA: Harvard University Press, 1983), pp. 80–6, 96–100, 122–3, 132.

20 extremely high out-marriage rates: Alixa Naff, "Lebanese Immigration into the United States: 1880 to the Present," in Albert Hourani and Nadim Shehadi, eds., *The Lebanese in the World: A Century of Emigration* (London: Centre for Lebanese Studies and I.B. Tauris & Co., 1992), pp. 159–60; Andrzej Kulczycki and Arun Peter Lobo, "Patterns, Determinants, and Implications of Intermarriage Among Arab Americans," *Journal of Marriage and Family* 64, no. 1 (2002), pp. 202–10.

20 WASP economic dominance in the United States declined: See, e.g., Jerome Karabel, *The Chosen: The Hidden History of Admission and Exclusion at Harvard, Yale, and Princeton* (Boston and New York: Houghton Mifflin Company, 2005); Robert C. Christopher, *Crashing the Gates: The De-WASPing of America's Power Elite* (New York: Simon & Schuster, 1989), pp. 17–8; Peter Schrag, *The Decline of the WASP* (New York: Simon & Schuster, 1971); Ronald W. Schatz, "The Barons of Middletown and the Decline of the North-Eastern Anglo-Protestant Elite," *Past & Present* 219 (2013), pp. 165, 197–200; James D. Davidson et al., "Persistence and Change in the Protestant Establishment, 1930–1992," *Social Forces* 74, no. 1 (1995), pp. 157–75; E. Digby Baltzell, "The Protestant Establishment Revisited," *The American Scholar* 45, no. 4 (1976), pp. 499–518.

20 Jews may feel less insecure: See, e.g., Charles E. Silberman, *A Certain People: American Jews and Their Lives Today* (New York: Summit Books, 1985), p. 23 ("Americans Jews now live in a freer, more open society than that of any Diaspora community in which Jews have ever lived before").

20 precipitous drop: See Unz, "The Myth of American Meritocracy" (noting that the number of Jewish top scorers on the high school Math Olympiad appears to have dropped from roughly 40 percent in the 1970s to 2.5 percent today).

21 2.6 times out of 10: Clay Davenport, "Baseball Prospectus Basics – About EqA," Feb. 24, 2004, http://www.baseballprospectus.com/article.php?articleid=2596 (referring to "the all-time major-league batting average of .262"). Williams hit .406 in 1941. Bill Pennington, "Ted Williams's .406 Is More Than a Number," *New York Times*, Sept. 17, 2011.

22 narcissistic personality disorder: Walter Isaacson, *Steve Jobs* (New York: Simon & Schuster, 2011), pp. 265–6 (quoting Tina Redse).

22 "chip on his shoulder": Jeffrey S. Young and William L. Simon, *iCon Steve Jobs: The Greatest Second Act in the History of Business* (Hoboken, NJ: John Wiley & Sons, 2005), p. 7.

22 an especially strong parent or even grandparent: Oprah Winfrey has described her grandmother as filling just such a role. "I am what I am because of my grandmother. My strength. My sense of reasoning. Everything. All that was set by the time I was six. . . . I was raised with an outhouse, no plumbing. Nobody had any clue that my life could be anything but working in some factory or a cotton field in Mississippi. Nobody—nobody. . . . Thank goodness I was raised by my grandmother the first six years." Janet Lowe, *Oprah Winfrey Speaks: Insight from the World's Most Influential Voice* (New York: John Wiley & Sons, 1998) (inner quotations omitted), pp. 6, 8.

22 stereotype threat and stereotype boost: The seminal article on stereotype threat is Claude M. Steele and Joshua Aronson, "Stereotype Threat and the Intellectual Test Performance of African Americans," *Journal of Personality and Social Psychology* 69 (1995), but there is now a voluminous literature on the topic. For an excellent comprehensive treatment, see Michael Inzlicht and Toni Schmader, eds., *Stereotype Threat: Theory, Process, and Application* (New York: Oxford University Press, 2012), which includes a discussion of the converse phenomenon, stereotype boost. Both stereotype threat and stereotype boost will be discussed in more detail in chapter 3 of this book.

23 "ethnic armor": see Zhou, "The Ethnic System of Supplementary Education," p. 232; Ogbu and Simons, "Voluntary and Involuntary Minorities," pp. 170–2.

23 groundswell of studies: See chapter 8 and especially p. 213.

23 two of the leading twentieth-century studies: Victor Goertzel and Mildred George Goertzel, *Cradles of Eminence* (Boston and Toronto: Little, Brown and Company, 1962), pp. vii–viii, 55, 60–7, 131 (stating that the authors' study of 400 eminent individuals confirms theory that "contented," "untroubled" people tend to have "low aspirations for themselves and for their children"), 174, 183, 214–6, 272–3, 302–49; Howard Gardner (in collaboration with Emma Laskin), *Leading Minds: An Anatomy of Leadership* (New York: BasicBooks, 1995), pp. 1–9, 32–3, 249–53. See also Angela L. Duckworth, David Weir, Eli Tsukayama, and David Kwok, "Who Does Well in Life? Conscientious Adults Excel in Both Objective and Subjective Success," *Frontiers in Psychology* 3 (Sept. 2012), pp. 1, 5–6 (nationwide study of 9,646 American adults found that "more agreeable adults actually earned and saved less money").

23 "the twinge of adversity": Gardner, *Leading Minds*, p. 33 (quoting Winston Churchill).

23 "willpower" and "grit": Baumeister and Tierney, *Willpower*, pp. 1–17; Duckworth, "The Significance of Self-Control," pp. 2639–40; Duckworth, Peterson, Matthews, and Kelly, "Grit: Perseverance and Passion for Long-Term Goals," pp. 1087–1101; Kelly McGonigal, *The Willpower Instinct: How Self-Control Works, Why It Matters, and What You Can Do to Get More of It*, (New York: Avery, 2012), p. 12; Sullivan and Thompson, *The Plateau Effect*, pp. 57–74; Angela L. Duckworth, Patrick D. Quinn, and Eli Tsukayama, "What *No Child Left Behind* Leaves Behind: The Roles of IQ and Self-Control in Predicting Standardized Achievement Test Scores and Report Card Grades," *Journal of Educational Psychology* 104, no. 2 (2012), pp. 439–51.

23 "marshmallow test": For the original studies, see Walter Mischel, Ebbe B. Ebbeson, and Antonette Raskoff Zeiss, "Cognitive and Attentional Mechanisms in Delay of Gratification," *Journal of Personality and Social Psychology* 21, no. 2 (1972), pp. 204–18; Yuichi Shoda, Walter Mischel, and Philip K. Peake, "Predicting Adolescent Cognitive and Self-Regulatory Competencies from Preschool Delay of Gratification: Identifying Diagnostic Conditions," *Developmental Psychology* 26, no. 6 (1990), pp. 978–86.

23 Kids with more impulse control go on: Baumeister and Tierney, *Willpower*, pp. 10–13 (summarizing studies); Duckworth, "The Significance of Self-Control," p. 2639 ("self-control . . . during the first decade of life predicts income, savings behavior, financial security, occupational prestige, physical and mental health, substance use, and (lack of) criminal convictions . . . in adulthood. Remarkably, the predictive power of self-control is comparable to that of either general intelligence or family socioeconomic status"). For more detail, see chapter 5.

23 better predictors of grades and future success: Baumeister and Tierney, *Willpower*, p. 11; Duckworth, Peterson, Matthews, and Kelly, "Grit," p. 1098; Duckworth, "The Significance of Self-Control," p. 2639; Duckworth, Quinn, and Tsukayama, "What *No Child Left Behind*," p. 445; Angela L. Duckworth and Martin E. P. Seligman, "Self-Discipline Outdoes IQ in Predicting Academic Performance of Adolescents," *Psychological Science* 16 (2005), no. 12, pp. 939, 941.

24 Syrian Jewish enclave in Brooklyn: Jeffrey S. Gurock, ed., *American Jewish Life 1920–1990, American Jewish History* (New York and London: Routledge, 1998) vol. 4, pp. 112–3 (noting that most Syrian Jews in America live in Brooklyn, that they "are more business-oriented, and fewer attend college," and that "[w]hen the young marry, they often choose to resettle in the community"); see also Morris Gross, *Learning Readiness in Two Jewish Groups* (New York: Center for Urban Education, 1967), pp. 32–5; Hayyim Cohen, "Sephardi Jews in the United States: Marriage with Ashkenazim and Non-Jews," *Dispersion and Unity* (Jerusalem) 13-14 (1971/72), pp. 151, 152–3; Zev Chafets, "The Sy Empire," *New York Times*, Oct. 14, 2007.

25 relatively closed to intellectual and scientific inquiry: Bowman, *The Marmon People,* p. 180 (mid-twentieth-century Mormonism was dominated by a "theological conservatism rejecting enthusiasm for science and reason in favor of a strict emphasis on scripture"); Gregory A. Prince and Wm. Robert Wright, *David O. McKay and the Rise of Modern Mormonism* (Salt Lake City: University of Utah Press, 2005), p. 160 (quoting LDS Church president David McKay saying "Character is the aim of true education; and science, history, and literature are but means used to accomplish this desired end").

25 "men of God first": President Spencer W. Kimball, "Education for Eternity" (address delivered at Brigham Young University, Sept. 12, 1967) (in part quoting then-BYU president Ernest L. Wilkinson), http://education.byu.edu/edlf/archives/prophets/eternity.html.

25 knew nothing of Jewish learning: Nathan Glazer, "Disaggregating Culture," in Lawrence E. Harrison and Samuel P. Huntington, *Culture Matters: How Values Shape Human Progress* (New York: Basic Books, 2000), pp. 219, 224.

25 "I went to Yale much against my father's wishes": "Stanley Schachter," in Gardner Lindzey and William M. Runyan, eds., *A History of Psychology in Autobiography* (Washington, DC: American Psychological Association, 2007), quoted in Steven Pinker, "The Lessons of the Ashkenazim: Groups and Genes," *The New Republic,* June 26, 2006.

26 ultra-Orthodox Satmar community: Sam Roberts, "A Village with the Numbers, Not the Image, of the Poorest Place," *New York Times,* Apr. 20, 2011.

27 "longing to rise": Alexis de Tocqueville, *Democracy in America,* trans. George Lawrence, ed. J. P. Mayer (Garden City, NY: Anchor Books, 1969), vol. 2, part III, chap. 19, p. 627.

CHAPTER 2: WHO'S SUCCESSFUL IN AMERICA?

30 more than three hundred definitions in the literature: John R. Baldwin et al., eds., *Redefining Culture: Perspectives Across Disciplines* (Mahwah, NJ: Lawrence Erlbaum, 2006), pp. 139–226; see A. L. Kroeber and Clyde Kluckhohn, *Culture: A Critical Review of Concepts and Definitions* (New York: Vintage, 1952).

Many scholars have defined culture solely in terms of mental states, a view associated with American archeologist Walter Taylor, who wrote, "Culture consists of ideas . . . . [It] is a mental phenomenon . . . not material objects or observable behavior." W. R. Taylor, *A Study of Archeology* (American Anthropological Association, 1948), pp. 98–110. This conception, which in fact predated Taylor, see Albert Blumenthal, *Culture Consists of Ideas* (Marietta, OH: Marietta College Press, 1937), remains highly influential today. See, e.g., Samuel P. Huntington, "Foreword: Cultures Count," in Lawrence E. Harrison and Samuel P. Huntington, eds., *Culture Matters: How Values Shape Human Progress* (New York: Basic Books, 2000), p. xv ("we define culture in purely subjective terms as the values, attitudes, beliefs, orientations, and underlying assumptions prevalent among people in a society"); Dan Sperber, *Explaining Culture: A Naturalistic Approach* (Oxford: Blackwell Publishers, 1996), p. 1 ("Culture is made up first and foremost of contagious ideas").

The problem with this view is what it leaves out. Food isn't an idea, attitude, or belief, but most would agree that what people eat (sushi or falafel), how they eat it (hands, fork, chopsticks), and how much they eat are part of their culture. Clothing isn't a "mental phenomenon" either, but whether a person wears a sari or a tank top can surely be of cultural importance. In short, culture has to include not only what people think, but what they do. See Roy G. D'Andrade, "Cultural Meaning Systems," in Richard Schweder and Richard Alan LeVine, eds., *Culture Theory: Essays on Mind, Self, and Education* (Cambridge, UK: Cambridge University Press, 1984), pp. 88, 90, 96, 115 (criticizing the view that culture consists of "knowledge and symbol," "knowledge and belief," or "conceptual structures," "rather than habit and behavior").

Perhaps for this reason, some anthropologists, like Clifford Geertz, have opted for a "thick" definition of culture as "the entire way of life of a society: its values, practices, symbols, institutions, and human relationships." Huntington, "Foreword," p. xv. But this approach, as Huntington points out, is too inclusive; everything becomes culture. The United States Congress is certainly an "institution." Is it culture?

Sociologist Talcott Parsons classified all human action in terms of four hierarchically arranged "systems": behavioral or physiological; personality; social; and cultural. The first was to be the realm of biology; the second, psychology; the third (which included institutions), sociology; the fourth, anthropology. See, e.g., Talcott Parsons, *On Institutions and Social Evolution*, Leon Mayhew, ed. (Chicago: University of Chicago Press, 1985), pp. 157–58; Talcott Parsons, *Social Systems and the Evolution of Action Theory* (New York: Free Press, 1977), p. 86. Parsons's classification may have been useful to assign domains of expertise to different academic disciplines, but in real life psychology, institutions, and culture don't come in hermetically sealed packages; they constantly intermix and codetermine one another.

For just this reason, we don't try to define culture in this book; some things we treat as cultural could just as easily be described as psychological or social. Generally speaking, taking a cue from Orlando Patterson, we include in culture anything—artwork or commodity, behavior or attitude—that expresses a group's sense of who it is and how its members should live. See Orlando Patterson, "Taking Culture Seriously: A Framework and an Afro-American Illustration," in Harrison and Huntington, *Culture Matters*, pp. 202, 208. Culture, on this view, is a collective effort to answer certain fundamental human questions. Parents impart answers to these questions to their children; so do schools; so does television. It's for this reason that family, education, and TV are so intensely cultural (in addition to everything else they are).

Of particular importance to this book, cultures orient people in *time*, urging them for example to venerate the past, live for the moment, or direct themselves to the future. Writers across a wide range of genres have recognized and elaborated on this phenomenon. For example, Weber's *Protestant Work Ethic* is one such elaboration; another is Milan Kundera's novel *Slowness*; another is organizational anthropologist Geert Hofstede's concept of "long-term orientation," said to be empirically testable and to vary from country to country. See Geert Hofstede, *Culture's Consequences: Comparing Values, Behaviors, Institutions, and Organizations Across Nations* (2d ed.) (Thousand Oaks, CA, London, and New Delhi: Sage Publications, 2001), pp. 351–72. These temporal orientations, as we'll see, can have a powerful impact on individuals' and groups' economic outcomes.

30 The term "Mormon": Claudia L. Bushman, *Contemporary Mormonism: Latter-day Saints in Modern America* (Westport, CT: Praeger, 2006), p. xiii; Matthew Bowman, *The Mormon People: The Making of an American Faith* (New York: Random House, 2012), pp. ix, xiv–xv.

30 Concentrated in Utah: Neal R. Peirce, *The Mountain States of America: People, Politics, and Power in the Eight Rocky Mountain States* (New York: W. W. Norton, 1972), p. 192.

30–31 discriminated against blacks: Bushman, *Contemporary Mormonism*, pp. 91–3, 97–98; Armand L. Mauss, *The Angel and the Beehive: The Mormon Struggle with Assimilation* (Chicago: University of Illinois Press, 1994), pp. 51–3.

31 still a rarity on Wall Street and in Washington: James Crabtree, "The Rise of a New Generation of Mormons," *Financial Times*, July 9, 2010; Dan Gilgoff, "With or Without Romney, DC a Surprising Mormon Stronghold," CNN.com, May 12, 2012, http://religion.blogs.cnn.com/2012/05/12/hfr-with-or-without-romney-d-c-a-surprising-mormon-stronghold.

31 apple-pie variety: See Bushman, *Contemporary Mormonism*, pp. 187–8 ("Mormons do well in finance, business, and law. . . . More Mormons are moving into performance, scholarship, and the arts.") For a fascinating history of Mormon music, literature, theater, and visual arts, see Terryl L. Givens, *People of Paradox: A History of Mormon Culture* (Oxford and New York: Oxford University Press, 2007), chaps. 7–10.

31 5 to 6 million Mormons: The LDS Church officially claims about 6 million U.S. members, which would make Mormons about 2 percent of the population. Independent researchers, however, believe the true figure to be lower. See, e.g., Rick Phillips and Ryan T. Cragun, *Mormons* *in the United States 1990–2008: Socio-demographic Trends and Regional Differences* (Hartford, CT: Trinity College, 2011), p. 1 (estimating Mormons at 1.4 percent of adult population); The Pew Forum on Religion and Public Life, *A Portrait of Mormons in the U.S.* (July 24, 2009) (estimating Mormons at 1.7 percent of adult population and Protestants at 51.3 percent), http://www .pewforum.org/2009/07/24/a-portrait-of-mormons-in-the-us.

31 $230 million: Edwin Durgy, "What Mitt Romney Is Really Worth: An Exclusive Analysis of His Latest Finances," *Forbes*, June 4, 2012.

31 Other leading Mormon politicians include: Richard N. Ostling and Joan K. Ostling, *Mormon America: The Power and the Promise* (New York: HarperOne, 2007), pp. xiii, xv, 137, 140; Caroline Winter, "God's MBAs: Why Mormon Missions Produce Leaders," *Business Week*, June 9, 2011.

31 prominent Mormons include: Jeff Benedict, *The Mormon Way of Doing Business: How Nine Western Boys Reached the Top of Corporate America* (New York and Boston: Business Plus, 2007); Winter, "God's MBAs"; Crabtree, "The Rise of a New Generation of Mormons"; Larissa MacFarquhar, "When Giants Fail: What Business Can Learn from Clayton Christensen," *New Yorker*, May 12, 2012; "The Mormon Way of Business," *The Economist*, May 5, 2012; "The List: Famous Mormons," *Washington Times*, Oct. 21, 2011; "Mormons in Business," Businessweek .com, June 9, 2011.

32 just the tip of the iceberg: "Mormons in Business"; Brent Schlender, "Incredible: The Man Who Built Pixar's Animation Machine," CNN Money.com, Nov. 15, 2004, http://money .cnn.com/magazines/fortune/fortune_archive/2004/11/15/8191082; Max B. Knudson, "2 More Utahns Join Forbes' 400," Deseret News, October 10, 1990, http://www.deseretnews.com/ article/126660/2-MORE-UTAHNS-JOIN-FORBES-400.html?pg=all.

32 overrepresented in the CIA: Grace Wyler, "11 Surprising Things You Didn't Know About Mormons," Business Insider, June 24, 2011, http://www.businessinsider.com/11-surprising-things -you-didnt-know-about-mormons-2011-6.

32 "out of nowhere": Jon Mooallem, "When Hollywood Wants Good, Clean Fun, It Goes to Mormon Country," *New York Times*, May 23, 2013.

33 Ken Jennings: Ostling and Ostling, *Mormon America*, p. 149.

33 Mormon senior executives in any Fortune 500 company: Data compiled for authors in May 2012 (on file with authors). For all the companies ever listed on the Fortune 500 since it began in 1955, see CNN Money, "Fortune 500," http://money.cnn.com/magazines/fortune/ fortune500_archive/full/1955.

33 Goldman Sachs announced the addition: See Winter, "God's MBAs"; Julie Steinberg, "Goldman Looks West for New Flock," Fins Finance, Mar. 12, 2012, http://www.fins.com/ Finance/Articles/SBB0001424052970204781804577271884170878826/Goldman-Looks-West -for-New-Flock.

33 thirty-one of its graduates: Winter, "God's MBAs."

33 correlation of faith and money: Lisa A. Keister, *Faith and Money: How Religion Contributes to Wealth and Poverty* (Cambridge, UK: Cambridge University Press, 2011), pp. 181–8.

33–34 somewhat more likely to make: The Pew Forum on Religion and Public Life, *A Portrait of Mormons in the U.S*, Table: Education and Income.

34 Mormon women who describe themselves as housewives: Phillips and Cragun, *Mormons in the United States 1990–2008*, pp. 5–6.

34 "secret" to their success: Benedict, *The Mormon Way of Doing Business*, pp. 144–6.

34 most are relatively poor: Pew Forum on Religion and Public Life, *A Portrait of Mormons in the U.S.*, Table: A Demographic Portrait of Converts to Mormonism (40 percent of Mormon converts earn less than $30,000 a year, compared to only 21 percent of non-converts).

34 Non-convert Mormons: Ibid.; see also Phillips and Cragun, *Mormons in the United States 1990–2008*, p. 8 (58 percent of Mormons in Utah earn $50,000 or more, compared to 45 percent of the U.S. population as a whole).

34 not considered Mormon by: Ostling and Ostling, *Mormon America*, p. 57.

34 "prophet" Rulon Jeffs: Jon Krakauer, *Under the Banner of Heaven: A Story of Violent Faith* (New York: Doubleday, 2003), pp. 12–13.

35 the church's highest leadership council: Bushman, *Contemporary Mormonism*, pp. 31–3, 111–3.

35 Church holdings: Ostling and Ostling, *Mormon America*, pp. 124–6; Winter, "God's MBAs."

35 Church of England: The Church of England, "Funding the Church of England," http://www.churchofengland.org/about-us/facts-stats/funding.aspx; William C. Symonds, "The Economic Strain on the Church," Apr. 15, 2002, http://www.businessweek.com/magazine/content/02_15/b3778001.htm.

36 U.S. Catholic Church: "The U.S. Catholic Church: How It Works," *Business Week*, April 14, 2002, http://www.businessweek.com/stories/2002-04-14/table-the-u-dot-s-dot-catholic-church-how-it-works; United States Conference of Catholic Bishops, "Catholic Information Project," August 2006, http://old.usccb.org/comm/cip.shtml (estimating total parish annual revenues of $7.35 billion); Pew Research Center, "U.S. Catholics: Key Data from Pew Research," Feb. 25, 2013 ("about 75 million Catholics in the United States"), http://www.pewresearch.org/key-data-points/u-s-catholics-key-data-from-pew-research.

36 "no other religion comes close": Ostling and Ostling, *Mormon America*, pp. 118, 408; Winter, "God's MBAs."

36 Goizueta name in a lot of places in Atlanta: See "Distinguished Faculty," Georgia Tech Office of Hispanic Initiatives, http://www.hispanicoffice.gatech.edu/goizueta-fellowship/faculty-staff (discussing "The Goizueta Foundation Faculty Chair" and the "Roberto C. Goizueta Chair for Excellence in Chemical Engineering"); "Atlanta Ballet Receives $2 Million Gift from The Goizueta Foundation," The Atlanta Ballet, Jan. 4, 2012, http://www.atlantaballet.com/press/goizueta-foundation-gift-010412. Since its establishment in 1992, the Goizueta Foundation has given over $370 million to support education, family services, and the arts.

36 Roberto Goizueta: David Greising, *I'd Like to Buy the World a Coke: The Life and Leadership of Roberto Goizueta* (New York: John Wiley & Sons, 1998), pp. xvi–xviii; "Goizueta Remembered for Service to Church," *The Georgia Bulletin*, Oct. 23, 1997.

36 "Cuban Exiles": See Miguel Gonzalez-Pando, *The Cuban Americans* (Westport, CT, and London: Greenwood Press, 1998), pp. 20–1, 33–4, 44–6; María Cristina García, *Havana USA: Cuban Exiles and Cuban Americans in South Florida, 1959–1994* (Berkeley and Los Angeles: University of California Press, 1996), chap. 1; see generally Richard D. Alba and Victor Nee, *Remaking the American Mainstream: Assimilation and Contemporary Immigration* (Cambridge, MA; and London: Harvard University Press, 2003), pp. 189–92; Guillermo J. Grenier and Lisandro Pérez, *The Legacy of Exile: Cubans in the United States* (Boston, MA: Pearson Education, 2003), pp. 23–7.

36 four major waves of post-Castro Cuban immigration: Grenier and Pérez, *The Legacy of Exile*, pp. 23–5; see Susan Eva Eckstein, *The Immigrant Divide: How Cuban Americans Changed the U.S. and Their Homeland* (New Haven, CT, and London: Routledge, 2009), pp. 2-3; U.S. Census, American Community Survey, Table S0201: Selected Population Profile in the United States (2010 3-year dataset) (population group code 403 – Cuban).

37 middle and upper class: García, *Havana USA*, p. 15; Gonzalez-Pando, *The Cuban Americans*, p. 33. Specifically, of the so-called Golden Exiles, who arrived in the first wave of immigration to the U.S., about 25 percent had been judges, lawyers, professionals, and semiprofessionals; 12 percent had managerial and executive positions in Cuba; and 31 percent had clerical or sales jobs. Thomas D. Boswell and James R. Curtis, *The Cuban-American Experience: Culture, Images, and Perspectives* (Totowa, NJ: Rowman & Allanheld, 1984), pp. 45–6.

37 "the psychological impetus": Gonzalez-Pando, *The Cuban Americans*, p. 47.

37 Miami was still largely a resort: García, *Havana USA*, pp. 5, 86–7.

37 more than 1,100 multinational corporations: WorldCity, "The 2009 Who's Here Multinational Economic Impact Study" (2009), p. 2, http://www.bus.miami.edu/_assets/files/news-media/recent-news/WhosHere09.pdf.

37 eleventh-highest gross metropolitan product: The U.S. Conference of Mayors, *U.S. Metro Economies: Outlook—Gross Metropolitan Product, and Critical Role of Transportation Infrastructure* (2012), appendix Table 1; see also García, *Havana USA*, pp. 86–7; Grenier and Pérez, *The Legacy of Exile*, pp. 48–50.

37 Alfonso Fanjul: See Patrick Verel, "Fordham Alumnus Recounts Rise to Top of Sugar Industry," *Inside Fordham*, Feb. 25, 2013, http://www.fordham.edu/campus_resources/newsroom/inside_fordham/february_25_2013/news/fordham_alumnus_reco_90307.asp; Marci McDonald, "A Sweet Deal for Big Sugar's Daddies, *U.S. News & World Report*, Aug. 6, 2001.

37 Bacardis: Tom Gjelten, *Bacardi and the Long Fight for Cuba: The Biography of a Cause* (New York: Viking, 2008), pp. 194–5, 205–6, 208, 235–6; W. Blake Grey, "Bacardi, and Its Yeast, Await a Return to Cuba," *Los Angeles Times*, Oct. 6, 2011.

38 As of 1961: Gonzalez-Pando, *The Cuban Americans*, p. 21.

38 Gedalio Grinberg: Douglas Martin, "G. Grinburg, Watch Baron, Dies at 77," *New York Times*, Jan. 6, 2009; "Movado Group, Inc.," 2013 Form 10-K, filed with the Securities and Exchange Commission, http://www.sec.gov/Archives/edgar/data/72573/000119312513126699/d444274d10k.htm#tx444274_3.

38 Carlos Gutierrez: Kateri Drexler, "Carlos Gutierrez," in *Icons of Business: An Encyclopedia of Mavericks, Movers, and Shakers* (Westport, CT: Greenwood Press, 2007), vol.1, pp. 203–21.

38 Ralph de la Vega . . . "Spam-like meat": Ralph de la Vega, *Obstacles Welcome: How to Turn Adversity into Advantage in Business and Life* (Nashville: Thomas Nelson, 2009), pp. 1–3, 10; Robert Reiss, "A Journey from Cuba to Corporate Leadership," *Forbes*, April 5, 2010.

38 "no idea . . . where to take their families": García, *Havana USA*, p. 18.

38–39 Former executives parked cars . . . "I was determined that my children": Gonzalez-Pando, *The Cuban Americans*, pp. 33–4, 36, 47–8 (quoting an interview with anthropologist Mercedes Sandoval); García, *Havana USA*, p. 20; see also "Marco Rubio Won't Be V.P.," *New York Times Magazine*, Jan. 26, 2012.

39 "Cubans more than most Latin American immigrants": Susan Eckstein, "Cuban Émigrés and the American Dream," *Political Science and Politics* 4, no. 2 (June 2006), p. 297.

39 wealthiest Hispanic American: Miguel de la Torre, *La Lucha for Cuba: Religion and Politics on the Streets of Miami* (Berkeley and Los Angeles: University of California Press, 2003), p. 35.

39 managerial and professional positions: U.S. Census Bureau, American Community Survey Reports, *The American Community—Hispanics: 2004* (February 2007), p. 16 and fig. 12.

39 enrolled full-time in college: Marcela Muñiz, *Latinos and Higher Education: Snapshots from the Academic Literature* (College Board, 2006), p. 17, http://advocacy.collegeboard.org/sites/default/files/latinos-and-highered_snapshots.pdf.

39 total revenue of Cuban American businesses: Eckstein, "Cuban Émigrés," p. 297 (calculated using the unofficial, de facto exchange rate).

39 About a third of Miami's population: U.S. Census, American Community Survey, Table S0201: Selected Population Profile in the United States (2011 1-year dataset) (county: Miami-Dade, Florida) (population group codes 001 – total population; 403 – Cuban) (Cubans represent about 900,000 of Miami-Dade County's 2.6 million residents).

39 also dominate Miami politics: Eckstein, *The Immigrant Divide*, p. 94; Miami-Dade County—Office of the Mayor, http://www.miamidade.gov/mayor/; see also de la Torre, *La Lucha for Cuba*, p. 19.

39 all three Latinos elected in 2012: "Latino Congress Members: 2012 Election Sets a New Record with the Most Latinos Elected to U.S. Senate, House in History," Huffington Post, Nov. 7, 2012, http://www.huffingtonpost.com/2012/11/07/latino-congress-members_n_2090311.html.

39 Cuban actors: Tom Seymour, "Andy Garcia on Coppola, Corleone and City Island," Empire, http://www.empireonline.com/interviews/interview.asp?IID=1235; Robert Sullivan, "Sunshine Superwoman," *Vogue*, June 2009, p. 97; Cindy Pearlman, "Eva Mendes Relies on What's Inside to Figure Things Out," *Chicago Sun-Times*, Mar. 31, 2013.

39 later waves of Cuban immigrants: Alba and Nee, *Remaking the American Mainstream*, pp.

189–92; Gonzalez-Pando, *The Cuban Americans*, pp. 20–1, 33–4, 122; Grenier and Pérez, *The Legacy of Exile*, pp. 23–27; Emily H. Skop, "Race and Place in the Adaptation of Mariel Exiles," *International Migration Review* 35, no. 2 (2001), pp. 449–71.

40 mostly white, whereas a substantial fraction: Grenier and Pérez, *The Legacy of Exile*, pp. 38–9; García, *Havana USA*, p. xi; Skop, "Race and Place in the Adaptation of Mariel Exiles," p. 450.

40 cold shoulder from the Exiles: Gonzalez-Pando, *The Cuban Americans*, pp. 68–9; Interview with Jose Pico.

40 absent from Miami's power elite: Eckstein, *The Immigrant Divide*, pp. 88–9.

40 New Cubans . . . other Hispanics: Ibid.; Pew Hispanic Center, *Fact Sheet: Cubans in the United States*, Aug. 25, 2006, http://pewhispanic.org/files/factsheets/23.pdf.

40 the Exiles' U.S.-born children: Kevin A. Hill and Dario Moreno, "Second-Generation Cubans," *Hispanic Journal of Behavioral Science* 18, no. 2 (1996), pp. 177–8.

40 majority of the nonwhite . . . outside Miami: Skop, "Race and Place in the Adaptation of Mariel Exiles," p. 461.

41 3.2 million black immigrants: U.S. Census, American Community Survey, Table S0201: Selected Population Profile in the United States (2010 3-year dataset) (population group codes 001 – total population; 004 – Black or African American) As of 2009, approximately one million of America's black immigrants came from African countries; the principal countries of origin were Nigeria (19 percent), Ethiopia (13 percent), Ghana (10 percent), and Kenya, Somalia, and Liberia (6 percent each). Randy Capps, Kristen McCabe, and Michael Fix, *Diverse Streams: African Migration to the United States* (Washington, DC: Migration Policy Institute, 2012), p. 4. The same year, there were approximately 1.7 million black West Indian immigrants, with Jamaica (36 percent) and Haiti (31 percent) by far the dominant countries of origin. Kevin J.A. Thomas, *A Demographic Profile of Black Caribbean Immigrants in the United States* (Washington, DC: Migration Policy Institute, 2012), p. 5.

41 41 percent of black Ivy League freshmen: Douglas S. Massey, Margarita Mooney, Kimberly C. Torres, and Camille Z. Charles, "Black Immigrants and Black Natives Attending Selective Colleges and Universities in the United States," *American Journal of Education* 113, no. 2 (Feb. 2007), pp. 243, 249, 267. Note that Massey's study covered only four Ivy League schools (Columbia, Princeton, Penn, and Yale).

41 Gates Jr. and Lani Guinier: Sara Rimer and Karen W. Arenson, "Top Colleges Take More Blacks, but Which Ones?," *New York Times*, June 24, 2004.

42 Yale Law, 18 students: Study commissioned by authors, June 2012, based on personal interviews with members of the Yale Black Law Students Association (on file with authors).

42 260,000 Nigerians in the U.S.: U.S. Census, American Community Survey, Table S0201: Selected Population Profile in the United States (2010 3-year dataset) (population group code 567 – Nigerian).

42 Harvard Business School: Study commissioned by authors, January 2013, based on the Harvard Business School Intranet, the HBS African-American Student Union Facebook group, and personal interviews (on file with authors).

42 factor of about ten: Massey found that 27 percent of black freshmen at America's selective schools in 1999 were of "immigrant origin" (either immigrants themselves or the children of immigrants) and that 17 percent of these were Nigerian, meaning that about 4.6 percent of the black students were Nigerian. Massey et al., "Black Immigrants," pp. 248, 251. At that time, Nigerians represented about .48 percent of America's black population (roughly 165,000 out of 34,000,000), so their percentage of the black student body was approximately ten times their percentage of the U.S. black population as a whole. U.S. Census, Table PCT001: Total Population (year: 2000) (population group codes 004 – Black or African American; 567 – Nigerian).

42 particularly well in medicine: In Harvard Medical School's class of 2015, there are 13 black medical students (out of 170 overall). Of those 13, reportedly 4 had at least one Nigerian parent. Study commissioned by authors, January 2013 (on file with authors). The AMA counted

34,000 black physicians in the U.S. in 2008. American Medical Association, "Total Physicians by Race/Ethnicity – 2008," http://www.ama-assn.org/ama/pub/about-ama/our-people/member -groups-sections/minority-affairs-section/physician-statistics/total-physicians-raceethnicity .page. Of these, around 3,000 to 3,500 appear to be Nigerian, including more than 50 in Charlotte, NC, alone. See Ronald H. Baylor, ed., *Multicultural America: An Encyclopedia of the Newest Americans* (Santa Barbara, CA: Greenwood, 2011), vol. 1, p. 1610; Mike Stobbe, "Nigerian Physicians' 'Phenomenal' Influx into Charlotte, North Carolina," *Charlotte Observer*, Dec. 27, 2004, http://naijanet.com/news/source/2004/dec/27/1000.html.

42 "Nigerians dominate" investment banking: Patricia Ngozi Anekwe, *Characteristics and Challenges of High Achieving Second-Generation Nigerian Youths in the United States* (Boca Raton, FL: Universal Publishers, 2008), p. 129 (quoting interviewee).

43 overrepresented at America's top law firms: Complete information about the number of black lawyers at America's top law firms, or their national origins, is not readily available. Using a well-known national ranking of U.S. law firms (the "Vault Law 100"), we examined the five top-ranked firms in the country, all of which are headquartered in New York City, and five of the top-ranking law offices in Washington, DC, as well. Study commissioned by authors, July/ August 2013 (on file with authors). The results of our study, based on interviews, firm Web sites, and the NALP Directory of Legal Employers, suggest that, in the aggregate, upwards of 5 percent of the black lawyers (including partners, counsel, and associates, but excluding staff attorneys) at these ten firms are of Nigerian origin, which is an at least sevenfold overrepresentation as compared to Nigerian Americans' percentage (0.7) of the U.S. black population as a whole.

43 West Indian by birth or descent: See Alba and Nee, *Remaking the American Mainstream*, p. 198; Violet M. Showers Johnson, "What, Then, Is the African American? African and Afro-Caribbean Identities in Black America," *Journal of American Ethnic History* 28 (2008), p. 101 n. 26.

43 only Jamaica sent: Massey et al., "Black Immigrants," p. 251.

43 Colin Powell: Colin Powell, *My American Journey* (New York: Ballantine, 2003), pp. 7–8.

43 Clifford Alexander: Catherine Reef, *African Americans in the Military* (New York: Facts on File, 2004), p. 3.

43 "business, professional, and political elites": Alba and Nee, *Remaking the American Mainstream*, p. 198.

43 among the nation's poorest: Capps, McCabe, and Fix, *Diverse Streams*, p. 17; Sam Roberts, "Government Offers Look at Nation's Immigrants," *New York Times*, Feb. 20, 2009 ("Somalis are the youngest and poorest" immigrants in America). According to one researcher, African immigrants as of the late 1990s were doing worse overall than other black immigrants, F. Nii Amoo-Dodoo, "Assimilation Differences Among Africans in America," *Social Forces* 76, no. 2 (1997), pp. 527, 528, and another says that Haitians and "black immigrants from Africa do not . . . have stronger outcomes than African Americans." Suzanne Model, *West Indian Immigrants: A Black Success Story?* (New York: Russell Sage Foundation, 2008), p. 2. These authors do not, however, break out African subgroups like Nigerians or Ghanaians.

43 Nigerian Americans dramatically outperform black Americans: U.S. Census, American Community Survey, Table DP03: Selected Economic Characteristics (2010 5-year dataset) (population group codes 004 – Black or African American; 567 – Nigerian).

43 and also outperform Americans: U.S. Census, American Community Survey, Table S0201: Selected Population Profile in the United States (2010 3-year dataset) (population group codes 001 – total population; 567 – Nigerian). About 21 percent of American households as a whole earn over $100,000 a year, with 4.2 percent earning over $200,000. U.S. Census, American Community Survey, Table DP03: Selected Economic Characteristics (2010 5-year dataset).

44 includes a significant number of new arrivals: Over 25 percent of America's Nigerians entered the country in or after 2000. U.S. Census Bureau, American Community Survey, Table S0201: Selected Population Profile in the United States (2010 3-year dataset) (population group code 567 – Nigerian); Dennis D. Cordell and Manuel Garcia y Griego, "The Integration of Ni-

gerian and Mexican Immigrants in Dallas/Fort Worth Texas," Working Paper, http://iussp2005
.princeton.edu/papers/51068, p. 20 (noting contrast between high incomes of Nigerian immigrants overall and low incomes of recent arrivals, and explaining that "many of the Nigerians who
have arrived since 2000 . . . are students living on a restricted budget").

44 end up in unskilled jobs: Capps, McCabe, and Fix, *Diverse Streams*, pp. 16–9; Patrick L.
Mason and Algernon Austin, "The Low Wages of Black Immigrants: Wage Penalties for U.S.-
Born and Foreign-Born Black Workers," EPI Briefing Paper No. 298 (Washington, DC: Eco-
nomic Policy Institute, Feb. 25, 2011).

44 "dysfunctional" culture: Dinesh D'Souza, *The End of Racism: Principles for a Multicultural
Society* (New York: Free Press, 1996), p. 24.

44 product of racism: See Richard Thompson Ford, Book Review, "Why the Poor Stay
Poor," *New York Times*, Mar. 6, 2009 (describing these arguments); see also Mary C. Waters,
Black Identities: West Indian Immigrant Dreams and American Realities (New York: Russell Sage
Foundation, 1999), pp. 7–8, 13; Ezekiel Umo Ette, *Nigerian Immigrants in the United States. Race,
Identity, and Acculturation* (Plymouth, UK: Lexington Books, 2012), pp. 129–30.

45 "highest-income, best-educated": Pew Research Center, *The Rise of Asian Americans*
(Washington, DC: Pew Research Center, Apr. 4, 2013) (updated edition), p. 1.

45 "Asian American": Ibid., p. 22.

45 several of the poorest: Asian American Center for Advancing Justice, *A Community of
Contrasts: Asian Americans in the United States: 2011* (2011), pp. 9–10, 34, 36, 38–40, 45 (citing
data from the U.S. Census Bureau, 2007–9 American Community Survey, 3-year estimates).

45 The six largest U.S. Asian groups: Pew Research Center, *The Rise of Asian Americans*,
pp 2, 7, 57; Asian American Center for Advancing Justice, *A Community of Contrasts*, pp. 9 10.

46 twenty-three of the fifty top prizewinners: Society for Science and the Public, "Science
Talent Search Through the Years," http://student.societyforscience.org/science-talent-search
-through-years.

46 "Super Bowl of science": Chris Higgins, "The Super Bowl of Science," mental_floss, last
updated Mar. 26, 2010, http://mentalfloss.com/article/24306/super-bowl-science.

46 The résumés of these Intel winners: Society for Science and the Public, Intel Science
Talent Search 2012 Finalists, http://apps.societyforscience.org/sts/71sts/finalists.asp (bios for
Amy Chyao, Jack Li, and Anirudh Prabhu).

46 "synthesized a nanoparticle": Kenneth Chang, "Nanotechnology Gets Star Turn at
Speech," *New York Times*, Jan. 25, 2011, http://www.nytimes.com/2011/01/26/science/26light
.html.

46 30–50 percent of the student bodies: Grace Wang, "Interlopers in the Realm of High
Culture: 'Music Moms' and the Performance of Asian and Asian American Identities," *American
Quarterly* 61, no. 4 (2009), p. 882.

46 Tchaikovsky Competition: E-mail from Ivan Scherbak, International Projects Director
for the Association of Tchaikovsky Competition Stars, Aug. 6, 2013 (confirming that Jennifer
Koh [1992], Emily Shie [1992], Sirena Huang [2009], and Noah Lee [2012] are the only Americans to have won the International Tchaikovsky Youth Competition) (on file with authors).

47 spelling bees: Visi R. Tilak, "Why Indian Americans Are Best at Bees," *Wall Street Journal*, June 2, 2012, http://blogs.wsj.com/indiarealtime/2012/06/02/why-indian-americans-are
-best-at-bees/.

47 Presidential Scholars: U.S. Department of Education, U.S. Presidential Scholars Program, http://www2.ed.gov/programs/psp/2012/scholars.pdf and http://www2.ed.gov/programs/
psp/2011/scholars.pdf.

47 hypersuccessful Asian American teens: The statistics on Asian representation at Harvard,
Yale, Princeton, and Stanford are from those universities' sites at https://bigfuture.collegeboard.
org. The SAT statistics are from the College Board, cited in Scott Jaschik, "SAT Scores Drop
Again," Inside Higher Ed, Sept. 25, 2012, http://www.insidehighered.com/news/2012/09/25/
sat-scores-are-down-and-racial-gaps-remain.

47–48 "anti-Asian admissions bias" . . . CalTech: See Ron Unz, "The Myth of American Meritocracy," *The American Conservative*, Nov. 28, 2012.

48 Indian Americans . . . Taiwanese Americans: U.S. Census, American Community Survey, Table S0201: Selected Population Profile in the United States (2010 3-year dataset) (population group codes 001 – total population; 013 – Asian Indian; 018 – Taiwanese) (median household income and individual income); ibid., Table DP03: Selected Economic Characteristics (2010 5-year dataset) (population group codes 001 – total population; 013 – Asian Indian) (income level percentages).

48 relatively conventional and prestige-oriented: Bandana Purkayastha, *Negotiating Ethnicity: Second Generation South Asian Americans Traverse a Transnational World* (New Brunswick, NJ: Rutgers University Press, 2005), p. 91; Mei Tang, Nadya A. Fouad, and Philip L. Smith, "Asian Americans' Career Choices: A Path Model to Examine Factors Influencing Their Career Choices," *Journal of Vocational Behavior* 54 (1999), pp. 142–5; S. Alvin Leung, David Ivey, and Lisa Suzuki, "Factors Affecting the Career Aspirations of Asian Americans," *Journal of Counseling & Development* 72 (March/April 1994), pp. 404, 408; see also Min Zhou, "Assimilation the Asian Way," in Tamar Jacoby, ed., *Reinventing the Melting Pot: The New Immigrants and What It Means to Be American* (New York: Basic Books, 2003), pp. 139, 140–2.

48 "respectable," and "impressive": See Purkayastha, *Negotiating Ethnicity*, p. 91; Yuki Okubo et al. "The Career Decision-Making Process of Chinese American Youth." *Journal of Counseling & Development* 85, no. 4 (2007), pp. 440–1; Pei-Wen Winnie Ma and Christine J. Yeh, "Factors Influencing the Career Decision Status of Chinese American Youths," *The Career Development Quarterly* 53, no. 4 (2005), pp. 337, 338; Philip Guo, "Understanding and Dealing with Overbearing Asian Parents," December 2009, http://www.pgbovine.net/understanding-asian-parents.htm; see also Min Zhou, "Negotiating Culture and Ethnicity: Intergenerational Relations in Chinese Immigrant Families," in Ramaswami Mahalingam, ed., *Cultural Psychology of Immigrants* (Mahwah, NJ: Lawrence Erlbaum Associates Publishers, 2006), pp. 315, 323 (noting that Chinese immigrant parents often have three main goals: "To live in your own house, to be your own boss, and to send your children to the Ivy League").

48 a disproportionate number of Asian Americans: Mitchell J. Chang, Julie J. Park, Monica H. Lin, Oiyan A. Poon, and Don T. Nakanishi, *Beyond Myths: The Growth and Diversity of Asian American College Freshman, 1971-2005* (Los Angeles: UCLA Higher Education Research Institute, 2004), pp. 4, 17–20; Tang et al., "Asian Americans' Career Choices," pp. 142–3.

48 Nobel prizes: The Indian American Nobel laureates are: Har Gobind Khorana (medicine, 1968), Subramanyan Chandrasekhar (physics, 1983), Amartya Sen (economics, 1998), and Venkatraman Ramakrishnan (chemistry, 2009). Amartya Sen, an Indian citizen and longtime U.S. resident, won the Nobel Prize for economics in 1998. See Surekha Vijh, "Indian Americans Become a Force to Reckon With," *News East West*, Oct. 8, 2012, http://newseastwest.com/how-indian-americans-became-a-force-to-reckon-with. The post-1965 Chinese American Nobel laureates are: Samuel Ting (physics, 1976), Yuan Tseh Lee (chemistry, 1986), Steven Chu (physics, 1997), Daniel Chee Tsui, (physics, 1998), Roger Yonchien Tsien (chemistry, 2008), and the British-American Charles Kao (physics, 2009). See China Whisper, "The 10 Ethnic Chinese Nobel Prize Winners," Mar. 18, 2013, http://www.chinawhisper.com/the-10-ethnic-chinese-nobel-prize-winners.

48 Zappos founder Tony Hsieh: Tony Hsieh, *Delivering Happiness: A Path to Profits, Passion, and Purpose* (New York and Boston: Business Plus, 2010). Hsieh writes that his parents wanted him to attend medical school or get a PhD, but "I always fantasized about making money, because to me, money meant that later on in life I would have the freedom to do whatever I wanted." Ibid., pp. 9–10. See also "Jerry Yang," Forbes.com, http://www.forbes.com/profile/jerry-yang/; "Steve Chen," CrunchBase, http://www.crunchbase.com/person/steve-chen; Derek Andersen, "How the Huang Brothers Bootstrapped Guitar Hero to a Billion Dollar Business," TechCrunch, Dec. 30, 2012, http://techcrunch.com/2012/12/30/how-the-huang-brothers-bootstrapped-guitar-hero-to-a-billion-dollar-business/; Michael Arrington, "Alfred Lin Has the Midas Touch,"

TechCrunch, July 28, 2009, http://techcrunch.com/2009/07/28/alfred-lin-has-the-midas-touch
-the-man-with-2-billion-in-acquisitions-under-his-belt/.

49 The Indian American list: Katherine Jacobsen, "Amar Bose, Inventor of Bose Speak-
ers, Dies, *Christian Science Monitor,* July 15, 2013; "Amar Bose," Forbes.com, Mar. 2011,
http://www.forbes.com/profile/amar-bose; "Top 6 Indian CEOs in America," SiliconIndia,
Mar. 15, 2012, http://www.siliconindia.com/news/usindians/Top-6-Indian-CEOs-in-America
-nid-109397-cid-49.html; "Boot Camp for Engineers," Forbes.com, Apr. 16, 2001, http://www
.forbes.com/global/2001/0416/088.html; Julia Werdigier, "Two Executives Are Ousted at
HSBS," New York Times, Feb. 23, 2007.

49 Indian American Silicon Valley entrepreneurs: "Vinod Khosla—Partner Emeritus,"
Kleiner Perkins Caufield Byers, http://www.kpcb.com/partner/vinod-khosla; Vivek Wadhwa,
"The Face of Success, Part I: How the Indians Conquered Silicon Valley," Inc., Jan. 13, 2012,
http://www.inc.com/vivek-wadhwa/how-the-indians-succeeded-in-silicon-valley.html.

49 Bobby Jindal: Jill Konieczko, "10 Things You Didn't Know About Bobby Jindal," *U.S.
News & World Report,* May 22, 2008.

49 Nikki Haley: Jonathan Martin, "Democratic South Carolina Chair Defends Nikki Haley
Barb," Politico, May 14, 2013, http://www.politico.com/story/2013/05/dick-harpootlian-nikki
-haley-90918.html.

49 cover of *Time* magazine: See Massimo Calabresi and Bill Saporito, "The Street Fighter,"
Time, Feb. 13, 2012.

50 "I thought I had": Remark made by presenter Atul Gawande at Boston Public Library
"Literary Lights" award ceremony, Apr. 15, 2012 (attended by Amy Chua).

50 underrepresented at the top levels: See New York City Bar, *2010 Law Firm Diversity
Benchmarking Report: A Report to Signatory Law Firms* (New York City Bar Association, 2010),
p. 15. A study of the largest Bay Area corporations concludes that while Asians are well-
represented in the workforce, they are underrepresented in the boardroom and at the top executive levels. See Buck Gee and Wes Hom, "The Failure of Asian Success in the Bay Area: Asians
as Corporate Executive Leaders" (Corporate Executive Initiative, Mar. 28, 2009), p. 3.

50 five Indian American CEOs . . . no Chinese Americans: "Where's the Diversity in Fortune 500 CEOs?," DiversityInc, http://www.diversityinc.com/diversity-facts/wheres-the-diversity
-in-fortune-500-ceos.

50 discrimination . . . quarterback mentality: Gee and Hom, "The Failure of Asian Success
in the Bay Area," p. 3; Wesley Yang, "Paper Tigers," *New York Magazine,* May 8, 2011 (quoting
Columbia Professor Tim Wu).

50 Over 65 percent of Chinese Americans . . . over 90 percent: Zhou, "Negotiating Culture
and Ethnicity, pp. 315, 317.

50 87 percent of Indian American adults: Pew Research Center, *The Rise of Asian Americans,*
p. 44.

50 about 15 percent of American Jewish adults: Jonathon Ament, *Jewish Immigrants in the
United States* (United Jewish Communities Report Series on the National Jewish Population Survey 2001-01, October 2004), p. 4, n.3.

51 a 3 percent Chinese minority: Amy Chua, *World on Fire: How Exporting Free Market
Democracy Breeds Ethnic Hatred and Global Instability* (New York: Doubleday, 2003), p. 43.

51 Census is barred by law: U.S. Census Bureau, "Religion Statistics and Publications,"
http://www.census.gov/prod/www/religion.htm.

51 antiquity—but that's about the last time they weren't: Chua, *World on Fire,* pp. 79–82,
202–4, 217–21; Thomas Sowell, *Migrations and Culture: A World View* (New York: Basic Books,
1996), pp. 236–7 (noting that "in ancient times" Jews were not yet middlemen and "most Jews
were in fact poor"); see generally Michael Grant, *The Jews in the Roman World* (New York: Charles
Scribner's Sons, 1973).

51 about 43 percent: See "The Jewish Population of the World (2010)," Jewish Virtual Library, http://www.jewishvirtuallibrary.org/jsource/Judaism/jewpop.html. Estimates of the

American Jewish population vary. See Ira Sheskin and Arnold Dashefsky, *Jewish Population in the United States, 2012* (Storrs, CT: North American Jewish Data Bank, 2012), pp. 7–10, 15 (discussing methodological difficulties and estimating likely U.S. Jewish population at 6.0–6.4 million).

51 roughly zero were Jewish: Jeremy Fine, "No Jewish Masters," *Jewish Journal*, Apr. 11, 2010, http://www.jewishjournal.com/the_great_rabbino/item/no_jewish_masters_ 20100411. In general, Jews are underrepresented in American sports, but according to one entertaining recent book, their contributions have not been inconsequential. See Franklin Foer and Marc Tracy, eds., *Jewish Jocks: An Unorthodox Hall of Fame* (New York: Twelve, 2012).

51 highest-paid CEOs: "America's Highest Paid Chief Executives," *Forbes*, Mar. 25, 2011, http://www.forbes.com/lists/2011/12/ceo-compensation-11_rank.html.

52 hedge fund managers: Nathan Vardi, "The 40 Highest-Earning Hedge Fund Managers," *Forbes*, Mar. 1, 2012, http://www.forbes.com/lists/2012/hedge-fund-managers-12_land .html#fulllist.

52 Jews on the Forbes 400: Jacob Berkman, "At Least 139 of the Forbes 400 are Jewish," Oct. 5, 2009, http://blogs.jta.org/philanthropy/article/2009/10/05/1008323/at-least-139-of-the-forbes -400-are-jewish.

52 fortunes in real estate: "The Forbes 400: The Richest People in America," *Forbes*, http:// www.forbes.com/forbes-400/list/#page:1_sort:0_direction:asc_search:_filter:Real%20Estate_ filter:All%20states_filter:All%20categories (as of March 27, 2012) (religious affiliations compiled for and on file with authors); see also Steven L. Pease, *The Golden Age of Jewish Achievement* (Sonoma, CA: Deucalion, 2009), p. 238.

52 pledged to leave half their estates to charity: Marc Tracy, "Generous Jews: Nearly Half of 'Pledge' Billionaires are MOTs," *Tablet Magazine*, Aug. 5, 2010, http://www.tabletmag.com/ scroll/41829/generous-jews; see also Kent McGroarty, *Biography on Diane von Furstenberg* (Hyperink, 2012), p. 5.

52 the Tony: Stewart F. Lane, *Jews on Broadway: An Historical Survey of Performers, Playwrights, Composers, Lyricists and Producers* (Jefferson, NC: McFarland & Company, 2011), p. 190.

52 Many of America's best-loved comics: Lane, *Jews on Broadway*, pp. 42–3, 111; David K. Israel, "The Top 20 Jewish Comedians of All Time," Mental Floss, http://mentalfloss.com/ article/22596/top-20-jewish-comedians-all-time.

52 Academy Award . . . Pulitzer Prize winners: Pease, *The Golden Age*, pp. viii–ix, Table 1.

52 psychiatrists . . . physicians: See Will Dunham, "Psychiatrists Least Religious of U.S. Doctors: Study," Reuters, Sept. 3, 2007, http://www.reuters.com/article/2007/09/03/us -psychiatrists-religion-idUSN0228386620070903.

52 New York City's law offices: See Steven Silbiger, *The Jewish Phenomenon: Seven Keys to the Enduring Wealth of a People* (Atlanta: Longstreet Press, 2001), p. 29.

53 justices . . . Architecture superstars: Pease, *The Golden Age of Jewish Achievement*, pp. 76–7, 134–5.

53 Advice-giving: Ruth Andrew Ellenson, "Jewish Advice Columnists from Miss Manners to Dear Abby," *Jewish Women*, Fall 2009.

53 United Jewish Association report: Jacob B. Ukeles, Steven M. Cohen, and Ron Miller, *Jewish Community Study of New York: 2011 Special Report on Poverty* (rev. ed) (New York: UJA-Federation of New York, June 2013), pp. 15, 72. The report defines "poor" as "less than 150% of the 2010 federal poverty guideline." Ibid., p. 21. Cf. Keister, *Faith and Money*, p. 71 (stating that Jewish poverty rates are "notably low" compared to other religions).

53 Jewish median household income . . . $97,000 to $98,000: This estimate is based on data from the Pew Forum on Religious and Public Life, *U.S. Religious Landscape Survey—Religious Affiliation: Diverse and Dynamic* (Washington, DC: Pew Research Center, 2008) (reporting 2006 data). No overall household income figure is given, but the study indicates that as of 2006, 46 percent of American Jewish households had incomes of at least $100,000. Ibid., pp. 60, 78. From this, we conservatively extrapolated a 2006 median income of $90,000–90,900. We then converted that range to 2010 dollars (as the Census Bureau does for its multi-year estimates of other groups' income, such as the $90,500 figure mentioned in the text for Indian Americans' income), using the U.S. Department of Labor, Bureau of Labor Statistics conversion tool, http://www.bls .gov/data/ inflation_calculator.htm, producing a range from $97,346 to 98,320. The 2010 median household income for the U.S. population as a whole was about $51,000. U.S. Census, American Community Survey, Table S0201: Selected Population Profile in the United States (2010 3-year dataset). Throughout this book, we measure group income as of 2010; data from 2013 suggest that Jewish income may be declining slightly. See Pew Research Center, *A Portrait of Jewish Americans: Findings from a Pew Research Center Survey of U.S. Jews* (Washington, DC: Pew Research Center, 2013), p. 42 (reporting that 42 percent of American Jews had household incomes of at least $100,000).

53 Indian Americans . . . $90,500: U.S. Census, American Community Survey, Table S0201: Selected Population Profile in the United States (2010 3-year dataset) (population group code 013 – Asian Indian).

53 exceeds that of Protestants by up to 246 percent: Keister, *Faith and Money*, p. 171. One twenty-five-year tracking study found almost half of all Jews to have a net worth of over $600,000, as compared to 6.8 percent of white conservative Protestants, 0.8 percent of black conservative Protestants, 14.2 percent of "mainline" Protestants, 13.7 percent of Catholics, and 10.6 percent of all respondents. What about low net worth? Defining low net worth as zero or less, 11.2 percent of white conservative Protestants fell in that category, as did 32.5 percent of black conservative Protestants, and 8 percent of Catholics. Among all respondents in the survey, 13.3 percent had zero or lower net worth. For Jews, the figure was 2.1 percent. Ibid., p. 90 (Table 4.2).

53 According to Pew: Pew Forum on Religion and Public Life, *U.S. Religious Landscape Survey*, p. 78.

53 Reform Jews, 55 percent . . . nation as a whole . . . 18 percent: Ibid.

53 New Left and the neoconservative movement: See, e.g., Murray Friedman, *The Neoconservative Revolution: Jewish Intellectuals and the Shaping of Public Policy* (Cambridge, UK: Cambridge University Press, 2005), chaps. 2, 4–6.

54 American academia . . . "disproportionate role": Pease, *The Golden Age of Jewish Achievement*, pp. 29–39, 52–7, 64–71.

54 "When the studio chiefs": Joel Stein, "How Jewish Is Hollywood?," *Los Angeles Times*, Dec. 19, 2008, http://www.latimes.com/news/opinion/commentary/la-oe-stein19-2008dec19,0, 4676183.column.

54 a fifth of all Nobel laureates: Pease, *Golden Age of Jewish Achievement*, p. ix (23 percent); Raphael Patai, *The Jewish Mind* (Detroit: Wayne State University Press, 1996), pp. 342, 548 (between 1901 and 1994 Jews, won 35 percent of the Nobel Prizes for Economics, 27 percent for Physiology and Medicine, 22 percent for Physics, and 20 percent of Nobel Prizes overall).

54 Nobel Prize for economics: Pease, *The Golden Age of Jewish Achievement*, p. 68; "All Prizes in Economic Sciences," *Nobelprize.org*, http://www.nobelprize.org/nobel_prizes/economic-sciences/ laureates.

54 36 percent of all Nobel Prizes ever awarded to Americans: Ari Ben-Menahem, *Historical Encyclopedia of Natural and Mathematical Sciences* (New York: Springer, 2009), vol. 1, part 3, p. 2891, n. 39.

55 race for the atom bomb: See Pease, *The Golden Age of Jewish Achievement*, pp. 44–5, 53; Ronald W. Clark, *The Birth of the Bomb* (New York: Horizon Press, 1961), pp. 1–3, 8–13; Martin J. Sherwin, *A World Destroyed: The Atomic Bomb and the Grand Alliance* (New York: Alfred A. Knopf, 1975), pp. 49–50; C. P. Snow, *The Physicists* (Boston: Little, Brown & Co., 1981), pp. 79–80, 103, 131, 159.

55 Iranian immigration: Mehdi Bozorgmehr and Daniel Douglas, "Success(ion): Second-Generation Iranian Americans," *Iranian Studies* 44, no. 1 (January 2011), pp. 10–13; see also Mitra K. Shavarini, *Educating Immigrants: Experiences of Second-Generation Iranians* (New York:

LFB Scholarly Publishing, 2004), pp. 35–41; U.S. Census, American Community Survey, Table S0201: Selected Population Profile in the United States (2010 3-year dataset) (population group code 540 – Iranian).

55 numerous ethnic and religious groups: Shavarini, *Educating Immigrants*, pp. 2, 47–51.

55 relatively low profile: Ibid., pp. 2–6.

56 "vulgar, materialistic show-offs": Mike Hale, "The Children of Old Tehran Go Hollywood," *New York Times*, Mar. 9, 2012; Jon Michaud, "Life in Irangeles," *The New Yorker*, Mar. 14, 2012.

56 "the most highly educated ethnic group in the United States": Ali Mostashari and Ali Khodamhosseini, "An Overview of Socioeconomic Characteristics of the Iranian-American Community, based on the 2000 U.S. Census" (Iranian Studies Group at MIT), February 2004, http://www .isgmit.org/projects-storage/census/socioeconomic.pdf; see Phyllis McIntosh, "Iranian-Americans Reported Among Most Highly Educated in U.S.," *Payvand Iran News*, Jan. 26, 2004, http://www .payvand.com/news/04/jan/1191.html.

56 over 17 percent lived in houses: U.S. Census Bureau, American Community Survey, Table DP04: Selected Housing Characteristics (2010 5-year dataset) (population group codes 001 – total population; 540 – Iranian).

56 median household income . . . One in three: U.S. Census, American Community Survey, Tables S0201: Selected Population Characteristics in the United States (2010 3-year dataset) (population group code 540 – Iranian); DP03: Selected Economic Characteristics (2010 5-year dataset) (population group codes 001 – total population; 540 – Iranian).

56 Carlos Slim: "The World's Billionaires," *Forbes*, Mar. 2013; Chua, *World on Fire*, pp. 60–3.

56 Lebanese minorities: Chua, *World on Fire*, pp. 66–7, 115–20; see generally Albert Hourani and Nadim Shehadi, eds., *The Lebanese in the World: A Century of Emigration* (London: The Centre for Lebanese Studies, 1992).

56 Lebanese immigrants: Alixa Naff, "Lebanese Immigration into the United States: 1880 to the Present," in Hourani and Shehadi, *The Lebanese in the World*, pp. 141–2, 145–8, 161–3; Philippe W. Zgheib and Abdulrahim K. Kowatly, "Autonomy, Locus of Control, and Entrepreneurial Orientation of Lebanese Expatriates Worldwide," *Journal of Small Business and Entrepreneurship* 24, no. 3 (2011), pp. 345, 347; Zeinab Fawaz, *Success Factors of Lebanese Small Businesses in the United States* (Bloomington, IN: AuthorHouse, 2012), pp. 3–5, 19–20.

57 Lebanese American population . . . income numbers: U.S. Census, American Community Survey, Table S0201: Selected Population Profile in the United States (2010 3-year dataset) (population group code 509 – Lebanese).

57 foreign-born: U.S. Census, American Community Survey, Table S0201: Selected Population Profile in the United States (2010 3-year dataset) (population group codes 509 – Lebanese; 540 – Iranian).

57 probably only a half: See Nalf, "Lebanese Immigration into the United States," pp. 159–60.

57 Sununu . . . Shaheen: William Saletan, "Lawrence of Nashua: New Hampshire's All-Arab-American Senate Race," Slate, Oct. 16, 2002, http://www.slate.com/articles/news_and_ politics/ballot_box/2002/10/lawrence_of_nashua.html.

57 Japanese Americans: See U.S. Census, American Community Survey, Table S0201: Selected Population Profile in the United States (2010 3-year dataset) (population group code 022 – Japanese) (median household income of $65,573).

57 Greek Americans: See U.S. Census, American Community Survey, Table S0201: Selected Population Profile in the United States (2010 3-year dataset) (population group code 536 – Greek) (median household income of $62,552); Charles C. Moskos, *Greek Americans: Struggle and Success* (2d ed.) (New Brunswick, NJ, and London: Transaction Publishers, 1989), pp. 52, 112–15.

58 arguably the five most successful: No authoritative metric and no conclusive ranking exists for disproportionate group economic success in America. In terms of 2010 median household income, the Jewish figure of about $97,000 is probably the highest of any group; Indians (about $91,000) would be second-highest, with Iranians ($68,000), Chinese ($67,000), and Lebanese ($67,000) very close to the top as well. These five groups also rank at or near the top in percent of households earning more than $100,000 per year and percent of households earning more than $200,000 per year. (For median household incomes of all Census-tracked groups, see U.S. Census, American Community Survey, Table S0201 [2010 3-year dataset]; for percentages earning over $100,000 and $200,000, see ibid., Table DP03 [2010 5-year dataset]; for Jews, see the note on Jewish income above.) But income measures capture only a piece of economic success. Academic attainment among the young is a critical predictor of success in the next generation; corporate and business achievement is an important marker of conventional success in the American economy; and intergenerational mobility—the ability of children born to lower-income parents to rise to relative affluence—is another key measure of disproportionate group economic success in the United States as well. We tried to take into account all these factors in deciding which groups to focus on in this book.

But our selection was necessarily restricted by time, space, and information limits. As mentioned, our book measures income only as of 2010. We also restricted our scope to groups with significant populations in America; we chose 100,000 as the cut-off. In addition, we ruled out certain Census-generated group classifications because of the way the Census Bureau acquires its ancestry information. The decennial census no longer asks all respondents about their ethnic or national origins (although it does ask about their "race"). Instead, the Census Bureau takes data from the American Community Survey (ACS), an annual sample population survey (also administered by the Bureau), which asks individuals open-endedly to state their "ancestry or ethnic origin." More than one answer is permitted; an individual's first two responses are recorded. As a result, some Census-generated classifications are simply too amorphous for our purposes (e.g., "European"), while others are of uncertain validity. For example, Census data show about 1.1 million "British" Americans with a median income among the highest in the country; but Census data also show 27 million "English" Americans with a significantly lower median income. See U.S. Census, American Community Survey, Table S0201: Selected Population Profile in the United States (2010 3-year dataset) (population group codes 520 – British; 529 – English). While there is some reason to think that self-identified "British" Americans may form a distinct cultural group (quite possibly a Triple Package group), this hypothesis could not be verified, and we ultimately chose to exclude "British" Americans from our analysis.

Two Asian groups we decided not to include were Korean and Filipino Americans, although in certain respects both groups are quite successful. Koreans are an extremely interesting but bimodal case, displaying both extraordinary achievement and relatively high poverty levels. See Renee Reichl Luthra and Roger Waldinger, "Intergenerational Mobility," in David Card and Steven Raphael, eds., *Immigration, Poverty, and Socioeconomic Inequality* (New York: Russell Sage Foundation, 2013), pp. 169, 182 (Korean Americans have high percentages of both educational attainment and households in poverty). Overall, by standard income metrics, Korean Americans are not nearly as disproportionately successful as the groups we study; for example, their median household income is barely distinguishable from that of the U.S. population as a whole (both being around $51,000). See U.S. Census, American Community Survey, Table S0201: Selected Population Profile in the United States (2010 3-year dataset) (population group codes 001 – total population; 023 – Korean). Unlike the Koreans, Filipino Americans are more homogeneous and have a very high median household income. But on individual income measures, Filipino Americans rank surprisingly low; for example, working Filipino men earn a median income of about $45,000—actually below the national average. See ibid. (population group code 019 – Filipino). By contrast, Jewish, Indian, Chinese, Iranian, and Lebanese Americans are all in the top tier in terms of both household and individual income.

CHAPTER 3: THE SUPERIORITY COMPLEX

59 "one's own group (the in-group) as virtuous and superior": Ross A. Hammond and Robert Axelrod, "The Evolution of Ethnocentrism," *Journal of Conflict Resolution* 50, no. 6 (2006), p. 926. Sumner wrote that "[e]ach group nourishes its own pride and vanity, boasts itself superior, exalts its own divinities, and looks with contempt on outsiders." William Graham Sumner, *Folkways: A Study of the Sociolological Importance of Usages, Manners, Customs, Mores, and Morals* (Boston: The Athenaeum Press, 1907), p. 13; see also Paul C. Rosenblatt, "Origins and Effects of Group Ethnocentrism and Nationalism," *Journal of Conflict Resolution* 8, no. 2 (1964), p. 131.

59 For Adler . . . a "superiority complex" was in every case: Alfred Adler, *The Science of Living* (Garden City, NY: Garden City Publishing, 1929), p. 79 ("if we inquire into a superiority complex . . . we can always find a more or less hidden inferiority complex"), p. 97 ("the superiority complex is one of the ways which a person with an inferiority complex may use as a method of escape from his difficulties").

59 coining the term: Adler apparently first used the terms "inferiority complex" and "superiority complex" in 1925 or 1926. See Heinz L. Ansbacher and Rowena R. Ansbacher, eds., *The Individual Psychology of Alfred Adler* (New York: Harper & Row, 1964), pp. 256. But both terms were already in use in 1922. See, e.g., "Calls Our Women Superior: French Philosopher Declares Americans 'Will Go Far,'" *New York Times*, Jan. 30, 1922 ("In Europe Freud declared women are handicapped by an inferiority complex . . . but it is certainly not true here. They have what I would describe as the superiority complex").

60 New England . . . Cromwell: Todd Gitlin and Liel Leibovitz, *The Chosen Peoples: America, Israel, and the Ordeals of Divine Election* (New York: Simon & Schuster, 2010), pp. xiii, 65–72; Harry S. Stout, *The New England Soul: Preaching and Religious Culture in Colonial New England* (Oxford, UK: Oxford University Press, 1986), p. 8.

60 the New Testament: 1 Peter 2:9 ("Ye are a chosen generation, . . . an holy nation, a peculiar people").

60 "did not set his love upon you, nor choose you": Gitlin and Leibovitz, pp. 15–16.

61 "Blessed art Thou": Hayim Halevy Donin, *To Pray as a Jew: A Guide to the Prayer Book and the Synagogue Service* (New York: Basic Books, 1991), p. 326 (prayer for festival days); see also ibid., p. 52 (Torah blessing) ("chosen us from among the nations"), p. 325 (Sabbath blessing) ("chosen us and sanctified us above all nations").

61 "contaminated": See Jon D. Levenson, "Chosenness and Its Enemies," *Commentary*, December 2008, p. 26 (quoting José Saramago); see also Paul Berman, "Something's Changed: Bigotry in Print. Crowds Chant Murder," in Ron Rosenbaum, ed., *Those Who Forget the Past: The Question of Anti-Semitism* (New York: Random House, 2004), pp. 17–9.

61 "the root of evil": Levenson, "Chosenness and Its Enemies," p. 26; The Associated Press and Haaretz Service, "Israel Complains About Greek Composer's Anti-Semitic Remarks," *Haaretz*, Nov. 12, 2003; Jeff Weintraub, "Theodorakis's Jewish Problem (2004)," July 15, 2005, http://jeffweintraub.blogspot.com/2005/07/theodorakiss-jewish-problem-2004.html. Theodorakis has stated that his remarks were taken out of context (and that his later declaration, "I am an anti-Semite," was a "slip of the tongue"). Mikis Theodorakis, Letter to the Central Board of Jewish Communities in Greece, May 16, 2011, http://www.kis.gr/en/index.php?option=com_cont ent&view=article&id=421:mikis-theodorakis.

61 Jewish philosophers: See, e.g., Menachem Kellner, "Chosenness, Not Chauvinism: Maimonides on the Chosen People," in Daniel H. Frank, ed., *A People Apart: Chosenness and Ritual in Jewish Philosophical Thought* (Albany: State University of New York Press, 1993), pp. 51–76; David Novak, *The Election of Israel: The Idea of the Chosen People* (Cambridge, UK: Cambridge University Press, 1995), pp. 31–108.

61 Spinoza: Benedictus de Spinoza, *Tractatus Theologico-Politicus*, trans. Samuel Shirley (Boston: E.J. Brill, 1991) [1677], p. 100; Steven Nadler, *Spinoza: A Life* (Cambridge, UK: Cambridge University Press, 1999), pp. 120–32.

61 Mendelssohn: See Paul Johnson, *A History of the Jews* (New York: Harper & Row, 1987),

p. 301 ("tightrope"); Moses Mendelssohn, *Jerusalem: A Treatise on Ecclesiastical Authority and Judaism*, trans. M. Samuels (London: Longman, Orme, Brown, and Longmans, 1838), vol. 2, pp. 89, 102 ("Judaism boasts of no exclusive revelation of immutable truths . . . no revealed religion in the sense in which that term is usually taken. Revealed *religion* is one thing; revealed *legislation* is another"); but cf. Allan Arkush, *Moses Mendelssohn and the Enlightenment* (Albany: State University of New York Press, 1994), pp. 218–9 (suggesting that Mendelssohn nevertheless embraced a concept of Jewish election).

62 "To abandon the claim to chosenness": Arnold M. Eisen, *The Chosen People in America: A Study in Jewish Religious Ideology* (Bloomington: Indiana University Press, 1983), pp. 3–4.

62 Reconstructionist Judaism: See Avi Beker, *The Chosen: The History of an Idea, and the Anatomy of an Obsession* (New York: Palgrave MacMillan, 2008), pp. 23–4, 72–3. Reconstructionism is by far the smallest of the four major denominations of American Jews. As of 2013, only 1 percent of American Jews identified themselves as Reconstructionist; 10 percent identified themselves as Orthodox, 18 percent as Conservative, 35 percent as Reform, and 27 percent as "just Jewish." Pew Research Center, *A Portrait of Jewish Americans: Findings from a Pew Research Center Survey of U.S. Jews* (Washington, DC: Pew Research Center, 2013), p. 48.

62 "mission" . . . "witnesses to God's presence": Pittsburgh Platform, Union of American Hebrew Congregations (November 1885); "A Statement of Principles for Reform Judaism," Central Conference of American Rabbis (Pittsburgh, May 1999), http://ccarnet.org/rabbis-speak/platforms/statement-principles-reform-judaism.

62 deemphasizing if not rejecting: For a highly influential Reform rabbi's declaration that Reform Jews "believe we have a mission to perform," but "reject" the concept that "[Jews] are *the* Chosen People," see Jacob R. Marcus, "Genesis: College Beginnings (1978)," in Gary Phillip Zola, ed., *The Dynamics of American Jewish History: Jacob Rader Marcus's Essays on American Jewry* (Lebanon, NH: Brandeis University Press, 2004), p. 144. Other Reform leaders have retained the idea of chosenness. See Beker, *The Chosen*, p. 77.

62 "There is no doubt": Sigmund Freud, "Moses and Monotheism," in James Strachey, ed. and trans. *The Standard Edition of the Complete Psychological Works of Sigmund Freud* (London: Hogarth Press, 1986), vol. 23, pp. 105–6.

62 "nobler past": Louis Dembitz Brandeis, "The Jewish Problem, How to Solve It" (speech delivered in June 1915), reprinted in Steve Israel and Seth Forman, eds., *Great Jewish Speeches Throughout History* (Northvale, NJ: Jason Aronson Inc., 1994), p. 74.

62 "Persecution . . . deepened the passion for righteousness": Brandeis, "The Jewish Problem, How to Solve It," p. 77; see also Adam Garfinkle, *Jewcentricity: Why Jews Are Praised, Blamed, and Used to Explain Just About Everything* (Hoboken, NJ: Wiley, 2009), p. 58 (some Jews became "convinced of their own moral superiority to non-Jews, in rough proportion to their suffering at their hands").

63 "our role in history is actually unique": Martin Buber, "Hebrew Humanism," in *Israel and the World: Essays in a Time of Crisis* (Syracuse, NY: Syracuse University Press, 1997), pp. 240, 250.

63 "foundational ambiguity": Michael Chabon, "Chosen, but Not Special," *New York Times*, June 6, 2010.

63 "perverse sacralization": Peter Novick, *The Holocaust in American Life* (Boston: Houghton Mifflin, 1999), p. 280.

63 "two sides of the same coin": Office of the Chief Rabbi, "Faith Lectures: Jewish Identity: The Concept of a Chosen People," May 8, 2001, http://oldweb.chiefrabbi.org/ReadArtical.aspx?id=454.

63 "extravagant" Jewish overrepresentation: Charles Murray, "Jewish Genius," *Commentary*, April 2007, p. 29.

63 70 percent of Israeli Jews: Nir Hasson, "Survey: Record Number of Israeli Jews Believe in God," *Haaretz*, Jan. 27, 2012, http://www.haaretz.com/jewish-world/survey-record-number-of-israeli-jews-believe-in-god-1.409386; Harris Interactive, The Harris Poll No. 59, Oct. 15,

2003, p. 2, http://www.harrisinteractive.com/vault/Harris-Interactive-Poll-Research-While
-Most-Americans-Believe-in-God-Only-36-pct-A-2003-10.pdf.

64 locate this exceptionality: See, e.g., Raphael Patai, *The Jewish Mind* (Detroit: Wayne
State University Press, 1996), pp. 8–9, 324, 339; Brandeis, "The Jewish Problem, How to Solve
It," p. 74.

64 "it was up to you to invent your specialness": S. Leyla Gürkan, *The Jews as a Chosen People:
Tradition and Transformation* (London and New York: Routledge, 2009), p. 124 (quoting Philip
Roth); Charles E. Silberman, *A Certain People: American Jews and Their Lives Today* (New York:
Summit Books, 1985), pp. 80–1.

64 "a psychology without content": Michael P. Kramer, "The Conversion of the Jews and
Other Narratives of Self-Definition: Notes Towards the Writing of Jewish American Literary
History; Or, Adventures in Hebrew School," in Emily Miller Budick, ed., *Ideology and Jewish
Identity in Israeli and American Literature* (Albany: State University of New York Press, 2001),
p. 191.

64 "Israel" . . . "Zion" . . . "New Jerusalem" . . . exodus: Matthew Bowman, *The Mormon
People: The Making of an American Faith* (New York: Random House, 2012), p. 94; Claudia L.
Bushman, *Contemporary Mormonism: Latter-day Saints in Modern America* (Westport, CT: Praeger, 2006), pp. 15–16, 18; Terryl L. Givens, *People of Paradox: A History of Mormon Culture* (Oxford and New York: Oxford University Press, 2007), pp. xii, xvi.

64 had their Moses: Leonard J. Arrington, *Brigham Young: American Moses* (New York: Alfred A. Knopf, 1985).

64 "extermination order": Bowman, *The Mormon People*, p. 62.

64 "a religious genius": See, e.g., Harold Bloom, *The American Religion: The Emergence of the
Post-Christian Nation* (New York: Simon & Schuster, 1992), p. 82; Bowman, *The Mormon People*,
p. 24.

64 America's providential place in the world: see Givens, *People of Paradox*, p. xvi ("Mormons have long identified their faith with America's providential role in history").

65 Garden of Eden: On Smith's "sacralization" of the American continent, see Bowman, *The
Mormon People*, pp. xv–xvi, 26–7, 32–8; Bushman, *Contemporary Mormonism*, pp. 24–5.

65 "The whole of America is Zion": Richard Lyman Bushman, *Joseph Smith: Rough Stone
Rolling* (New York: Alfred A. Knopf, 2005), p. 519; see also pp. 94–7.

65 "confident amateurism": Bowman, *The Mormon People*, p. xv.

65 rejected, ostracized: See Givens, *People of Paradox*, pp. xiii, xvi (describing Mormonism's
"history of persecution and alienation from the American mainstream").

65 crossed the country in their covered wagons: Bowman, *The Mormon People*, pp. 98–100;
George W. Givens, *The Language of the Mormon Pioneers* (Springville, UT: Bonneville Books,
2003), p. 215; see also Arrington, *Brigham Young*, chap. 9.

65 "quintessentially American religion": Givens, *People of Paradox*, p. 59; see also Bowman,
The Mormon People, p. xv. ("quintessential American faith").

65 only strengthened Mormons' belief in their divine election: Jan Shipps, *Mormonism: The
Story of a New Religious Tradition* (Urbana and Chicago: University of Illinois Press, 1985), p. 125
("living in the kingdom in the nineteenth century was the sign of citizenship in God's elect nation").

65 "Great Apostasy" . . . true church: Bushman, *Contemporary Mormonism*, p. 3; Givens,
People of Paradox, pp. xiii, xvi; Jon Krakauer, *Under the Banner of Heaven: A Story of Violent Faith*
(New York: Doubleday, 2003), pp. 5, 69.

65 end-of-days: Bowman, *The Mormon People*, p. xviii ("Mormons believe that their church
has been given . . . the mandate to prepare the earth for the Second Coming of Jesus Christ"). For
a fascinating take on the Mormon end-of-days mentality, see Joanna Brooks, *The Book of Mormon
Girl: A Memoir of an American Faith* (New York: Free Press, 2012), pp. 30–1, 35.

65 divine communications and revelations: Bowman, *The Mormon People*, p. 33; Bushman, *Contemporary Mormonism*, pp. 3, 18; Richard Lyman Bushman, *Mormonism: A Very Short Introduction* (New York: Oxford University Press, 2008), pp. 27–31.

65 "priesthood": See Bowman, *The Mormon People*, pp. 46, 140–1; Bushman, *Contemporary Mormonism*, p. 30 ("The priesthood is divided between the lower or Aaronic Priesthood for those twelve to eighteen and the upper or Melchizedek Priesthood for men nineteen and up").

65–66 "power of God": Bowman, *The Mormon People*, p. 46.

66 "missions": Ibid., pp. 188–90; Bushman, *Contemporary Mormonism*, p. 4.

66 temples: Bushman, *Contemporary Mormonism*, pp. 2, 75. Temples are closed not only to non-Mormons, but also to Mormons who do not live by LDS Church teachings. "To enter a temple, members must have been baptized and confirmed and must be privately interviewed in searching discussions by two levels of ecclesiastical authority every two years." Ibid., p. 79.

66 Although some Christians argue: Ibid., pp. 3, 23.

66 Mormonism departs on key theological points: Bowman, *The Mormon People*, pp. 165–7.

66 "free from the original sin that degraded mankind": Bushman, *Contemporary Mormonism*, pp. x, 19, 242.

66 God is a corporeal, essentially man-like person: Bowman, *The Mormon People*, pp. 166–7, 230; Bushman, *Contemporary Mormonism*, pp. 19, 23; see also Givens, *People of Paradox*, p. xv ("With God an exalted man" and "man a God in embryo," Smith collapsed the conventional Christian dualism); Blake T. Ostler, "Worshipworthiness and the Mormon Concept of God," *Religious Studies* 33, no. 3 (1997), pp. 315, 319–20.

66 "it is no robbery to be equal to God": Bowman, *The Mormon People*, p. 167.

66 "godhood of their own": Ibid., p. 230; Ostler, "Worshipworthiness and the Mormon Concept of God," p. 320.

66 families as divinely ordained units: Bowman, *The Mormon People*, p. 126; Bushman, *Contemporary Mormonism*, pp. 42–3; Givens, *People of Paradox*, p. 57; Richard N. Ostling and Joan K. Ostling, *Mormon America: The Power and the Promise* (New York: HarperOne, 2007), p. 338; Brooks, *The Book of Mormon Girl*, p. 136.

67 early Mormon theology seems to have made the number of wives: In the 1850s, Brigham Young and the apostle Orson Pratt "declared that the procreative power that marriage legitimized persisted into the eternities, where glory was measured in the number of relations." Bowman, *The Mormon People*, pp. 125–6.

67 between twenty-seven and fifty-five wives: Bloom, *American Religion*, pp. 108.

67 Smith had perhaps thirty: Bowman, *The Mormon People*, p. 82; see also Bloom, *American Religion*, p. 108.

67 renounced polygamy: Bushman, *Contemporary Mormonism*, p. xi; Ostling and Ostling, *Mormon America*, p. xxiv; Shipps, *Mormonism*, p. 125.

67 abstemiousness, strong families, and clean-cut children: Tim B. Heaton, Kristen L. Goodman, and Thomas B. Holman, "In Search of a Peculiar People: Are Mormon Families Really Different?," in Marie Cornwall, Tim B. Heaton, and Lawrence A. Young, eds., *Contemporary Mormonism: Social Science Perspectives* (Urbana and Chicago: University of Illinois Press, 1994), pp. 87, 113 (data confirm that "[m]ore so than the average American" "Mormonism is a family-focused religion and that its members adhere to traditional family values"); see also Bushman, *Contemporary Mormonism*, p. 184 (noting that in the 1960s the "shift back to some aspects of old-style Mormonism took place against the cultural change of the Civil Rights Movement, an expansion of tolerance, a general loosening of traditional morality, and substance abuse"); Ostling and Ostling, *Mormon America*, p. xxiv ("More than anyone on the street, [the Mormon] might seem honest, reliable, hardworking, and earnest—all the Boy Scout virtues. His children are obedient, his family close-knit").

67 "an island of morality in a sea of moral decay": Bushman, *Contemporary Mormonism*, p. 35.

67 *I am a child of God*: Children's Songbook of The Church of Jesus Christ of Latter-day Saints, (Salt Lake City: The Church of Jesus Christ of Latter-day Saints, 2002), pp. 2–3.

67 "For nearly six thousand years": Ezra Taft Benson, "In His Steps," Mar. 4, 1979, www .speeches.byu.edu/?act=viewitem&id=89; Thomas S. Monson, "Dare to Stand Alone," October 2011, http://www.lds.org/general-conference/2011/10/dare-to-stand-alone?lang=eng.

68 "I don't know about you": Pepe Billete, "I'm Not a Latino, I'm Not a Hispanic, I'm a Cuban American!" *Miami New Times*, July 6, 2012, http://blogs.miaminewtimes.com/cultist/2012/07/ im_not_a_latino_im_not_a_hispa.php.

69 the post provoked outrage: Pepe Billete, "I'm Not a Latino, I'm Not a Hispanic, I'm a Cuban American! (Part Dos)," *Miami New Times*, July 13, 2012, http://blogs.miaminewtimes. com/cultist/2012/07/pepe_billete_im_not_a_latino_i.php ("Last week I managed to upset a pretty big segment of Miami's non-Cuban Spanish speaking population").

69 "Cubans are a different breed": Interview with José Pico, director and president, JPL Investments Corp., in Miami, Fla. (conducted by Eileen Zelek on Jan. 6, 2012) (on file with authors); see also María Cristina García, *Havana USA: Cuban Exiles and Cuban Americans in South Florida, 1959–1994* (Berkeley and Los Angeles: University of California Press, 1996), p. 84 (many Cuban Exiles saw themselves as "martyrs" and resisted the label "immigrant" because *"Immigrant* implies a choice" and they "believed that they had no choice; they had been pushed out of their country. . . . Preserving *cubanidad* became . . . a political responsibility"); Miguel Gonzalez-Pando, *The Cuban Americans* (Westport, CT, and London: Greenwood Press, 1998), p. 31 (quoting one Exile as saying, "We didn't come here looking for a better life. We already had a good life in Cuba, but our lives were stopped because of the political situation").

69 "unique and privileged position in the world": Guillermo J. Grenier and Lisandro Pérez, *The Legacy of Exile: Cubans in the United States* (Boston: Pearson Education, 2003), p. 33.

69–70 "most Spanish" . . . most similar to the United States: Ibid. pp. 29–31 (inner quotation marks omitted). For a detailed discussion of early Cuban "proto-nationalists," including those who envisioned a "white Cuba" peopled by "superior beings," see Richard Gott, *Cuba: A New History* (New Haven, CT, and London: Yale University Press, 2004), pp. 54–7.

70 "Looking the teacher straight in the eye": Carlos Eire, *Waiting for Snow in Havana: Confessions of a Cuban Boy* (New York: The Free Press, 2003), pp. 24–5.

71 The Exiles: Gonzalez-Pando, *The Cuban Americans*, pp. 31–4, 45–6; García, *Havana USA*, pp. xi, 13–4.

71 "the *crème de la crème* of Cuban society": Eire, *Waiting for Snow in Havana*, p. 27.

71 Murillo . . . slaves: Ibid., pp. 13–4.

71 never identified themselves with . . . other . . . Hispanic communities: Miguel de la Torre, *La Lucha for Cuba: Religion and Politics on the Streets of Miami* (Berkeley and Los Angeles: University of California Press, 2003), p. 128; see also ibid., p. 18 (referring to the "Exilic Cubans' attempt to identify themselves with the Euroamerican dominant culture, and thus against other Hispanic groups"); Grenier and Pérez, *The Legacy of Exile*, p. 34 (describing bumper stickers in Miami that read: *Yo no soy hispano . . . yo soy cubano*).

71 "the second largest Cuban city in the world": Grenier and Pérez, *The Legacy of Exile*, p. 48.

72 "only globally transplanted population": Sheyla Paz Hicks, trans., "Article about Cubans Written by Mexican Journalist Victor Mona," http://www.spanish-tvtucanal.com/beta/index .php?option=com_content&view=article&id=199:cuba-y-los-cubanos&catid=1:latest&Itemid=2 (translation edited by authors). Although the article is usually attributed to "Mexican journalist Victor Mona," some say it was actually written by a Cuban-American journalist.

73 "not one or ten or ten thousand things": Elizabeth Alexander, "Today's News," in Arnold Rampersad and Hilary Herbold, eds., *The Oxford Anthology of African-American Poetry* (New York: Oxford University Press, 2006), p. 1.

73 black "aristocracies": Lawrence Otis Graham, *Our Kind of People: Inside America's Black Upper Class* (New York: HarperCollins, 1999).

73 the second-wealthiest African American: See "Richest African Americans—Forbes Wealthiest Black Americans," Forbes.com, Apr. 20, 2012, http://www.therichest.com/rich-list/nation/wealthiest-african-americans.

73 "The thing that really shocked people": Timothy Greenfield-Sanders and Elvis Mitchell, *The Black List* (New York: Atria Books, 2008), pp. 130–1 (quoting Sean Combs).

74 "I was raised to think": E-mail to Amy Chua, Oct. 15, 2013 (on file with authors).

74 do not . . . grow up with a group superiority complex: See, e.g., Jayanti Owen and Scott M. Lynch, "Black and Hispanic Immigrants' Resilience Against Negative-ability Racial Stereotypes at Selective Colleges and Universities in the United States," *Sociology of Education* 85, no. 4 (2012), pp. 304–5. Empirical evidence of the power of these negative stereotypes can be found in the "stereotype threat" literature, discussed below.

74 "You are born to a single mother": Nicholas Powers, "Jim Crow America: Why Our Society's Racial Caste System Still Exists," *The Indypendent* (July 24, 2012).

75 "sheer force of will": Greenfield-Sanders and Mitchell, *The Black List*, p. 4.

75 a third of young black men: David J. Harding, *Living the Drama: Community, Conflict, and Culture Among Inner-City Boys* (Chicago: University of Chicago Press, 2010), p. 18; see generally Michelle Alexander, *The New Jim Crow: Mass Incarceration in the Age of Colorblindness* (New York: The New Press, 2010).

77 fostering the kind of collective pride: Gaetane Jean-Marie and Anthony H. Normore, "A Repository of Hope in Social Justice: Black Women Leaders at Historically Black Colleges and Universities," in Anthony H. Normore, ed., *Leadership for Social Justice: Promoting Equity and Excellence Through Inquiry and Reflective Practice* (Charlotte, NC: Information Age Publishing, 2008), pp. 7–8 (part of threefold "mission" of the HBCUs in their first century was "provid[ing] a service to the Black community and the country by aiding in the development of leadership [and] racial pride").

77 better academic and economic outcomes: Kassie Freeman, ed., *African American Culture in Higher Education Research and Practice* (Westport, CT: Praeger, 1998), p. 9 ("studies suggest that African American students who attend HBCUs experience higher intellectual gains and have a more favorable psychosocial adjustment, a more positive self-image, stronger racial pride, and higher aspirations"); Mikyong Minsun Kim and Clifton F. Conrad, "The Impact of Historically Black Colleges and Universities on the Academic Success of African-American Students," *Research in Higher Education* 47, no. 4 (2006), pp. 399–400, 402 (finding that African American students have a "similar probability of obtaining a BA degree whether they attended [HBCUs] or a historically White college or university (HWCU)," but "among African-American college graduates, a disproportionately high percentage of political leaders, lawyers, doctors, and Ph.D. recipients have graduated from HBCUs," and summarizing other studies that have found that African American students, especially females, exhibit greater cognitive growth at HBCUs, "receive higher grades and have higher degree aspirations than their counterparts at HWCUs") (citations omitted); Walter R. Allen, "The Color of Success: African-American College Student Outcomes at Predominantly White and Historically Black Public Colleges and Universities," *Harvard Educational Review* 62, no. 1 (1992), p. 39 (analyzing the National Study on Black College Students data set and concluding that "the college experience was most successful [in terms of academic achievement, social involvement, and occupational aspirations] for African-American students on campuses with Black majority student populations").

77 attended historically black colleges: Cynthia L. Jackson, *African American Education: A Reference Handbook* (Santa Barbara, CA: ABC-CLIO, 2001), p. 255; "Prominent Alumni," Morehouse College, https://www.morehouse.edu/about/prominent_alumni.html.

77 reclaim black history: Kwame Ture and Charles V. Hamilton, *Black Power: The Politics of*

Liberation in America (New York: Vintage Books, 1992), pp. 34–35; Cornel West, *Race Matters* (New York: Vintage Books, 1994), pp. 98–99.

77 Malcolm X: Diane C. Fujino, *Heartbeat of Struggle: The Revolutionary Life of Yuri Kochiyama* (Minneapolis: University of Minnesota Press, 2005), p. 155.

78 Louis Farrakhan: Richard W. Leeman, *African-American Orators: A Bio-Critical Sourcebook* (Westpost, CT: Greenwood Press, 1996), p. 128.

78 Martin Luther King: Martin Luther King, "Speech at the Great March on Detroit," June 23, 1963, http://mlk-kpp01.stanford.edu/index.php/encyclopedia/documentsentry/doc_speech_at_the_great_march_ on_detroit; see also Louis A. DeCaro Jr., *Malcolm and the Cross: The Nation of Islam, Malcolm X, and Christianity* (New York: New York University Press, 1998), p. 160.

78 pioneering "stereotype threat": See Claude M. Steele and Joshua Aronson, "Stereotype Threat and the Intellectual Test Performance of African Americans," *Journal of Personality and Social Psychology* 69, no. 5 (1995), pp. 797–811. For a comprehensive review, see Michael Inzlicht and Toni Schmader, eds., *Stereotype Threat: Theory, Process, and Application* (New York: Oxford University Press, 2012).

78 identify their race or gender: See Mary C. Murphy and Valerie Jones Taylor, "The Role of Situational Cues in Signaling and Maintaining Stereotype Threat," in Inzlicht and Schmader, *Stereotype Threat*, p. 21; Paul R. Sackett et al., "On Interpreting Stereotype Threat as Accounting for African American-White Differences on Cognitive Tests," *American Psychologist* 59, no. 1 (2004), p. 8.

78 black students score lower: See Steele and Aronson, "Stereotype Threat"; Claude M. Steele et al., "Contending with Group Image: The Psychology of Stereotype and Social Identity Threat," in Mark P. Zanna, ed., *Advances in Experimental Social Psychology* (New York: Academic Press, 2002), vol. 34, pp. 379–440; Sackett et al., "On Interpreting Stereotype Threat," pp. 7–8 (finding has been "well replicated").

78 White male Stanford students: Joshua Aronson et al., "When White Men Can't Do Math: Necessary and Sufficient Factors in Stereotype Threat," *Journal of Experimental Social Psychology* 35 (1999), pp. 29–46.

78 Women chess players: Anne Maass et al., "Checkmate? The Role of Gender Stereotypes in the Ultimate Intellectual Sport," *European Journal of Social Psychology* 38 (March/April 2008), pp. 231–5. Women's math scores have been repeatedly shown to fall when gender stereotypes are cued. Patricia M. Gonzales, Hart Blanton, and Kevin J. Williams, "The Effects of Stereotype Threat and Double-Minority Status on the Test Performance of Latino [sic] Women," *Personality and Social Psychology Bulletin* 28, no. 5 (2002), pp. 659–70; Toni Schmader, "Gender Identification Moderates Stereotype Threat Effects on Women's Math Performance," *Journal of Experimental Social Psychology* 38 (2002), pp. 194–201.

79 visual rotation: Margaret J. Shih, Todd L. Pittinsky, and Geoffrey C. Ho, "Stereotype Boost: Positive Outcomes from the Activation of Positive Stereotypes," in Inzlicht and Schmader, *Stereotype Threat*, pp. 145–6. White students did better on a test when it was described as measuring intellectual capacity than they did when the same test was also described as "fair for different racial groups." Gregory M. Walton and Priyanka B. Carr, "Social Belonging and the Motivation and Intellectual Achievement of Negatively Stereotyped Students," in Inzlicht and Schmader, *Stereotype Threat*, p. 97.

79 miniature golf experiment: Jeff Stone, Aina Chalabaev, and C. Keith Harrison, "The Impact of Stereotype Threat on Performance in Sports," in Inzlicht and Schmader, *Stereotype Threat*, pp. 220–1.

79 Asian undergraduates . . . "strongly identified": Brian E. Armenta, "Stereotype Boost and Stereotype Threat Effects: The Moderating Role of Ethnic Identification," *Cultural Diversity and Ethnic Minority Psychology* 16, no. 1 (2010), pp. 94–8.

79 Outside the laboratory: Jennifer Lee and Min Zhou, "Frames of Achievement and Opportunity Horizons," in David Card and Steven Raphael, eds., *Immigration, Poverty, and Socioeconomic Inequality* (New York: Russell Sage Foundation, 2013), pp. 206–31.

80 "Because of that choice": Helene Cooper, *The House at Sugar Beach: In Search of a Lost African Childhood* (New York: Simon & Schuster, 2008), p. 29.

81 stereotype of Nigerians: See e.g., Helen Fogarassy, *Mission Improbable: The World Community on a UN Compound in Somalia* (Lanham, MD: Lexington Books, 1999), p. 34; Okechukwu Jones Asuzu, *The Politics of Being Nigerian* (Lulu.com, 2006), p. 63.

81 Igbo or Yoruba: See Amy Chua, *World on Fire: How Exporting Free Market Democracy Breeds Ethnic Hatred and Global Instability* (New York: Doubleday, 2003), pp. 108–9; Donald L. Horowitz, *Ethnic Groups in Conflict* (Berkeley, Los Angeles, and London: University of California Press, 1985), pp. 27–8, 154–5, 250–1.

81 an illustrious royal lineage: See Robert S. Smith, *Kingdoms of the Yoruba* (3rd ed.) (Madison: University of Wisconsin Press, 1988); Babatunde Lawal, "Reclaiming the Past: Yoruba Elements in African American Arts," in Toyin Falola and Matt D. Childs, eds., *The Yoruba Diaspora in the Atlantic World* (Bloomington: Indiana University Press, 2004), pp. 291, 292.

81 Igbo are often called: See Thomas D. Boston, ed., *A Different Vision: African American Economic Thought* (London and New York: Routledge, 1997), pp. 177–8; see also Chua, *World on Fire*, pp. 108–9.

81 the "dangers of hubris": Chinua Achebe, *There Was a Country: A Personal History of Biafra* (New York: Penguin Press, 2012), pp. 76–8. A hint of the Igbos' exclusiveness can also be found in the new fascinating novel *Americanah* by (Igbo) Nigerian writer Chimamanda Ngozi Adichie, where a Senegalese hair braider insists that "Igbo marry Igbo always." See Chimamanda Ngozi Adichie, *Americanah* (New York: Alfred A. Knopf, 2013), p. 15.

81 recent study of more than 1,800 students: Owens and Lynch, "Black and Hispanic Immigrants' Resilience Against Negative-ability Racial Stereotypes," pp. 303–25.

81 the more strongly black immigrant students identify: Yaw O. Adutwum, "The Impact of Culture on Academic Achievement Among Ghanaian Immigrant Children" (unpublished dissertation) (USC Digital Library 2009), Kay Deaux et al., "Becoming American: Stereotype Threat Effects in Afro-Caribbean Immigrant Groups," *Social Psychology Quarterly* 70, no. 4 (2007), pp. 384–404; Alwyn D. Gilkes, "Among Thistles and Thorns: West Indian Diaspora Immigrants in New York City and Toronto" (unpublished dissertation) (ProQuest Dissertations and Theses: 2005).

81 what they perceive as defeatism: Mary C. Waters, *Black Identities: West Indian Immigrant Dreams and American Realities* (New York and Cambridge, MA: Russell Sage Foundation and Harvard University Press, 1999), pp. 65–6.

82 "Nigerians . . . feel they are capable of anything": Patricia Ngozi Anekwe, *Characteristics and Challenges of High Achieving Second-Generation Nigerian Youths in the United States* (Boca Raton, FL: Universal Publishers, 2008), p. 128. West Indian immigrants often express a similar view. According to a Jamaican American school teacher:

> I can't help them [African Americans] because they're so wrapped up in racism, and they act it out so often, they interpret it as such so often that sometimes they are not even approachable. If they're going to teach anything and it's not black, black, all black, they are not satisfied, you know. . . . Sometimes I feel sorry for them, but you find that you just can't change their attitude because they just tell you that you don't understand.

Waters, *Black Identities*, pp. 171–2.

82 "an ethnic armor": Min Zhou, "The Ethnic System of Supplementary Education: Nonprofit and For-Profit Institutions in Los Angeles' Chinese Immigrant Community," in Marybeth Shinn and Hirokazu Yoshikawa, eds., *Toward Positive Youth Development: Transforming Schools and Community Programs* (New York: Oxford University Press, 2008), p. 232.

82 "Congo people": Cooper, *The House on Sugar Beach*, p. 6.

82 "Honorables": Ibid., p. 11.

82 "a white girl from Seagrove": Helene Cooper, "Author Interview," http://authors.simonandschuster.com/Helene-Cooper/18871279/interviews/91.

CHAPTER 4: INSECURITY

85 "Old World" . . . "secret restlessness": Alexis de Tocqueville, *Democracy in America*, trans. George Lawrence, ed. J. P. Mayer (Garden City, NY: Anchor Books, 1969) vol. 2, part II, chap. 13, pp. 535–8.

85 "All are constantly bent": Ibid., vol. 2, part III, chap. 19, p. 627.

85–86 "never stop thinking" . . . "cloud": Ibid., vol. 2, part II, chap. 13, pp. 536, 538.

86 "longing to rise": Ibid., vol. 2, part III, chap. 19, p. 627.

86 "in the midst of their prosperity": Ibid., vol. 2, part II, chap. 13, p. 535.

87 "Hell hath no fury": What William Congreve actually has his character Zara say in *The Mourning Bride* is "Heav'n has no rage like love to hatred turn'd/Nor hell a fury like a woman scorn'd" (Act III, scene VIII). See "The Mourning Bride: A Tragedy," in John Bell, *Bell's British Theatre Consisting of the Most Esteemed English Plays* (London: George Cawthorn, British Library, 1797), vol. 19, p. 63.

87 Everything can be borne but contempt: Arthur O. Lovejoy, *Reflections on Human Nature* (Baltimore: The Johns Hopkins University Press, 1961), p. 181 (quoting Voltaire's *Traité de métaphysique*) ("To be an object of contempt to those with whom one lives is a thing that none ever has been, or ever will be, able to endure").

87 about a third: Thomas D. Boswell and James R. Curtis, *The Cuban-American Experience: Culture, Images, and Perspectives* (Totowa, NJ: Rowman & Allanheld, 1984), p. 46.

87 forced to take any work: María Cristina García, *Havana USA: Cuban Exiles and Cuban Americans in South Florida, 1959–1994* (Berkeley and Los Angeles: University of California Press, 1996), p. 20; Miguel Gonzalez-Pando, *The Cuban Americans* (Westport, CT, and London: Greenwood Press, 1998), p. 36.

87 "less-advertised corollary": Tad Friend, *Cheerful Money: Me, My Family, and the Last Days of Wasp Splendor* (New York: Little, Brown and Company, 2009), p. 153; see also E. Digby Baltzell, "The Protestant Establishment Revisited," *The American Scholar* 45, no. 4 (1976), pp. 499, 505 (noting that third-generation WASPS went to "Harvard, Yale, and Princeton, where they joined the best clubs and graduated usually . . . with 'gentleman Cs'"), pp. 506, 512.

88 A culture that "once valued education": Peter Sayles, "Report from Newport RI: American WASPs—Dispossessed, Degenerate . . . Or Both?" VDARE.com, Jan. 20, 2013; see also Jerome Karabel, *The Chosen: The Hidden History of Admission and Exclusion at Harvard, Yale, and Princeton* (Boston and New York: Houghton Mifflin Company, 2005), pp. 115, 131.

88 "got lazy": Robert Frank, "That Bright, Dying Star, the American WASP," *Wall Street Journal*, May 15, 2010; see also David Brooks, "Why Our Elites Stink," *New York Times*, July 12, 2012.

88 "the scum of the Earth": Gonzalez-Pando, *The Cuban Americans*, pp. 46–7.

88 "an ideological quest": Ibid.

88 contempt of discrimination: García, *Havana USA*, pp. 18–20, 40; Gonzalez-Pando, *The Cuban Americans*, p. 37; Guillermo J. Grenier and Lisandro Pérez, *The Legacy of Exile: Cubans in the United States* (Boston: Pearson Education, 2003), p. 52.

88 "When we first arrived in Miami": Interview with José Pico, director and president, JPL Investments Corp., in Miami, Fla. (conducted by Eileen Zelek on Jan. 6, 2012) (on file with authors); see also Gonzalez-Pando, *The Cuban Americans*, p. 37.

88 plummet in status: Susan Eva Eckstein, *The Immigrant Divide: How Cuban Americans Changed the US and Their Homeland* (New Haven, CT, and London: Routledge, 2009), p. 83; David Rieff, *The Exile: Cuba in the Heart of Miami* (New York: Simon & Schuster, 1993), p. 48; Gonzalez-Pando, *The Cuban Americans*, pp. 34–6.

88 "my father would run into people who knew him": Telephone interview with Professor Domitila Fox, Florida International University (conducted by Eileen Zelek on Mar. 17, 2012) (on file with authors); see also García, *Havana USA*, pp. 18–20.

89 "the one dependable emotional motive": Robert C. Solomon, "Nietzsche and the Emotions," in Jacob Golomb, Weaver Santaniello, and Ronald Lehrer, eds., *Nietzsche and Depth Psychology* (Albany, NY: State University of New York Press, 1999), pp. 127, 142.

89 Ancient Persia is seen, if at all, through a Greek lens: Much of this paragraph is taken from Amy Chua, *Day of Empire: How Hyperpowers Rise to Global Dominance—and Why They Fall* (New York: Doubleday, 2007), pp. 4, 6–7; see also Pierre Briant, *From Cyrus to Alexander: A History of the Persian Empire*, trans. Peter T. Daniels (Winona Lake, IN: Eisenbrauns, 2002), pp. 5–7; Hooman Majd, *The Ayatollah Begs to Differ: The Paradox of Modern Iran* (New York: Doubleday, 2008), p. 163.

89 "I just can't get over the humiliation": "Xerxes and the Persian Army: What They Really Looked Like," A Persian's Perspective, Mar. 18, 2007, http://persianperspective.wordpress.com/2007/03/18/xerxes-and-the-persian-army-what-they-really-looked-like.

90 Iran *is* Persia: Kenneth M. Pollack, *The Persian Puzzle: The Conflict Between Iran and America* (New York: Random House, 2004), pp. 3, 30–1.

90 "a superpower like nothing the world had ever seen": Pollack, *The Persian Puzzle*, p. 3; see Chua, *Day of Empire*, p. 4.

90 larger even than Rome's . . . 42 million people: see Chua, *Day of Empire*, p. 4.

90 Persian "superiority complex": See, e.g., Majd, *The Ayatollah Begs to Differ*, p. 164; Robert Graham, *Iran: The Illusion of Power* (London: St. Martin's Press, 1978), p. 192.

90 "All Iranians": see Majd, *The Ayatollah Begs to Differ*, p. 163.

90 "a widely remarked sense of superiority": Pollack, *The Persian Puzzle*, p. 3; see also Graham, *Iran*, pp. 190–2 (describing Iran's "sense of superiority" and sense of "uniqueness," which "derives from a somewhat romanticised view of their history, but centres round the suppleness with which they have been able to survive different waves of conquest and absorb cultural influences without having their own identity submerged"), Kathryn Babayan, *Mystics, Monarchs, and Messiahs: Cultural Landscapes of Early Modern Iran* (Cambridge, MA: Harvard University Press, 2002), p. 492–3 (noting how twentieth-century nationalist movements in Iran "underscored Persian superiority").

90 Alexander the Great: Chua, *Day of Empire*, p. 27.

90 "such a brute": Majd, *The Ayatollah Begs to Differ*, p. 164; see also ibid., p. 12 (mentioning the author's grandfather, "who also happened to be an Ayatollah"); Chua, *Day of Empire*, pp. 24–6.

91 "savage bedouins" . . . "camel's milk and lizards": Joya Blondel Saad, *The Image of Arabs in Modern Persian Literature* (Lanham, MD: University Press of America, 1996), pp. 6–7.

91 "constantly fight among themselves": Ibid., p. 8 (quoting from a classic eleventh-century work by Nâser Khosrow).

91 most famous modern author, Sâdeq Hedâyat: Ibid., p. 29.

91 "locusts and plague" . . . "black, with brutish eyes": Ibid., p. 37 (quoting Sâdeq Hedâyat).

91 "equated Arab domination of Iran": Janet Afary, *The Iranian Constitutional Revolution, 1906–1911: Grassroots Democracy, Social Democracy, and the Origins of Feminism* (New York: Columbia University Press, 1996), p. 25.

91 "Iranians don't like being called Arabs": Shadi Akhavan, "Close Enough" (op-ed), Iranian.com, Aug. 25, 2003, http://iranian.com/Opinion/2003/August/Close/index.html.

91 "tremendous sense of insecurity": Graham, *Iran*, p. 194; see also Graham E. Fuller, *The "Center of the Universe": The Geopolitics of Iran* (Boulder, CO: Westview Press, 1991), p. 20.

92 "with its immediate neighbors": Graham, *Iran*, p. 192.

92 "national pursuit of empowerment": Abbas Amanat, "The Persian Complex," *New York Times*, May 25, 2006.

92 insecurity, too, was part of the cultural inheritance: Nima Tasuji, "Reconstructing a New Identity," in Tara Wilcox-Ghanoonparvar, ed., *Hyphenated Identities: Second-Generation Iranian-Americans Speak* (Costa Mesa, CA: Mazda Publishers, 2007), p. 7.

92 Status loss, anxiety, resentment, and even trauma: Mohsen M. Mobasher, *Iranians in Texas: Migration, Politics, and Ethnic Identity* (Austin: University of Texas Press, 2012), p. 8.

92 status collapse: See ibid., p. 8; Tara Bahrampour, "Persia on the Pacific," *The New Yorker*, Nov. 10, 2003.

92 *House of Sand and Fog . . . Crash*: See Andre Dubus III, *House of Sand and Fog* (New York: W. W. Norton & Company, 1999); Bahrampour, "Persia on the Pacific"; Carol Gerster, "CRASH: A Crash Course on Current Race/Ethnicity Issues," *The Journal of Media Literacy* 55, nos. 1 and 2.

92 Iranian flags were burned in public, and demonstrators carried signs: See, e.g., Mobasher, *Iranians in Texas*, p. 34.

93 fled to the United States precisely to escape: Mitra K. Shavarini, *Educating Immigrants: Experiences of Second-Generation Iranians* (New York: LFB Scholarly Publishing, 2004), pp. 38–41.

93 "God, please don't let them be Muslim or Iranian": Shadi Akhavan, "Take It from a Good Girl: Fight Back!," *The Iranian*, Dec. 27, 2002, http://iranian.com/Features/2002/December/Tough/index.html.

93 "stigma to be hidden . . . insecurity and even feelings of self-hatred": Tasuji, "Reconstructing a New Identity," p. 6; see also Shavarini, *Educating Immigrants*, pp. 7, 113–4; Maryam Daha, "Contextual Factors Contributing to Ethnic Identity Development of Second-Generation Iranian American Adolescents," *Journal of Adolescent Research* 26, no. 5 (2011), pp. 543, 554, 563; Mohsen Mobasher, "Cultural Trauma and Ethnic Identity Formation Among Iranian Immigrants in the United States," *American Behavioral Scientist* 50 (2006), pp. 100, 108.

93 branded part of the "axis of evil": Mobasher, *Iranians in Texas*, pp. 45–7.

93 Some Iranian parents: Shavarini, *Educating Immigrants*, p. 5; Mobasher, "Cultural Trauma and Ethnic Identity Formation Among Iranian Immigrants in the United States," pp. 103, 113.

93 self-parodying Internet video: "Iranian Census 2010 PSA with Maz Jobrani," http://www.youtube.com/watch?v=kgoLjFJ0rVg.

93 "stigmatized and humiliated": Mobasher, *Iranians in Texas*, p. 8.

93 survey of second-generation Iranian Americans: Daha, "Contextual Factors," pp. 543, 547, 552–4.

93 intense need to distinguish themselves: Shavarini, *Educating Immigrants*, p. 6.

94 status-conscious: See, e.g., Daha, "Contextual Factors," pp. 560–1; Mehdi Bozorgmehr and Daniel Douglas, "Success(ion): Second-Generation Iranian Americans," *Iranian Studies* 44, no. 1 (2011), pp. 5, 7.

94 "every Mercedes you see belongs to an Iranian person": Shavarini, *Educating Immigrants*, p. 150; see also ibid., pp. 147–51; Daha, "Contextual Factors," pp. 560–1.

94 "The American ideal": Neema Vedadi, "And Iran, Iran's So Far Away," in Wilcox-Ghanoonparvar, *Hyphenated Identities*, pp. 24–5.

94 95 percent of second-generation adolescent Iranian Americans: Daha, "Contextual Factors," pp. 550, 554–8.

94 taught that Persian culture is older, richer: Shavarini, *Educating Immigrants*, p. 113.

94 "confused as to what exactly constitutes": Ibid., pp. 112–3; see also Ali Akbar Mahdi, "Ethnic Identity among Second-Generation Iranians in the United States," *Iranian Studies* 31, no. 1 (Winter 1998), p. 91 (quoting a young Iranian American respondent in a study), ("I really do not know who an Iranian is. To me Iranian probably means *contradiction*").

95 "All Iranians are successful": Daha, "Contextual Factors," p. 560.

95 "If you don't get an A": Ibid., p. 561; see also Mahdi, "Ethnic Identity Among Second-Generation Iranians in the United States," p. 88 (quoting a respondent in a study of 48 Iranian-American youths as saying "My parents expect too much. They want me to be exceptional").

95 "that extra effort to have the bigger house": Daha, "Contextual Factors," p. 561; see also

Bozorgmehr and Douglas, "Success(ion): Second-Generation Iranian Americans," p. 5 (second-generation Iranians internalize their parents' values and become "very motivated to excel in school and to choose professional occupations which will garner them respect").

95 "We have to prove it": Shavarini, *Educating Immigrants*, p. 6.

95 bewildering array of caste, regional, ethnic: See Kanti Bajpai, "Diversity, Democracy, and Devolution in India," in Michael E. Brown and Šumit Ganguly, eds., *Government Policies and Ethnic Relations in Asia and the Pacific* (Cambridge, MA, and London: MIT Press, 1997), pp. 34–8; see also Donald L. Horowitz, *Ethnic Groups in Conflict* (Berkeley, Los Angeles, and London: University of California Press, 1985), p. 37; Nicholas B. Dirks, *Castes of Mind: Colonialism and the Making of Modern India* (Princeton, NJ: Princeton University Press, 2001), pp. 5–6.

95 great majority of Indian immigrants in America: See Eric Mark Kramer, "Introduction: Assimilation and the Model Minority Ideology," in Eric Mark Kramer, ed., *The Emerging Monoculture: Assimilation and the "Model Minority"* (Westport, CT: Praeger, 2003), pp. xi, xxi ("Most Indians in the United States are upper-class, upper-caste Indians"); see also Gita Rajan and Shailja Sharma, eds., *New Cosmopolitanisms: South Asians in the U.S.* (Stanford, CA: Stanford University Press, 2006), p. 13; Vijay Prasad, *Uncle Swami: South Asians in America Today* (New York: The New Press, 2012), pp. 10–1.

95 A few immigrants from East India: Ramaswami Mahalingam, Cheri Philip, and Sundari Balan, "Cultural Psychology and Marginality: An Explorative Study of Indian Diaspora," in Ramaswami Mahalingam, ed., *Cultural Psychology of Immigrants* (Mahwah, NJ: Lawrence Erlbaum Associates Publishers, 2006), pp. 151, 152.

95 first significant Indian community . . . was made up of Punjabi Sikhs: Madhulika S. Khandelwal, *Becoming American, Being Indian: An Immigrant Community in New York City* (Ithaca and London: Cornell University Press, 2002), pp. 9–10.

95 initial wave of post-1965 Indian immigrants. Ibid., p. 6; Bandana Purkayastha, *Negotiating Ethnicity: Second Generation South Asian Americans Traverse a Transnational World* (New Brunswick, NJ: Rutgers University Press, 2005), pp. 15–6.

95 as of 1975, 93 percent: Aparna Rayaprol, *Negotiating Identities: Women in the Indian Diaspora* (Delhi: Oxford University Press, 1997), p. 15.

95–96 Since 1990: Khandelwal, *Becoming American, Being Indian*, p. 6; Sunaina Marr Maira, *Desis in the House: Indian American Youth Culture in New York City* (Philadelphia: Temple University Press, 2002), p. 10 ("in 1989, 85 percent of Indian immigrants entered under family reunification categories, while only 1 pecent came with occupation-based visas [down from 18 percent in 1969]").

96 astonishing 87 percent of adults: Pew Research Center, *The Rise of Asian Americans* (Washington, DC: Pew Research Center, April 4, 2013) (updated edition), p. 44.

96 "ethnic anxiety": Maira, *Desis in the House*, p. 75; see also Purkayastha, *Negotiating Ethnicity*, p. 93.

96 the most successful Census-tracked ethnic group: U.S. Census, American Community Survey, Table S0201; Selected Population Profile in the United States (2010 3-year dataset) (population group code 013 – Asian Indian) (estimating median Indian household income of $90,525 as compared to $51,222 for U.S. population overall).

96 widely agreed that most Indian Americans: See Kramer, "Introduction," p. xxi; see also Mahalingam, Philip, and Balan, "Cultural Psychology and Marginality," p. 155 ("most of the current wave of immigrants are from the upper castes"); Prema A. Kurien, *A Place at the Multicultural Table: The Development of an American Hinduism* (New Brunswick, NJ: Rutgers University Press, 2007), p. 45 ("Given the elite nature of the migration, we can assume that most Indian Americans are from upper-caste backgrounds. Brahmins seem to be particularly overrepresented"); see also Peter F. Geithner, Paula D. Johnson, and Lincoln C. Chen, eds., *Diaspora Philanthropy and Equitable Development in China and India* (Cambridge, MA: Global Equity Initiative, Asia Center, Harvard University, 2004), p. 350; Himanee Gupta, "Hidden in Plain Sight: The Semiotics of Caste Among Hindu Indians in the United States" (August 2001) (paper presented at 2001 APARRI Conference).

96 traditional Hindu "castes": See Dirks, *Castes of Mind,* pp. 202–3, 221; Dietmar Rothermund, *India: The Rise of an Asian Giant* (New Haven, CT, and London: Yale University Press, 2008), p. 162; Rajendra K. Sharma, *Indian Society, Institutions, and Change* (New Delhi: Atlantic, 2004), p. 14.

96 For centuries, some say millennia: Although the caste system can be traced back thousands of years, some scholars emphasize that "[u]nder colonialism, caste was . . . made out to be far more— far more pervasive, far more totalizing, and far more uniform—than it had ever been before." Dirks, *Castes of Mind,* p. 13; see also Sharma, *Indian Society, Institutions, and Change,* pp. 11–2, 14–5.

96 "required to place clay pots": Narendra Jadhav, *Untouchables: My Family's Triumphant Escape from India's Caste System* (Berkeley and Los Angeles: University of California Press, 2005), pp. 1–4; see also Robert Deliège, *The Untouchables of India,* trans. Nora Scott (Oxford, UK: Berg Publishers, 1999), pp. 1–3.

96 "master symbol of their inferiority": David I. Kertzer, *Ritual, Politics & Power* (New Haven, CT, and London: Yale University Press, 1989), pp. 112–3.

96 Nehru . . . Indira Gandhi: John McLeod, *The History of India* (Westport, CT: Greenwood Publishing Group, 2002), pp. 185–6.

96 Tagore: The caste status of the celebrated Tagore, sometimes called India's Tolstoy, illustrates the arcane but potent role caste traditionally played in India. Although the Tagores were Brahman, they were so-called Pirali Brahman, "unmarriageable by orthodox Hindus" because (according at least to widely accepted lore) their ancestors had been "tainted by contact with Muslims" some time in the fifteenth century. In the early 1800s, a Brahman who merely ate a meal with a Pirali Brahman had to pay 50,000 rupees to regain caste. Krishna Dutta and Andrew Robinson, *Rabindranath Tagore: The Myriad-Minded Man* (London: Bloomsbury, 1995), pp. 17–8.

96 Mohandas Gandhi: Rajmohan Gandhi, *Gandhi: The Man, His People, and the Empire* (Berkeley and Los Angeles: University of California Press, 2008), p. 2.

97 about a third of the population: Bajpai, "Diversity, Democracy, and Devolution in India," p. 53; International Institute for Population Sciences, *India: National Family Health Survey (NFHS-3), 2005–06:* (Mumbai, 2007), vol. 1, chap. 3, http://hetv.org/india/nfhs/nfhs3/NFHS-3-India-Full-Report-Volume-I.pdf; 61st National Sample Survey (conducted by the Ministry of Statistics and Programme Implementation).

97 "Just try and check how many brahmins": "Arundhati Roy in Conversation with Venu Govindu," October 29, 2000, Friends of River Narmada, http://www.narmada.org/articles/arinterview.html. See also Ramesh Bairy T. S., *Being Brahmin, Being Modern: Exploring the Lives of Caste Today* (London, New York, and New Delhi: Routledge, 2010), pp. 85–6 (in the state of Karnataka, Brahmans, although just 4.28 percent of the total population, "continue to be disproportionately represented in the bureaucracy, spaces of higher education, judiciary," medicine, and engineering).

97 the Indian Constititution: Constitution of India (1949), art. 15, 17.

97 deeply ingrained source of superiority: Bairy, *Being Brahmin,* pp. 87, 280–1; Jadhav, *Untouchables,* p. 3; Louis Dumont, *Homo Hierarchicus: The Caste System and Its Implications* (Chicago and London: University of Chicago Press, 1980), pp. 79–80.

97 Bengalis pride themselves: Mark Magnier, "In India, Bengalis Seek to Recapture Their Glory as Intellectuals," *Los Angeles Times,* Sept. 8, 2012.

97 Bengali Brahman families: "Amartya Sen," Indians Abroad, indobase.com, http://www.indobase.com/indians-abroad/amartya-sen.html; Reshmi R. Dasgupta, "I Had No Idea a Pulitzer Was So Prestigious, Says Pulitzer Prize Winner Siddhartha Mukerjee's Father," *The Economic*

Times (India), Apr. 22, 2011; see also Maxine P. Fisher, *The Indians of New York City: A Study of Immigrants from India* (Columbia, MO: South Asia Books, 1980), p. 49 (noting that Bengali names ending with "ji" as in "Mukerji" or "Banerji" are "those of Brahmins").

97 Gujaratis: Pawan Dhingra, *Life Behind the Lobby: Indian American Motel Owners and the American Dream* (Stanford, CA: Stanford University Press, 2012), pp. 4–5, 14; Marjorie Howard, "A Motel of One's Own," *Tufts Now*, Nov. 27, 2012; Nimish Shukla, "16 Gujaratis in Forbes List," *The Times of India*, Mar. 8, 2008.

97 two of the top three: "Bhai is king: Gujaratis are ruling Forbes 40 richest Indians list," Nov. 14, 2008, DeshGujarat.com, http://deshgujarat.com/2008/11/14/bhai-is-king-gujaratis -are-ruling-forbes-40-richest-indians-list.

97 who number about 200,000: Pew Research Center, "How Many U.S. Sikhs?," Aug. 6, 2012, http://www.pewresearch.org/2012/08/06/ask-the-expert-how-many-us-sikhs/.

97 Sikhs . . . have their own superiority: Ved Mehta, *Rajiv Ghandi and Rama's Kingdom* (New Haven, CT, and London: Yale University Press, 1994), p. 65; see Dawinder S. Sidhu and Neha Singh Gohil, *Civil Rights in Wartime: The Post 9/11 Sikh Experience* (Farnham, UK: Ashgate, 2009), p. 51 (quoting Prabhjot Singh) (describing the Sikh turban as "a manifestation of the mission given to all Sikhs—to act as a divine prince or princess by standing firm against tyranny and protecting the downtrodden"); D. H. Butani, *The Third Sikh War? Towards or Away from Khalistan?* (New Delhi: Promilla & Co., 1986), p. 26.

97 competing north/south snobberies: See Fisher, *The Indians of New York City*, pp. 29–34; Rayaprol, *Negotiating Identities*, p. 75; Chetan Bhagat, *2 States: The Story of My Marriage* (New Delhi: Rupa & Co, 2009), pp. 13–4, 51; Steve Sailer, "Why Are South Indians So Smart?," Mar. 23, 2002, http://isteve.blogspot.com/2002/03/why-are-south-indians-so-smart.html. On the colonial origins of the north/south divide, see Dirks, *Castes of Mind*, pp. 140–3.

97 Indian Institutes of Technology: See Anita Raghavan, *The Billionaire's Apprentice: The Rise of the Indian-American Elite and the Fall of the Galleon Hedge Fund* (New York and Boston: Business Plus, 2013), pp. 40–1; Thomas L. Friedman, *The World Is Flat 3.0: A Brief History of the Twenty-First Century* (New York: Picador, 2007), p. 127; Matthew Schneeberger, "Why IIT grads abroad are returning to India," Rediff India Abroad, May 15, 2008, http://www.rediff .com/money/2008/may/15iit.htm (reporting that 35 percent of IIT graduates emigrated to America between 1965 and 2002, and 16 percent thereafter).

97 anti-Brahman movements: Bairy, *Being Brahmin*, p. 124; see also Rothermund, *India*, p. 164 (noting that long before the twentieth century "Buddhism and Jainism were social and religious movements founded on the quest for individual salavation" that "challenged [the] order based on caste and endogamy").

98 British colonial rule: See, e.g., Chua, *Day of Empire*, pp. 224–28; Niall Ferguson, *Empire: How Britain Made the Modern World* (London: Allen Lane, 2003), pp. 146–54, 210–15, 325–8.

98 "suffer a degradation not fit for human beings": Dutta and Robinson, *Rabindranath Tagore*, p. 216.

98 "I always felt so embarrassed by my name": Benjamin Anastas, "Inspiring Adaptation," *Men's Vogue*, March/April 2007, p. 113; see Julia Leyda, "An Interview with Jhumpa Lahiri," *Contemporary Women's Writing* 5, no. 1 (2011), p. 66.

98 The popular claim . . . "worshipping cows": Purkayastha, *Negotiating Ethnicity*, p. 39.

98 "smelling like curry": Rosalind S. Chou and Joe R. Feagin, *The Myth of the Model Minority: Asian Americans Facing Racism* (Boulder and London: Paradigm Publishers, 2008), p. 70.

99 Sikh men: Sidhu and Gohil, *Civil Rights in Wartime*, pp. xiii–xv, 48, 64–8.

99 "You fucking Arab rag-head": Prashad, *Uncle Swami*, pp. 4–5.

99 Indian cabdrivers: Sidhu and Gohil, *Civil Rights in Wartime*, p. 65; Palash Ghosh, "South Asian Taxi Drivers: Victims and Perpetrators of Racism," *International Business Times*, June 22, 2012, http://www.ibtimes.com/south-asian-taxi-drivers-victims-perpetrators-racism-705690.

99 although perhaps "Caucasian," was not "white": *United States v. Thind*, 261 U.S. 204, 210, 214–6 (1923).

99 Many Indian Americans attest: Prashad, *Uncle Swami*, pp. 3–7; Purkayastha, *Negotiating Ethnicity*, pp. 27–31, 38–41.

99 "whiten" their lobbies . . . "Even foreign is not a bad thing": Dhingra, *Life Behind the Lobby*, pp. 126–9.

100 Rajat Gupta came to the United States: Raghavan, *The Billionaire's Apprentice*, pp. 40–2, 46.

100 "oldest bloodlines" . . . "natural superiority": Ibid., p. 11.

100 passed over by every firm: Ibid., pp. 78, 82–4.

100 McKinsey's chief executive: Ibid., p. 123.

100 "'model minorities'" . . . "'real' minorities": Purkayastha, *Negotiating Ethnicity*, p. 93; see also Maira, *Desis in the House*, p. 72 ("[t]he anti-Black prejudices of South Asian immigrants are reinforced by the Black/White lines of American racial formations"); Vijay Prashad, *The Karma of Brown Folk* (Minneapolis and London: University of Minnesota Press, 2000), pp. 157, 178–9 (arguing that many Indian immigrants accept anti-black racism as part of a need to enhance their own foothold in their new country), pp. 97–8 (noting "obsession with skin color" prevalent in India); see also Mahalingam, Philip, and Balan, "Cultural Psychology and Marginality," pp. 160–62.

100 "better, smarter, more high-achieving": Purkayastha, *Negotiating Ethnicity*, p. 93.

100 Young South Asians who date African Americans: Maira, *Desis in the House*, pp. 71–2.

100 may be more insulated from American racism: See Prashad, *Uncle Swami*, p. 13.

100 "whiten" their complexion: See, e.g., Purkayastha, *Negotiating Ethnicity*, pp. 33–4.

101 "naked brown male bodies": E-mail to Amy Chua, Dec. 4, 2012 (on file with authors).

101 reconfigured through their confrontation with American society: Karen Leonard, "South Asian Religions in the United States: New Contexts and Configurations," in Rajan and Sharma, *New Cosmopolitanisms*, pp. 91, 94–6.

101 caste distinctions have become much less significant: John Y. Fenton, *Transplanting Religious Traditions: Asian Indians in America* (New York: Praeger, 1988), p. 34; Fisher, *The Indians of New York City*, pp. 52–3; Rhitu Chatterjee, "Beyond Class Part V—Indians in America— Caste Adrift," *The World*, Public Radio International, May 23, 2012, http://pri.org/ stories/2012-05-23/beyond-class-part-v-indians-america-caste-adrift?utm_source=rss&utm_ medium=rss&utm_campaign=india-caste-us; Joseph Berger, "Family Ties and the Entanglements of Caste," *New York Times*, Oct. 24, 2004. Even in India, the importance of caste varies greatly in different contexts and in different regions. It is very common for Indians in the United States (as in India) to interact and socialize across caste lines. For this reason, it would be both politically incorrect and inconsistent with common practice for an Indian American to openly express a belief that he or she is superior because of his or her caste background.

101 generally viewed as discriminatory: See, e.g., Chatterjee, "Beyond Class" (describing caste as "discriminatory and outdated").

101 across caste and regional divides: Rayaprol, *Negotiating Identities*, p. 76; Purkayastha, *Negotiating Ethnicity*, p. 97; Prashad, *Uncle Swami*, pp. 13–4; Fisher, *The Indians of New York City*, pp. 74–5.

101 "'superior culture' narrative": Purkayastha, *Negotiating Ethnicity*, pp. 89–90.

101 lowest out-marriage rate: Only 14 percent of Indian American newlyweds in 2008–10 married outside their ethnic group, compared to 64 percent Japanese Americans, 54 percent Filipinos, 39 percent Koreans, and 35 percent Chinese. See Pew Research Center, *The Rise of Asian Americans*, p. 106.

101 almost 70 percent of Indians: Ibid., p. 128.

102 emphasizing their Hinduism: Leonard, "South Asian Religions," p. 98; Purkayastha, *Negotiating Ethnicity*, pp. 96–9. Professor Purkayastha says that this Hindu emphasis is part of "a new ideology of superiority" among (Hindu) Indian Americans. Ibid., p. 97.

102 10 percent of Indian Americans who are Muslim: Percentages are from Pew Research

Center, *Asian Americans: A Mosaic of Faiths* (Washington, DC: Pew Research Center, 2012), p. 16. On Muslim Indian Americans, see also Aminah Mohammed-Arif, *Salaam America: South Asian Muslims in New York*, trans. Sarah Patey (London: Anthem Press, 2002), p. 33 (estimating that in 1990, there were approximately 80,000 Muslim Indian Americans out of a total Indian American population of 815,447). The Pew report found that 51 percent of Indian Americans identify themselves as Hindu and that 59 percent say they were raised Hindu. But see "So, How Many Hindus Are There in the US?," *Hinduism Today* (Jan., Feb., Mar., 2008), http://www.hinduism today.com/archives/2008/1-3/61_swadhyay%20pariwar.shtml (suggesting that 80 percent of Indian Americans are Hindu).

102 Hindu temple building . . . transformation: Purkayastha, *Negotiating Ethnicity*, p. 97.

102 also serve as social centers: Ibid., pp. 97–98.

102 feedback loop . . . "superior family/ethnic culture": Ibid., pp. 89, 91–7; see also Gupta, "Hidden in Plain Sight" (noting that Indian American culture is often self-depicted as the "'best of both worlds': American independence, determination and self-reliance coupled with Indian morals, religious beliefs and family values").

102 not the only Asians who experience racism: A 2012 Pew survey found that 21 percent of Chinese Americans, 20 percent of Korean Americans, 19 percent of Filipino Americans, 18 percent of Indian Americans, 14 percent of Vietnamese Americans, and 9 percent of Japanese Americans "have personally experienced discrimination." Pew Research Center, *The Rise of Asian Americans*, p. 114.

102 "Sometimes I'll glimpse": Wesley Yang, "Paper Tigers," *New York Magazine*, May 8, 2011.

103 "racially gendered stereotypes": Nancy Wang Yuen, "Performing Race, Negotiating Identity: Asian American Professional Actors in Hollywood," in Jennifer Lee and Min Zhou, eds., *Asian American Youth: Culture, Identity and Ethnicity* (New York and London: Routledge, 2004), pp. 251, 266.

103 "You're a quarterback": Yang, "Paper Tigers" (quoting Columbia Law Professor Tim Wu). On the male Asian American reaction to Jeremy Lin, see, e.g., Deanna Fei, "The Real Lesson of Linsanity," Huffington Post, Feb. 16, 2012, http://www.huffingtonpost.com/deanna-fei/jeremy-lin-asian-americans_b_1281916.html; Adrian Pei, "Jeremy Lin & Asian American Male Sexuality," Next Gener.Asian Church, Feb. 9, 2012, http://nextgenerasianchurch.com/2012/02/09/jeremy-lin-asian-american-male-sexuality.

103 Bullying of East Asian kids: See Jin Li, *Cultural Foundations of Learning: East and West* (New York: Cambridge University Press, 2012), pp. 214–16; Desiree Baolian Qin, Niobe Way, and Meenal Rana, "The 'Model Minority' and Their Discontent: Examining Peer Discrimination and Harassment of Chinese American Immigrant Youth," in Hirokazu Yoshikawa and Niobe Way, eds., *Beyond the Family: Contexts of Immigrant Children's Development*, no. 121 (2008), pp. 27, 29–36.

104 Yul Kwon's: Alexis Chiu and Cynthia Wang, "Master Strategist Yul Kwon Wins *Survivor*," *People*, Dec. 18, 2006; "Yul Kwon, from Bullying Target to Reality TV Star," NPR, May 15, 2012, http://www.npr.org/2012/05/16/152775069/yul-kwon-from-bullying-target-to-reality-tv-star.

104 "chink" or "gook" . . . "However improbable it might be": "Opinion: Red Chair Interview: Why Yul Kwon Ditched Law for TV," CNN in America, Nov. 16, 2011, http://inamerica.blogs.cnn.com/2011/11/16/red-chair-interview-yul-kwon.

104 "Fear of being persecuted and even murdered": Dennis Prager, "Explaining Jews, Part III: A Very Insecure People," Townhall Magazine, Feb. 21, 2006, http://townhall.com/columnists/dennisprager/2006/02/21/explaining_jews,_part_iii_a_very_insecure_people/page/full.

105 "You were born into anxiety": Daniel Smith, "Do the Jews Own Anxiety?," *New York Times*, Opinionator, May 26, 2012.

105 A history of persecution: Ancient Egypt enslaved its Jews (if the Bible is to be believed) and killed their newborn sons; Hadrian drove the Jews out of Jerusalem; England expelled its

Jews in 1290; France did so in 1306, 1322, and 1394; Germany's Jews were massacred in 1298, 1336–38, and 1348; Spain expelled them in 1492; Pope Pius V expelled them from the Papal States in 1569; over 50,000 Jews were killed in Poland between 1648 and 1654. And all this was before the rise of modern anti-Semitism. Chua, *Day of Empire*, pp. 38, 130, 134, 138; J. H. Elliott, *Imperial Spain, 1469–1716* (London: Edward Arnold, 1963), p. 98.

105 Franz Kafka: Daniel L. Medin, *Three Sons: Franz Kafka and the Fiction of J. M. Coetzee, Philip Roth, and W. G. Sebald* (Evanston, IL: Northwestern University Press, 2010), p. 22. In a similar vein, Walter Isaacson writes that "[l]iving as a Jew under the Nazis" left Henry Kissinger with an insecurity that fueled his ambition: "Confidence coexisting with insecurities, vanity with vulnerability, arrogance with a craving for approval: the complexities that were layered into Kissinger's personality as a young man would persist throughout his life." Walter Isaacson, *Kissinger: A Biography* (New York: Simon & Schuster, 1992), p. 56.

106 "That's a mother's prayer": Joyce Antler, *You Never Call! You Never Write! A History of the Jewish Mother* (New York: Oxford University Press, 2007), pp. 1, 135 (citing *An Evening with Mike Nichols and Elaine May*, original cast recording, Polygram Records, 1960).

107 "A Jewish girl becomes president": Paul Mazursky, in Abigail Pogrebin, ed., *Stars of David: Prominent Jews Talk About Being Jewish* (New York: Broadway Books, 2005), pp. 79–80, cited in Antler, *You Never Call!*, p. 5.

107 "construct developed by male writers": Martha A. Ravits, "The Jewish Mother: Comedy and Controversy in American Popular Culture," in Harold Bloom, ed., *Philip Roth's Portnoy's Complaint* (Philadelphia: Chelsea House, 2004), p. 163; Antler, *You Never Write!*, p. 145. For those who haven't read *Portnoy's Complaint*, Sophie Portnoy is the mother of the protagonist, Alexander Portnoy, and "the definitive article: she cleans up after the maid, worries endlessly about what goes into Alex and what comes out of him, and exists to protect him from gentiles and manhood." Christopher Lehmann-Haupt, "A Portrait of the Artist as a Young Jew," *New York Times*, Feb. 18, 1969.

107–8 "often the only thing [they] could clearly communicate": Robert Warshow, "Poet of the Jewish Middle Class," *Commentary*, May 1946, p. 20, quoted in Alexander Bloom, *Prodigal Sons: The New York Intellectuals & Their World* (New York and Oxford: Oxford University Press, 1986), p. 19.

108 "self-perception of failure": Bloom, *Prodigal Sons*, p. 18.

108 "forever disappointed in my father": Isaac Rosenfeld, *Passage from Home* (New York: Dial Press, 1946), p. 7.

108 "[W]hat *was* it with these Jewish parents": Philip Roth, *Portnoy's Complaint* (New York: Vintage Books, 1994), p. 119.

109 "worked for many years as a banquet bartender": Marco Rubio, *An American Son: A Memoir* (New York: Sentinel, 2012), p. 100; transcript of speech by Marco Rubio delivered at Republican National Convention, Aug. 30, 2012, http://www.politico.com/news/stories/0812/80493.html.

109 "Perceiving the sacrifices made by their parents": Rubén G. Rumbaut, "Children of Immigrants and Their Achievement: The Role of Family, Acculturation, Social Class, Gender, Ethnicity, and School Contexts," p. 1, http://www.hks.harvard.edu/inequality/Seminar/Papers/Rumbaut2.pdf.

110 almost a third: Ruth K. Chao, "Chinese and European American Mothers' Beliefs About the Role of Parenting in Children's School Success," *Journal of Cross-Cultural Psychology* 27, no. 4 (July 1996), pp. 403, 412.

110 "familial obligation and prestige": Vivian S. Louie, *Compelled to Excel: Immigration, Education, and Opportunity Among Chinese Americans* (Stanford, CA: Stanford University Press, 2004), p. 48. See also Peter H. Huang, "Tiger Cub Strikes Back: Memoirs of an Ex-Child Prodigy About Legal Education and Parenting," *British Journal of American Legal Studies* 1 (2012), pp.

297, 300–1. Quantitative psychological studies have confirmed that a sense of family obligation is more prevalent in Hispanic and Asian American families than in white American families and have found evidence that, in Asian American adolescents, a strong sense of family obligation acts as a buffer against the negative impact of low socioeconomic status on academic performance. See Lisa Kiang and Andrew J. Fuligni, "Ethnic Identity and Family Processes Among Adolescents from Latin American, Asian, and European Backgrounds," *Journal of Youth and Adolescence* 38, no. 2 (February 2009), pp. 228–41; Lisa Kiang et al., "Socioeconomic Stress and Academic Adjustment Among Asian American Adolescents: The Protective Role of Family Obligation," *Journal of Youth and Adolescence* 42, no. 6 (June 2013), pp. 837–47.

110 a child's best—perhaps only—protection: See, e.g., Louie, *Compelled to Excel*, pp. 54, 57, 60–1; Beloo Mehra, "Multiple and Shifting Identities: Asian Indian Families in the United States," in Clara C. Park, A. Lin Goodwin, and Stacey J. Lee, eds., *Asian American Identities, Families, and Schooling* (Greenwich, CT: Information Age Publishing, 2003), pp. 27, 45–6; Jamie Lew, "The Re(construction) of Second-Generation Ethnic Networks: Structuring Academic Success of Korean American High School Students," in Park et al., *Asian American Identities*, pp. 157, 166–7.

110 "Harvard #1!": See, e.g., Tony Hsieh, *Delivering Happiness: A Path to Profits, Passion, and Purpose* (New York and Boston: Business Plus, 2010), p. 8 ("Harvard yielded the most prestigious bragging rights"); Louie, *Compelled to Excel*, pp. 42, 107; Lew, "The Re(construction) of Second-Generation Ethnic Networks," p. 166; Tracy Jan, "Chinese Aim for the Ivy League," *New York Times*, Jan. 4, 2009.

110 "Why just an A": See, e.g., Jennifer Lee and Min Zhou, "Frames of Achievement and Opportunity Horizons," in David Card and Steven Raphael eds., *Immigration, Poverty, and Socioeconomic Inequality* (New York: Russell Sage Foundation, 2013), pp. 206, 216 ("Asian respondents described the value of grades on an Asian scale as 'A is for average, and B is an Asian fail' . . . the stakes have risen so that an A minus is now an Asian fail"); Louie, *Compelled to Excel*, p. 46 ("Why not 100?"), p. 109 ("she always expected 100s").

110 "Koreans . . . they take it to another level": Rebecca Y. Kim, *God's New Whiz Kids? Korean American Evangelicals on Campus* (New York and London: New York University Press, 2006), p. 79.

111 pointed comparisons: See, e.g., Lee and Zhou, "Frames of Achievement and Opportunity Horizons," p. 216; Louie, *Compelled to Excel*, pp. 97–8; see also Li, *Cultural Foundations of Learning*, p. 207; Jin Li, Susan D. Holloway, Janine Bempechat, and Elaine Loh, "Building and Using a Social Network: Nurture for Low-Income Chinese American Adolescents' Learning," in Yoshikawa and Way, *Beyond the Family*, pp. 9, 18.

111 "my parents thought I was a bad girl": Purkayastha, *Negotiating Ethnicity*, p. 92.

111 "average" but not "great" . . . "better if I was first or second": Lee and Zhou, "Frames of Achievement and opportunity Horizons," pp. 216, 217.

111 *lowest* self-esteem of any racial group: Douglas S. Massey et al., *The Source of the River: The Social Origins of Freshmen at America's Selective Colleges and Universities* (Princeton, NJ: Princeton University Press, 2003), pp. 120–1; Carl L. Bankston III and Min Zhou, "Being Well vs. Doing Well: Self-Esteem and School Performance Among Immigrant and Nonimmigrant Racial and Ethnic Groups," *International Migration Review* 36, no. 2 (2002), pp. 389–415.

112 "intensely proud" . . . "descendants of the ancient Phoenicians": Joseph J. Jacobs, *The Anatomy of an Entrepreneur: Family, Culture, and Ethics* (San Francisco: ICS Press, 1991), pp. 13–16, 21.

112 singled out the Phoenicians . . . alphabet, arithmetic, and glass: See Benjamin Isaac, *The Invention of Racism in Classical Antiquity* (Princeton, NJ: Princeton University Press, 2004), pp. 324–8; J.P.V.D. Balsdon, *Romans and Aliens* (London: Gerald Duckworth & Co., 1979), pp. 63, 67; Elsa Marston, *The Phoenicians* (New York: Benchmark Books, 2002), p. 60; Sabatino

Moscati, ed., *The Phoenicians* (New York: Abbeville Press, 1988), pp. 551–52, 558. In the elder Pliny's *Natural History*, a mammoth encyclopedia written in the 70s AD, he literally says, "The Phoenicians invented trade." See Isaac, *The Invention of Racism in Classical Antiquity*, pp. 324–8.

112 "greedy knaves" . . . "greed and luxury": Isaac, *The Invention of Racism in Classical Antiquity*, pp. 324–8.

112–3 "the original Christian disciples": Jacobs, *The Anatomy of an Entrepreneur*, p. 13–14.

113 "look up at the Christians": Kristine J. Ajrouch and Abdi M. Kusow, "Racial and Religious Contexts: Situational Identities Among Lebanese and Somali Muslim Immigrants," *Ethnic and Racial Studies* 30, no. 1 (2007), pp. 72, 83.

113 "Having no one to speak for us": Jacobs, *The Anatomy of an Entrepreneur*, pp. 16–7.

113 "need to please my mother": Ibid., pp. 29-35.

114 "I meant your *own* store!": Ibid., pp. 28–9.

114 "doubly driven to succeed": Ibid., p. 17.

114 self-esteem-centered . . . popular psychology: See, e.g., Lori Gottlieb, "How to Land Your Kid in Therapy," *Atlantic Monthly*, July/Aug. 2011 (questioning "self-esteem" that "comes from constant accommodation and praise rather than earned accomplishment"). We discuss the self-esteem movement in more detail in chapter 8.

CHAPTER 5: IMPULSE CONTROL

117 "I've missed more than 9,000 shots": Eric Zorn, "Without Failure, Jordan Would Be False Idol," *Chicago Tribune*, May 19, 1997.

117 large and growing body of research: See, e.g., Roy F. Baumeister and John Tierney, *Willpower: Rediscovering the Greatest Human Strength* (New York: Penguin Press, 2011); Carol S. Dweck, *Mindset: The New Psychology of Success* (New York: Random House, 2006); Angela L. Duckworth, Christopher Peterson, Michael D. Matthews, and Dennis Kelly, "Grit: Perseverance and Passion for Long-Term Goals," *Journal of Personality and Social Psychology* 92, no. 6 (2007), pp. 1087–1101. See also Kelly McGonigal, *The Willpower Instinct: How Self-Control Works, Why It Matters, and What You Can Do to Get More of It* (New York: Avery, 2012).

118 "marshmallow test": Walter Mischel, Ebbe B. Ebbeson, and Antonette Raskoff Zeiss, "Cognitive and Attentional Mechanisms in Delay of Gratification," *Journal of Personality and Social Psychology* 21, no. 2 (1972), pp. 204-18; Baumeister and Tierney, *Willpower*, pp. 9–11; Jonah Lehrer, "The Secret of Self-Control," *The New Yorker*, May 18, 2009.

118 Mischel followed up . . . doing much better academically: Walter Mischel, Yuichi Shoda, and Monica L. Rodriguez, "Delay of Gratification in Children," *Science* 244, no. 4907 (1989), pp. 933–8; Yuichi Shoda, Walter Mischel, and Philip K. Peake, "Predicting Adolescent Cognitive and Self-Regulatory Competencies from Preschool Delay of Gratification: Identifying Diagnostic Conditions," *Developmental Psychology* 26, no. 6 (1990), pp. 978–86; Inge-Marie Eigsti, Vivian Zayas, Walter Mischel et al., "Predicting Cognitive Control from Preschool to Late Adolescence and Young Adulthood," *Psychological Science* 17, no. 6 (2006), pp. 478–84.

118 numerous studies: Baumeister and Tierney, *Willpower*, pp. 10–3; McGonigal, *The Willpower Instinct*, p. 12.

118 better predictor than SAT scores: Baumeister and Tierney, *Willpower*, p. 11; Duckworth et al., "Grit," pp. 1098, 1099; Angela L. Duckworth and Martin E. P. Seligman, "Self-Discipline Outdoes IQ in Predicting Academic Performance of Adolescents," *Psychological Science* 16, no. 12 (2005), pp. 939–44; Angela L. Duckworth, Patrick D. Quinn, and Eli Tsukayama, "What *No Child Left Behind* Leaves Behind: The Roles of IQ and Self-Control in Predicting Standardized Achievement Test Scores and Report Card Grades," *Journal of Educational Psychology* 104, no. 2 (2012), pp. 439–45.

118 researchers in New Zealand tracked: Terrie E. Moffitt et al., "A Gradient of Childhood Self-Control Predicts Health, Wealth, and Public Safety," *Proceedings of the National Academy of*

Science of the United States of America 108, no. 7 (2011), pp. 2693–8. For a summary, see Baumeister and Tierney, *Willpower*, pp. 12–3.

119 Willpower and perseverance can be strengthened: See, e.g., Baumeister and Tierney, *Willpower*, pp. 11, 124–41; Dweck, *Mindset*, pp. 71–4; Mark Muraven, Roy F. Baumeister, and Dianne M. Tice, "Longitudinal Improvement of Self-Regulation Through Practice: Building Self Control Through Repeated Exercise," *Journal of Social Psychology* 139, no. 4 (1999), pp. 446–57; see also Heidi Grant Halvorson, *Succeed: How We Can Reach Our Goals* (New York: Plume, 2012), pp. xvii–xxi; M. Oaten and K. Cheng, "Improved Self-Control: The Benefits of a Regular Program of Academic Study," *Basic and Applied Social Psychology* 28, no. 1 (2006), pp. 1–16; Megan Oaten and Ken Cheng, "Longitudinal Gains in Self-Regulation from Regular Physical Exercise," *British Journal of Health Psychology* 11, no. 4 (2006), pp. 717–33; Lori Gottlieb, "How to Land Your Kid in Therapy," *The Atlantic*, July/August 2011.

119 Blaine's feats: John Tierney, "If He Starts Nodding Off, Try Another Million Volts," *New York Times*, Oct. 1, 2012; "David Blaine Nears Final Hours of 'Shocking' Stunt," *USA Today*, Oct. 8, 2012.

119 "[F]or once": Chris Britcher, "David Blaine's Latest Trick: Making Mountains Out of Molehills," Kentnews.co.uk, Oct. 8, 2012, http://www.kentnews.co.uk/blogs/david_blaine_s_latest_trick_making_mountains_out_of_molehills_1_1592494.

119 "hunger artist": Franz Kafka, *A Hunger Artist and Other Stories*, trans. Joyce Crick (New York: Oxford University Press, 2012).

120 two-thirds of today's Chinese Americans: U.S. Census, American Community Survey, Table S0201: Selected Population Profile in the United States (2010 3-year dataset) (population group code 016 – Chinese) (showing that 69 percent of Chinese Americans are foreign-born).

120 China's massive superiority complex: See, e.g., John K. Fairbank, "China's Foreign Policy in Historical Perspective," *Foreign Affairs* 47, no. 3 (1969), pp. 449, 456–63; Q. Edward Wang, "History, Space, Ethnicity: The Chinese Worldview," *Journal of World History* 10, no. 2 (1999), pp. 285, 287–8, 291. See also Yingjie Guo, *Cultural Nationalism in Contemporary China* (New York: RoutledgeCurzon, 2004), pp. 83–4 (tracing emergence of "Chinese superiority complex" at least as far back as Confucius); Larry Clinton Thompson, *William Scott Ament and the Boxer Rebellion: Heroism, Hubris and the "Ideal Missionary"* (Jefferson, NC: McFarland & Co., 2008), p. 67 (referring to the "overweening Chinese superiority complex"); Unryu Suganuma, *Sovereign Rights and Territorial Space in Sino-Japanese Relations: Irredentism and the Diaoyu/Senkaku Islands* (Honolulu: University of Hawaii Press, 2000), p. 15 ("Chinese superiority complex").

120 combined to make China an extreme case: see Amy Chua, *Day of Empire: How Hyperpowers Rise to Global Dominance—and Why They Fall* (New York: Doubleday, 2007), p. 62.

121 "Since ancient China began as a culture island": Fairbank, "China's Foreign Policy in Historical Perspective," p. 456; Wang, "History, Space, and Ethnicity," pp. 287–8.

121 "barbarians": Wang, "History, Space, and Ethnicity," pp. 287–8.

121 "tribute" to China: J. K. Fairbank and S. Y. Têng, "On the Ch'ing Tributary System," *Harvard Journal of Asiatic Studies* 6, no. 2 (1941), pp. 135, 148–57, 182–90; see also Chua, *Day of Empire*, pp. 79–81.

121 Mongols . . . ultimately "sinicized": Fairbank, "China's Foreign Policy in Historical Perspective," pp. 456–8; see also Paul Heng-chao Ch'en, *Chinese Legal Tradition Under the Mongols: The Code of 1291 as Reconstructed* (Princeton, NJ: Princeton University Press, 1979); Ping-Ti Ho, "In Defense of Sinicization: A Rebuttal of Evelyn Rawksi's 'Reenvisioning the Qing,'" *Journal of Asian Studies* 57, no. 1 (1998), pp. 123, 141.

121 China rose to heights unprecedented: Chua, *Day of Empire*, pp. 178–81; Paul Kennedy, *The Rise and Fall of the Great Powers* (New York: Vintage Books, 1989), pp. 4–9; Gavin Menzies, *1421: The Year China Discovered America* (New York: HarperCollins, 2003), pp. 45, 52, 63, 70; Leo Suryadinata, ed., *Admiral Zheng He and Southeast Asia* (Singapore: Institute of Southeast

Asian Studies, 2005), p. 150 (contemporary Ibn Battuta reported seeing "1000 men on board" vessels).

122 "plates" of stale bread: Chua, *Day of Empire*, p. 178; Menzies, *1421*, p. 63.

122 "barbarians that send tribute": Fairbank and Têng, "On the Ch'ing Tributary System," pp. 182–5; see also Chua, *Day of Empire*, p. 80.

122 "used to take every opportunity": Philip Jia Guo, *On the Move: An Immigrant Child's Global Journey* (New York: Whittier Publications, 2007), p. 93. In passing down a sense of Chinese identity and pride from one generation to the next, parents are not the only agents. Chinese language schools, which children attend after regular school or on weekends, also play a critical role. Sociologist Min Zhou has been a pioneer on the subject of such ethnic institutions as "mediating grounds" and "cultural centers" where "traditional values and ethnic identity are nurtured." Originally located mostly in West Coast Chinatowns, these Chinese schools have exploded in number over the last several decades. As of 2006, there were 95 such schools in the Los Angeles area alone. In addition to Mandarin language instruction, these schools teach young children Chinese history, geography, painting, and calligraphy, as well as badminton, Chinese chess, kung fu, cooking, and dragon dance. Children memorize Chinese sayings, recite Chinese poems, and quote Confucian sayings. They are taught to write phrases like "I am Chinese" and "My ancestral country is in China." Many kids will eventually drop out of Chinese school, protesting that it's boring and that their parents forced them to attend. Nevertheless, as Zhou writes, most children of Chinese immigrants have attended a Chinese language school "at some point in their preteen years," and it is often "a definitive ethnic affirming experience." It's important to note that Chinese school is different from—and usually piled on top of—academic tutoring or music lessons. A primary function of Chinese school is to "nurture ethnic identity and pride that may otherwise be rejected by the children because of the pressure to assimilate." See Min Zhou, "Negotiating Culture and Ethnicity: Intergenerational Relations in Chinese Immigrant Families in the United States," in Ramaswami Mahalingam, ed., *Cultural Psychology of Immigrants* (Mahwah, NJ: Lawrence Erlbaum Associates Publishers, 2006), pp. 328–32; Min Zhou, "The Ethnic System of Supplementary Education: Nonprofit and For-Profit Institutions in Los Angeles' Chinese Immigrant Community," in Marybeth Shinn and Hirokazu Yoshikawa, eds. *Toward Positive Youth Development: Transforming School and Community Programs* (New York: Oxford University Press, 2008), pp. 229–51, especially pp. 234–8, 242; the Los Angeles statistic is from p. 237.

123 "[I]n our house, everything important in life came from China": Andrea Jung, "International Conference Keynote," *Business Today* (Spring 2009), p. 20.

123 date to the fifteenth century: Kennedy, *The Rise and Fall of the Great Powers*, pp. 3–7.

123 shame and humiliation: See Orville Schell and John DeLury, *Wealth and Power: China's Long March to the Twenty-First Century* (New York: Random House, 2013), pp. 6–7; Suisheng Zhao, "'We are Patriots First and Democrats Second': The Rise of Chinese Nationalism in the 1990s," in Edward Friedman and Barrett L. McCormick, eds., *What If China Doesn't Democratize? Implications for War and Peace* (New York: M. E. Sharpe, 2000), pp. 25, 42.

123 NO DOGS AND CHINESE ALLOWED: See Stella Dong, *Shanghai: The Rise and Fall of a Decadent City* (New York: HarperCollins, 2000), p. 198. Dong adds, however, that "those with sharp memories say no such sign existed: instead, two notices were posted on the gate, one reading "No Dogs Allowed," and the other, "Only For Foreigners").

123 massacring . . . and raping: Iris Chang, *The Rape of Nanking: The Forgotten Holocaust of World War II* (New York: Basic Books, 1997), p. 6.

123 "Never will China be humiliated": Jean-Pierre Lehmann, "Learning from China's Past," Forbes.com (Oct. 1, 2009), http://www.forbes.com/2009/10/01/china-history-60-anniversary-opinions-contributors-jean-pierre-lehmann.html. See also Zhao, "'We Are Patriots First and Democrats Second,'" p. 42.

123 "twin burdens:" Vivian S. Louie, *Compelled to Excel: Immigration, Education, and Opportunity Among Chinese Americans* (Stanford, CA: Stanford University Press, 2004), p. 60.

124 "you won't have to work like me in the restaurant": Ibid., p. 54.

124 "we don't speak English that clearly": Ibid., pp. 58–9.

124 only 24 percent of Chinese Americans: Pew Research Center, *The Rise of Asian Americans* (Washington, DC: Pew Research Center, April 4, 2013) (updated edition), pp. 110, 114.

124 "work too hard": See Chiung Hwang Chen, "'Outwhiting the Whites': An Examination of the Persistence of Asian American Model Minority Discourse," in Rebecca Ann Lind, ed., *Race/Gender/Media: Considering Diversity Across Audiences, Content, and Producers* (Boston: Allyn & Bacon, 2004), pp. 146, 149; see also Suein Hwang, "The New White Flight," *Wall Street Journal*, Nov. 19, 2005, p. A1.

124 anti-Chinese animus: See Matthew Yglesias, "White People Think College Admissions Should Be Based on Test Scores, Except When They Learn Asians Score Better Than Whites," Slate.com, Aug. 13, 2013, http://www.slate.com/blogs/moneybox/2013/08/13/white_people_s_meritocracy_hypocrisy.html; Chen, "'Outwhiting the Whites,'" pp. 146–9; Lee Siegel, "Rise of the Tiger Nation," *Wall Street Journal*, Oct. 27, 2012, p. C1; Ian Lovett, "U.C.L.A. Student's Video Rant Against Asians Fuels Firestorm," *New York Times*, Mar. 16, 2011, p. A21; Hwang, "The New White Flight," p. A1.

124 "so driven to prove her wrong": E-mail to Amy Chua, Nov. 14, 2012 (on file with authors).

125 "I'm going to have to prove myself more": Ben Golliver, "Jeremy Lin: Bias Provides 'Chip on the Shoulder,'" CBS Sports, Feb. 24, 2012, http://www.cbssports.com/mcc/blogs/entry/22748484/34980303.

125 "you have to be smarter than other people": Louie, *Compelled to Excel*, p. 61.

125 "It's better that you're taught the truth": Anchee Min, *The Cooked Seed* (New York: Bloomsbury, 2013), p 343.

125 "You have to work harder": Louie, *Compelled to Excel*, p. 61.

125 Confucian "learning virtue": Jin Li, *Cultural Foundations of Learning: East and West* (Cambridge, UK: Cambridge University Press, 2012), pp. 49–50, 63–5, 91–2, 139–42; Nicholas D. Kristof, "China Rises, and Checkmates," *New York Times*, Jan. 9, 2011.

126 A cultural chasm separates "learning should be fun": See, e.g., Li, *Cultural Foundations of Learning*, pp. 111–2, 258–60, 267–8; Ruth K. Chao, "Chinese and European Mothers' Beliefs About the Role of Parenting in Children's School Success," *Journal of Cross-Cultural Psychology* 27, no. 4 (July 1996), pp. 403–23. Cf. Gish Jen, *Tiger Writing: Art, Culture, and the Interdependent Self* (Cambridge, MA, and London: Harvard University Press, 2013), p. 48 ("I am struck every year by how consistently President Drew Faust's addresses to Harvard freshmen emphasize free exploration and playfulness—an emphasis appropriate to a stable, egalitarian, individualistic society. . . . The traditional Chinese template . . . was geared toward attaining safety and social standing in a dangerous, interdependent, hierarchical world").

126 you'll find students sitting upright: See Li, *Cultural Foundations of Learning*, pp. 124–5, 129, 145–6; Carmina Brittain, "Sharing the Experience: Transnational Information About American Schools Among Chinese Immigrant Students," in Clara C. Park, A. Lin Goodwin, and Stacey J. Lee, eds., *Asian American Identities, Families, and Schooling* (Greenwich, CT: Information Age Publishing, 2003), pp. 177, 189–90; Hywel Williams, "When It Comes to Education Can Britain Be the Singapore of the West?," Mail Online, Jan. 23, 2012, http://www.dailymail.co.uk/debate/article-2089507/When-comes-education-Britain-Singapore-West.html.

126 Calligraphy: Li, *Cultural Foundations of Learning*, p. 131.

126 hours of additional study and tutoring: Ibid., pp. 167–70; Amy Chua, "Tigress Tycoons," *Newsweek*, Mar. 12, 2012, pp. 30, 34–5; Chuing Prudence Chou and James K. S. Yuan, "Buxiban in Taiwan," *The Newsletter*, no. 56 (Spring 2011), p. 15, http://www.iias.nl/sites/default/files/IIAS_NL56_15_0.pdf.

126 "You need to make more effort, not be so lazy": Li, *Cultural Foundations of Learning*, pp. 258–60.

126 "little emperors": "Thirty years into China's one-child policy, many are concerned about the prospect of letting a hundred spoiled brats bloom. . . . But China's 'little emperors' are coddled in a distinctly Chinese way. While doted on and catered to, they are also loaded up with the expectations of parents who have invested all their dreams—not to mention money—in their only child. These 'spoiled' children often study and drill from 7 a.m. to 10 p.m. every day." Chua, "Tigress Tycoons," *Newsweek*, pp. 30, 34–5.

126 Chinese immigrants parent far more strictly: See, e.g., Louie, *Compelled to Excel*, pp. 42–5; Chao, "Chinese and European Mothers' Beliefs," pp. 403–23; Paul E. Jose, Carol S. Huntsinger, Phillip R. Huntsinger, and Fong-Ruey Liaw, "Parental Values and Practices Relevant to Young Children's Social Development in Taiwan and the United States," *Journal of Cross-Cultural Psychology* 31, no. 6 (November 2000), pp. 677–702; Richard R. Pearce, "Effects of Cultural and Social Structural Factors on the Achievement of White and Chinese American Students at School Transition Points," *American Education Research Journal* 43, no. 1 (Spring 2006), pp. 75–101; Robert D. Hess, Chang Chih-Mei, and Teresa M. McDevitt, "Cultural Variations in Family Beliefs About Children's Performance in Mathematics: Comparisons Among People's Republic of China, Chinese American, and Caucasian American Families," *Journal of Educational Psychology* 79, no. 2 (1987), pp. 179–88; Wenfan Yan and Qiuyun Lin, "Parent Involvement and Mathematics Achievement: Contrast Across Racial and Ethnic Groups," *Journal of Educational Research* 99, no. 2 (2005), pp. 118, 120; Parminder Parmar et al., "Teacher or Playmate: Asian Immigrant and Euro-American Parents' Participation in Their Young Children's Daily Activities," *Social Behavior and Personality* 36, no. 2 (2008), pp. 163–76.

127 Juilliard Pre-College students: See Grace Wang, "Interlopers in the Realm of High Culture: 'Music Moms' and the Performance of Asian and Asian American Identities," *American Quarterly* 61, no. 4 (2009), pp. 894, 901 n. 6; Joseph Kahn and Daniel Wakin, "Increasingly in the West, the Players Are from the East," *New York Times*, Apr. 4, 2007. See also Mari Yoshihara, *Musicians from a Different Shore: Asians and Asian Americans in Classical Music* (Philadelphia: Temple University Press, 2007), pp. 2–3, 53.

127 "focused activity": Jose et al., "Parental Values and Practices," pp. 689 (Table 4), 690.

127 one-third less television: Ibid.; see also Pearce, "Effects of Cultural and Social Structural Factors," pp. 75, 81, 89 (table 4).

127 Asian kids are more likely: Laurence Steinberg, *Beyond the Classroom: Why School Reform Has Failed and What Parents Need to Do* (New York: Simon & Schuster, 1996), pp. 91–4.

127 extra work: Chao, "Chinese and European Mothers' Beliefs," p. 410; Louie, *Compelled to Excel*, pp. 42–5. See generally Zhou, "Negotiating Culture and Ethnicity," pp. 326–7; Min Zhou, "Assimilation the Asian Way," in Tamar Jacoby, ed., *Reinventing the Melting Pot: The New Immigrants and What It Means to Be American* (New York: Basic Books, 2003), pp. 146–51.

127 social skills and self-esteem: Chao, "Chinese and European Mothers' Beliefs," especially pp. 408–10.

128 "an hour per instrument per day": Tony Hsieh, *Delivering Happiness: A Path to Profits, Passion, and Purpose* (New York and Boston: Business Plus, 2010), pp. 7–9.

128 One recent study: Su Yeong Kim et al., "Does 'Tiger Parenting' Exist? Parenting Profiles of Chinese Americans and Adolescent Developmental Outcomes," *Asian American Journal of Psychology* 4, no. 1 (2013), pp. 7–18. "[I]t may be," the authors of the study acknowledge, "that the parents identified as supportive in the current study would no longer be identified as supportive if they were part of a sample that included European American families." Ibid., p. 16.

128 strict parenting is "uncommon" . . . "supportive" parenting is the norm: See, e.g., Lindsay Abrams, "The Queen Bee's Guide to Parenting," *The Atlantic*, Apr. 2013; Susan Adams, "Tiger Moms Don't Raise Superior Kids, Says New Study," *Forbes*, May 8, 2013, http://www.forbes.com/sites/susanadams/2013/05/08/tiger-moms-dont-raise-superior-kids-says-new-study.

The same study was further reported to have shown that "tiger parenting doesn't work"—that the "supportive" parents had children with better academic grades and fewer adjustment problems. See, e.g., Adams, "Tiger Moms." But once again, the study's methodology makes it difficult to draw conclusions. Because the study looked only at Chinese American households, the parents who were classified as "supportive" (and whose children were found to have better outcomes) could very well have been engaging in what was, relative to American norms, very strict parenting—strong on discipline, high in expectations, less concerned with raising self-esteem through praise, and so on. The study's lead author, Su Yeong Kim, acknowledged in an interview that the Chinese parents classified as "supportive" in the study engaged in practices that most white American "supportive" parents probably don't engage in (and might not consider supportive): for example, "shaming" their children, emphasizing "filial obligation," making their children feel that achieving is necessary to preserve family "honor," and constantly reminding them of parental "sacrifices." Kim also said that this "may be the key to why some of these kids are doing well scholastically." Jeff Yang, "Tiger Babies Bite Back," *The Wall Street Journal*, May 14, 2013, http://blogs.wsj.com/speakeasy/2013/05/14/tiger-babies-bite-back.

Moreover, notwithstanding media reports, Kim's paper does not in fact make the claim that "supportive" parenting *caused* better academic and psychological outcomes among her subjects, or that harsher parenting *caused* worse outcomes. The study had no way of measuring causation and carefully states its findings in terms of *"association,"* not causation. Kim et al., "Does 'Tiger Parenting' Exist?", pp. 7, 10, 12–3. In other words, the true cause-and-effect might have been the reverse of what the media reported: parents in the study may have adopted harsher disciplinary measures because their kids' grades dropped or because the kids began engaging in problematic behavior.

128 Studies also confirm: See Zhou, "Negotiating Culture and Ethnicity," 327–8; Louie, *Compelled to Excel*, p. 43; see also Rebecca Y. Kim, *God's New Whiz Kids? Korean American Evangelicals on Campus* (New York and London: New York University Press, 2006), pp. 79–80.

128 "sleepovers" . . . "for the longest time": Louie, *Compelled to Excel*, pp. 45–6.

128 "I have a friend": Ibid., p. 43.

129 intergenerational conflict is a frequent theme: See Zhou, "Negotiating Culture and Ethnicity," p. 328; Louie, *Compelled to Excel*, pp. 43, 45–6.

129 the consistent finding . . . "educational advantage": See, e.g., Steinberg, *Beyond the Classroom*, pp. 85–7; Pearce, "Effects of Cultural and Social Structural Factors," pp. 94–5; see also Louie, *Compelled to Excel*, p. 47 (quoting a respondent as saying "I felt I had an advantage because I was Chinese because I felt that I was coming from a better work ethic").

129 sense of the superiority of the Chinese work ethic: See, e.g., Louie, *Compelled to Excel*, pp. 39–40, 46–8; Wang, "Interlopers in the Realm of High Culture," pp. 892, 894, 896.

129 "American kids, they do not have the discipline": Wang, "Interlopers in the Realm of High Culture," p. 896; see also Louie, *Compelled to Excel*, pp. 47–8 (quoting a respondent as saying that unlike "Hispanics, blacks, whites" she could "force myself to study" even when she was tired).

129 "most American parents": Pew Research Center, "The Rise of Asian Americans," p. 135.

130 "American parents, if they too hard have to sacrifice": Wang, "Interlopers in the Realm of High Culture," p. 892.

130 "There wasn't much praise": E-mail to Amy Chua, July 2, 2012 (on file with authors). Zappos founder Tony Hsieh described a similar upbringing:

My parents were your typical Asian American parents . . . They had high expectations in terms of academic performance for myself as well as for my two younger brothers. . . . There weren't a lot of Asian families living in Marin County, but somehow my parents managed to find all ten of them, and we would have regular gatherings . . . The kids would watch TV while the adults were in a separate room socializing and bragging to each other about their kids' accomplishments. . . the children had to perform [either piano or violin] in front of the group of parents after dinner was over.

This was ostensibly to entertain the parents, but really it was a way for parents to compare their kids with each other.

Hsieh, *Delivering Happiness*, pp. 7–8; see also Li, *Cultural Foundations of Learning*, p. 207; Jin Li, Susan D. Holloway, Janine Bempechat, and Elaine Loh, "Building and Using a Social Network: Nurture for Low-Income Chinese American Adolescents' Learning," in Hirokazu Yoshikawa and Niobe Way, eds., *Beyond the Family: Contexts of Immigrant Children's Development* no. 121 (2008), pp. 9, 18.

131 conventional forms of prestigious achievement: Louie, *Compelled to Excel*, pp. 42, 107, 136–7; Zhou, "Negotiating Culture and Ethnicity, pp. 326–7.

131 first-generation Korean and Indian Americans: see Baumeister and Tierney, *Willpower*, pp. 194–5; Dr. Soo Kim Abboud and Jane Kim, *Top of the Class: How Asian Parents Raise High Achievers—and How You Can Too* (New York: Berkeley Books, 2006); see Clara C. Park, "Educational and Occupational Aspirations of Asian American Students," in Park, Goodwin, and Lee, *Asian American Identities, Families, and Schooling*, pp. 135, 148 (study of 978 white, Chinese, Japanese, Korean, and other high school students in southern California showed that "Korean students had the highest educational aspirations among all groups" as well as "the highest perceived parental influence").

131 Preet Bharara: Anita Raghavan, *The Billionaire's Apprentice: The Rise of the Indian-American Elite and the Fall of the Galleon Hedge Fund* (New York and Boston: Business Plus, 2013), p. 361.

131 "I remember when I was learning decimals": E-mail from South Asian young man to Amy Chua, Apr. 7, 2011 (on file with authors); see also Pew Research Center, *The Rise of Asian Americans*, p. 46 ("Indian Americans stand out from most other U.S. Asian groups in the personal importance they place on parenting. 78% of Indian Americans say being a good parent is one of the most important things to them personally"); Parag Khanna, "Confessions of a Tiger Dad," Huffington Post, July 31, 2013, http://www.huffingtonpost.com/parag-khanna/confessions-of-a -tiger-dad_b_3682869.html.

132 "My parents were obsessed": E-mail from Indian American young woman to Amy Chua, May 15, 2012 (on file with authors); see also Bandana Purkayastha, *Negotiating Ethnicity: Second Generation South Asian Americans Traverse a Transnational World* (New Brunswick, NJ: Rutgers University Press, 2005), pp. 91–3; Nitya Ramanan, "Raising an Indian American Teen," India Currents, June 4, 2012, http://www.indiacurrents.com/articles/2012/06/04/ raising-indian-american-teen, pp. 140–1.

132 "your job is to study" . . . "Going to the mall was forbidden": Patricia Ngozi Anekwe, *Characteristics and Challenges of High Achieving Second-Generation Nigerian Youths in the United States* (Boca Raton, FL: Universal Publishers, 2008), pp. 140–1 (quoting interviewees).

132 "At school you talk": Kim, *God's New Whiz Kids?*, p. 79; see also Ramanan, "Raising an Indian American Teen."

133 "place greater expectations on children": Carl L. Bankston III and Min Zhou, "Being Well vs. Doing Well: Self-Esteem and School Performance Among Immigrant and Nonimmigrant Racial and Ethnic Groups," *International Migration Review* 36, no. 2 (2002), pp. 389, 393, 395; Park, "Educational and Occupational Aspirations of Asian American Students," pp. 148–53; see also John U. Ogbu and Herbert D. Simons, "Voluntary and Involuntary Minorities: A Cultural-Ecological Theory of School Performance with Some Implications for Education," *Anthropology & Education Quarterly* 29, no. 2 (1998), pp. 155, 172–3, 176; Lingxin Hao and Melissa Bonstead-Bruns, "Parent-Child Differences in Educational Expectations and the Academic Achievement of Immigrant and Native Students," *Sociology of Education* 71, no. 3 (1998), pp. 175–98.

133 dramatically lower rates of drug use: U.S. Department of Health and Human Services, *Results from the 2011 National Survey on Drug Use and Health: Summary of National Findings*, NSDUH Series H-44, HHS Publication No. (SMA) 12-4713 (Rockville, MD: Substance Abuse and Mental Health Services Administration, 2012), figs. 2.11, 3.2; Li-Tzy Wu et al., "Racial/ Ethnic Variations in Substance-Related Disorders Among Adolescents in the United States," *Archives of General Psychiatry* 68, no. 11 (2011), p. 1179.

133 lowest rates of teenage childbirth: Centers for Disease Control and Prevention, "Birth Rates for U.S. Teenagers Reach Historic Lows for All Age and Ethnic Groups," NCHS Data Brief, April 2012, fig. 3.

133 so highly correlated with adverse economic outcomes: Rubén G. Rumbaut, "Paradise Shift: Immigration, Mobility and Inequality in Southern California," Working Paper No. 14 (Vienna: Austrian Academy of Sciences, KMI Working Paper Series, October 2008) p. 5.

133 Impulse control is like stamina: Baumeister and Tierney, *Willpower*, pp. 129–41; see also Mark Muraven, Roy F. Baumeister, and Dianne M. Tice, "Longitudinal Improvement of Self-Regulation Through Practice: Building Self-Control Strength Through Repeated Exercise," *Journal of Social Psychology* 139, no. 4 (1999), pp. 446–57; see also Heidi Grant Halvorson, *Succeed: How We Can Reach Our Goals* (New York: Plume, 2012), pp. xvii–xxi; Megan Oaten and Ken Cheng, "Improved Self-Control: The Benefits of a Regular Program of Academic Study," *Basic and Applied Social Psychology* 28, no. 1 (2006), pp. 1–16; Megan Oaten and Ken Cheng, "Longitudinal Gains in Self-Regulation from Regular Physical Exercise," *British Journal of Health Psychology* 11, no. 4 (2006), pp. 717–33.

134 "health code": Claudia L. Bushman, *Contemporary Mormonism: Latter-day Saints in Modern America* (Westport, CT, and London: Praeger, 2006), p. 20–2.

134 "sin exceeded in seriousness only by murder": Tim B. Heaton, Kristen L. Goodman, and Thomas B. Holman, "In Search of a Peculiar People: Are Mormon Families Really Different?," in Marie Cornwall, Tim B. Heaton, and Lawrence A. Young, eds., *Contemporary Mormonism: Social Science Perspectives* (Chicago: University of Illinois Press, 1994), pp. 87, 100; see also Bushman, *Contemporary Mormonism*, p. 22.

134 three hours at church . . . "seminary": Bushman, *Contemporary Mormonism*, p. 30.

134 National Study of Youth and Religion: Ibid., p. 47.

134 Missionary Training Center: Ibid., p. 62.

135 ten to fourteen hours a day: Ibid., pp. 59–61, 63.

135 constant rebuffs, rejections: Jeff Benedict, *The Mormon Way of Doing Business: How Nine Western Boys Reached the Top of Corporate America* (New York: Business Plus, 2007), pp. 14–5.

135 "The thing a mission does is teach you persistency": Ibid., p. 20 (quoting Gary Crittenden).

135 "companionship" requirement: Keith Parry, "The Mormon Missionary Companionship," in Cornwall, Heaton, and Young, *Contemporary Mormonism*, pp. 182–206; Bushman, *Contemporary Mormonism*, p. 64.

135 "The idea of having a companion": Parry, "The Mormon Missionary Companionship," p. 183.

135 not playing golf: Benedict, *The Mormon Way of Doing Business*, pp. 51, 60–1.

135 Family Home Evening: Bushman, *Contemporary Mormonism*, pp. 44–5.

136 "gods in embryo": Monte S. Nyman, *28 Truths Taught by the Book of Mormon* (San Clemente, CA: Sourced Media Books, 2011), p. 56.

136 "sea of moral decay": Bushman, *Contemporary Mormonism*, p. 35.

136 "creepy" (as Mitt Romney's sons were repeatedly described): See, e.g., "Which Romney Son Is Creepiest?," Gawker.com, http://gawker.com/5953005/which-romney-son-is-creepiest.

136 do "not regard the Mormon chuch": Bushman, *Contemporary Mormonism*, p. 23.

136 2006 South Carolina poll: Richard N. Ostling and Joan K. Ostling, *Mormon America: The Power and the Promise* (New York: HarperOne, 2007), p. xiv.

136 "peculiar people": Ostling and Ostling, *Mormon America*, p. 185; see also Heaton, Goodman, and Holman, "In Search of a Peculiar People," pp. 87–117.

136 "cognitive dissonance": Armand L. Mauss, "Refuge and Retrenchment: The Mormon Quest for Identity," in Cornwall, Heaton, and Young, *Contemporary Mormonism*, pp. 24, 36–7.

136 clean-cut, all-American: Mitt Romney's five sons are sometimes perceived as "too strapping, too wholesome, and too perfect somehow," particularly "in an age when complicated, messy families increasingly seem like the new normal." Ashley Parker, "Romney Times Four," *New York Times*, Jan. 6, 2012.

136 Joseph Smith ran for president: Ostling and Ostling, *Mormon America*, p. xiii.

136 Orrin Hatch: Bushman, *Contemporary Mormonism*, p. 16.

137 vacillated between assimilation and retrenchment: Armand L. Mauss, *The Angel and the Beehive: The Mormon Struggle with Assimilation* (Chicago: University of Illinois Press, 1994), p. 5; Ostling and Ostling, pp. xviii, xx–xxvi.

137 *Twilight* books: The wildly popular *Twilight* series was written by Mormon author Stephenie Meyer, and the Mormon symbolism, which the author says is overplayed, has been much discussed. See, e.g., Angela Aleiss, "Mormon Influence, Imagery Run Deep Through 'Twilight,'" Huffington Post, June 24, 2010, http://www.huffingtonpost.com/2010/06/24/mormon -influence-imagery_n_623487.html.

137 "As somebody who grew up in Utah": Walter Kirn, "Mormons Rock!," *Daily Beast*, June 5, 2011 (quoting Dave Checketts).

137 "A big part of my drive": Benedict, *The Mormon Way of Doing Business*, p. 19 (quoting Dave Checketts).

137 "one of the most misunderstood organizations": Ibid., p. xiii (quoting David Neeleman).

137 "It's all about doing better than everyone": Ibid., p. 24 (quoting David Neeleman).

137 divine favor: James Carroll, "The Mormon Arrival," *Boston Globe*, Aug. 7, 2011.

137 "Puritan anachronism": Harold Bloom, *The American Religion: The Emergence of the Post-Christian Nation* (New York: Simon & Schuster, 1992), p. 103.

138 Asceticism has never loomed large in Judaism: See, e.g., Louis Ginzberg, "Israel Salanter," in Jacob Neusner, ed., *Understanding Rabbinic Judaism, from Talmudic to Modern Times* (New York: KTAV Publishing, 1974), pp. 355, 378 ("there can be no doubt as to the correctness of the view that Judaism is not an ascetic religion"); George Robinson, *Essential Judaism: A Complete Guide to Beliefs, Customs, and Rituals* (New York: Pocket Books, 2000), p. 84 ("Judaism is most decidedly not an ascetic religion"); but for dissenting voices, see Eliezer Diamond, *Holy Men and Hunger Artists: Fasting and Asceticism in Rabbinic Culture* (New York: Oxford University Press, 2004), pp. 5–6 ("Though I have heard over and over again that Judaism is not an ascetic faith, experience teaches me otherwise"); James A. Montgomery, "Ascetic Strains in Early Judaism," *Journal of Biblical Literature* 51, no. 3 (1932), pp. 183–213.

138 613 to be precise: See Ronald L. Eisenberg, *The 613 Mitzvot: A Contemporary Guide to the Commandments of Judaism* (Rockville, MD: Schreiber Publishing, 2005), p. xxi; Robinson, *Essential Judaism*, pp. 196–219.

138 law to bind the Jewish people: Jews traditionally consider the revelation of the law—the Torah—to Moses on Mt. Sinai as the "defining event in the history of Judaism." Wayne D. Dosick, *Living Judaism: The Complete Guide to Jewish Belief, Tradition, and Practice* (New York: HarperCollins, 1995), p. 177. For one of the leading twentieth-century treatments of Jewish law, see Menachem Elon, *Jewish Law: History, Sources, Principles* (Philadelphia: Jewish Publication Society, 1994).

138 "Who is strong?": Erica Brown, *Spiritual Boredom: Rediscovering the Wonder of Judaism* (Woodstock, VT: Jewish Lights Publishing, 2009), p. 84 (quoting *Pirkei Avot*, 4:1). Josephus, the Jewish scholar of ancient Rome, "stresse[d] that gentile religious practices lead to a lack of self-control," as compared to the "great discipline the Jewish law requires." Stanley K. Stowers, *A Rereading of Romans: Justice, Jews, and Gentiles* (New Haven, CT: Yale University Press, 1994), p. 64.

139 "and the organs below the belly": Maren Niehoff, *Philo on Jewish Identity and Culture* (Tübingen: Mohr Siebeck, 2001), p. 94 (quoting Philo, 2 Spec. 195); Stowers, *A Rereading of Ro-*

mans, p. 58 (nothing that Philo and other Jewish writers of that era viewed Judaism as a "philosophy for the passions, a school for self-control"); Hans Svebakken, *Philo of Alexandria's Exposition of the Tenth Commandment* (Atlanta: Society of Biblical Literature, 2012), p. 14; for the Stoic influences on Philo's concept of *enkrateia*, see Carlos Lévy, "Philo's Ethics," in Adam Kamesar, ed., *The Cambridge Companion to Philo* (New York: Cambridge University Press, 2009), pp. 146, 150, 159.

139 the Shulchan Aruch: Written by Rabbi Joseph Karo, the *Shulchan Aruch* was a mere "handy reference" work, to which Karo turned only after spending twenty years on a multivolume, encyclopedic compilation of Talmudic law. Sol Scharfstein, *The Five Books of Moses: Translation, Rabbinic and Contemporary Commentary* (Jersey City, NJ: KTAV Publishing House, 2008), p. 547. For a translation of parts of the *Shulchan Aruch*, see Gersion Appel, *The Concise Code of Jewish Law: Compiled from the Kitzur Shulhan Aruch and Traditional Sources*, vols. 1–2 (New York: KTAV Publishing, 1977, 1989). For a list of other partial translations, see Phyllis Holman Weisbard and David Schonberg, eds., *Jewish Law: Bibliography of Sources and Scholarship in English* (Littleton, CO: Fred B. Rothman & Co., 1989), p. 19.

138 "Jews . . . do not get drunk": Immanuel Kant, *Anthropology from a Pragmatic Point of View*, trans. and ed. Robert B. Louden (Cambridge, UK: Cambridge University Press, 2006) [1798], p. 63.

139 largely German: Ofer Shiff, *Survival Through Integration: American Reform Jewish Universalism and the Holocaust* (Leiden and Boston: Brill, 2005), p. 33; Caryn Aviv and David Shneer, "From Diaspora Jews to New Jews," in Laurence J. Silberstein, ed., *Postzionism: A Reader* (New Brunswick, NJ: Rutgers University Press, 2008), p. 350 ("German Jews made up the bulk of Jewish immigrants to the United States in the nineteenth century" and imported the "Reform movement"). On the German origins of Reform Judaism, whose leading figure declared that the "Talmud must go," and "the Bible . . . as a divine work must also go," see Michael A. Meyer, *Response to Modernity: A History of the Reform Movement in Judaism* (New York: Oxford University Press, 1988), p. 91.

139 Circumcision . . . a "remnant of savage African life": Richard Rosenthal, "Without Milah and Tevilah," in Walter Jacob and Moshe Zemer, eds., *Conversion to Judaism in Jewish Law: Essays and Responsa* (Pittsburgh: Rodef Shalom Press, 1994), p. 110.

139 the bar mitzvah an obsolete ritual: Nathan Glazer, *American Judaism* (Chicago: University of Chicago Press, 1972), p. 55; Jerold S. Auerbach, *Rabbis and Lawyers: The Journey from Torah to Constitution* (Bloomington: Indiana University Press, 1990), pp. 79–80.

139 the menu included littleneck clams: John J. Appel, "The Trefa Banquet," *Commentary*, Feb. 1966, pp. 75–78. Other sources list the offending items at Hebrew Union's famous 1883 banquet as "oysters, shrimp and crabmeat." Auerbach, *Rabbis and Lawyers*, p. 78.

139 brought with them an orthodox Judaism: See, e.g., Irving Howe, *World of Our Fathers* (New York: Harcourt Brace Jovanovich, 1976), pp. 169–70, 193–94 (noting conflict between the orthodoxy of the new immigrants and the Reform Judaism of the "German Jews").

139 1914 book about Jewish life. Israel Cohen, *Jewish Life in Modern Times* (New York: Dodd, Mead & Co., 1914), p. 18.

139 "bound and shackled" . . . "tend even to forsake": Ba'al Makhshoves, "Mendele, Grandfather of Yiddish Literature," quoted in Irving Howe and Eliezer Greenberg, eds., *Voices from the Yiddish: Essays, Memoirs, Diaries* (Ann Arbor: University of Michigan Press, 1972), pp. 32, 37. Makhshoves was the pen name of the Kovno-born Israel Isidor Elyashev (1873–1924), a pioneer in Yiddish literary criticism.

139 Jewish Sabbath . . . a highly disciplined regimen: Appel, *The Concise Book of Jewish Law*, vol. 2, pp. 224, 239–81, 326; "The Shabbat Laws," Chabad.org, http://www.chabad.org/library/article_cdo/aid/95907/jewish/The-Shabbat-Laws.htm.

140 "I hadda chance to make a dollar": Budd Schulberg, *What Makes Sammy Run?* (New York: Random House, 1941), p. 237.

141 "We push our children too much" . . . "[A] piano in the front room": Howe, *World of Our Fathers*, p. 261 (quoting *The Forward*, Jan. 20, 1911, and July 6, 1903).

141–2 probably strengthened their impulse control: See Charles E. Silberman, *A Certain People: American Jews and Their Lives Today* (New York: Summit Books, 1985), p. 29 (describing the "fundamental rule" on which the author's generation of American Jews was raised in the 1930s as "Be quiet!—Do not call attention to yourself. . . . In talking about a Jewish subject in public we lowered our voices automatically, and we were careful never to read a Hebrew book or magazine . . . when riding on a subway or bus").

142 bar and bat mitzvahs: See Stefanie Cohen, "$1 Million Parties—Have NYC Bar Mitzvahs Gone Too Far?," *New York Post*, Apr. 18, 2010; Ralph Gardner Jr., "Bash Mitzvahs!," *New York Magazine*, Mar. 9, 1998, p. 20.

142 even the Cultural Revolution: See Li, *Cultural Foundations of Learning*, pp. 341–2. Professor Li writes that "[i]f the senseless Cultural Revolution did . . . manage[] to dent the family system, it did not destroy it permanently. It would be hard to imagine any force that would succeed in eradicating Confucian family relationships and child-rearing practices after they have survived for thousands of years."

142 traditional strict parenting . . . softens: Bryan Strong, Christine DeVault, and Theodore F. Cohen, *The Marriage and Family Experience: Intimate Relationships in a Changing Society* (11th ed.) (Belmont, CA: Wadsworth, 2011), p. 97.

142 expectations drop . . . sharp fall-off: Suet-ling Pong, Lingxin Hao, and Erica Gardner, "The Roles of Parenting Styles and Social Capital in the School Performance of Immigrant Asian and Hispanic Adolescents," *Social Science Quarterly* 86 (2005), pp. 928, 942, 944, 946; Yanwei Zhang, "Immigrant Generational Differences in Academic Achievement: The Case of Asian American High School Students," in Park, Goodwin, and Lee, *Asian American Identities, Families, and Schooling*, pp. 204, 209; see also Lingxin Hao and Han S. Woo, "Distinct Trajectories in the Transition to Adulthood: Are Children of Immigrants Advantaged?," *Child Development* 83, no. 5 (2012), pp. 1623, 1635.

143 junk-food corporations: Michael Moss, *Salt Sugar Fat: How the Food Giants Hooked Us* (New York: Random House, 2013), p. 341.

143 *The Wire*: Anmol Chaddha and William Julius Wilson, "Why We're Teaching 'The Wire' at Harvard," *Washington Post*, Sept. 12, 2010, p. B2.

143 *Breaking Bad*: Patrick Radden Keefe, "The Uncannily Accurate Description of the Meth Trade in 'Breaking Bad,'" *The New Yorker*, July 13, 2012.

CHAPTER 6: THE UNDERSIDE OF THE TRIPLE PACKAGE

146 Wittgenstein's paradoxical ladder: The ladder appears on the last page of Ludwig Wittgenstein, *Tractatus Logico-Philosophicus*, ed. C. K. Ogden (New York: Harcourt, Brace & Co. 1922), p. 189; one must climb the ladder of philosophy, Wittgenstein suggests, in order to see that philosophy is "senseless."

146 "The youth of America is their oldest tradition": Oscar Wilde, "A Woman of No Importance" (1894), Act 1.

146 America is a youth culture: See, e.g., Jon Savage, *Teenage: The Creation of Youth Culture* (New York: Viking, 2007); Patricia Cohen, *In Our Prime: The Invention of Middle Age* (New York: Scribner, 2012), p. 166 ("Our digital world incessantly assails us with artificially maintained images of youth"; in a 2005 Harris survey "[h]alf of those polled agreed that a youthful appearance is necessary for professional success and for personal happiness"); Martha Irvine and Lindsey Tanner, "Youthfulness an American Obsession at What Cost?," Associated Press, Dec. 7, 2008. Youth culture is of course not uniquely American. See Jed Rubenfeld, *Freedom and Time* (New Haven, CT, and London: Yale University Press, 2001), p. 34 ("Modernity adores youth because it imagines youth as exquisitely unburdened by temporal engagements . . . *To be young is to live in the present*").

146 Idealizing childhood: See, e.g., David Gettman, *Basic Montessori: Learning Activities for Under-Fives* (New York: St. Martin's Press, 1987), p. 37 (adults suffering "from their own adult pressures and anxieties" often "believe that early childhood should be fun and carefree"); Kay

Sambell, Mel Gibson, and Sue Miller, *Studying Childhood and Early Childhood: A Guide for Students* (2d ed.) (London and Thousand Oaks, CA: Sage Publications, 2010), pp. 18–9 (noting the tendency to idealize childhood, imagining it as "carefree," "fun," and "perpetually happy"); Christina Schwarz, "Leave Those Kids Alone," *The Atlantic*, Feb. 24, 2011 (describing childhood as "those first, fresh experiences of the world, unclouded by reason and practicality"; "[c]hildren have a knack for simply living that adults can never regain"); Linda F. Burghardt, "A Symbol of Carefree, Innocent Fun? Not in Oyster Bay," *New York Times*, May 28, 2006 (describing a Long Island carousel as "a potent symbol of the happy, carefree childhood that parents want to give their youngsters").

147 "I was burdened to excel": E-mail to Amy Chua, May 22, 2012 (on file with authors); see also Rubén G. Rumbaut, "Children of Immigrants and Their Achievement: The Role of Family, Acculturation, Social Class, Gender, Ethnicity, and School Contexts," p. 1, http://www.hks.harvard.edu/inequality/Seminar/Papers/Rumbaut2.pdf; Lisa Sun-Hee Park, "Ensuring Upward Mobility: Obligations of Children of Immigrant Entrepreneurs," in Benson Tong, ed., *Asian American Children: A Historical Handbook Guide* (Westport, CT: Greenwood Press, 2004), pp. 123, 129 (in a study of Chinese and Korean immigrants' children "[a]ll the respondents expressed a need to 'repay' their parents" and "believed that their parents purposely stunted their own growth so that their children might prosper").

147 "happiness has to take a back seat": See Park, "Ensuring Upward Mobility," pp. 125, 128.

147 Confucian expert Jin Li explains: Jin Li, *Cultural Foundations of Learning: East and West* (Cambridge, UK: Cambridge University Press, 2012), pp. 38, 73–4, 90–2.

148 "I've never felt ready to receive her": "Talk Asia: Interview with Fashion Designer Phillip Lim," CNN.com (aired Mar. 23, 2011), http://transcripts.cnn.com/TRANSCRIPTS/1103/23/ta.01.html.

148 "I always see where I didn't do things the right way": Elisa Lipsky-Karasz, "The Vera Wang Interview: Made of Honor," *Harper's Bazaar*, Mar. 24, 2011.

148 "I feel like I'm just an investment good": E-mail to Amy Chua, Mar. 14, 2013 (on file with authors); see also Vivian S. Louie, *Compelled to Excel: Immigration, Education, and Opportunity Among Chinese Americans* (Stanford, CA: Stanford University Press, 2004), p. 48; Ruth K. Chao, "Chinese and European Mothers' Beliefs About the Role of Parenting in Children's School Success," *Journal of Cross-Cultural Psychology* 27, no. 4 (1996), pp. 403, 412.

148 "they're only proud of me because they can boast": Louie, *Compelled to Excel*, p. 91; see also Tony Hsieh, *Delivering Happiness: A Path to Profits, Passion, and Purpose* (New York and Boston: Business Plus, 2010), p. 8.

148 "'Okay, well, then, I'm garbage'": Louie, *Compelled to Excel*, p. 93.

149 "[My parents] just didn't understand": Amy Tan interview, Academy of Achievement, June 28, 1996, http://www.achievement.org/autodoc/page/tan0int-5 (accessed Mar. 25, 2013); see also Kim Wong Keltner, *Tiger Babies Strike Back: How I Was Raised by a Tiger Mom but Could Not Be Turned to the Dark Side* (New York: William Morrow, 2013).

150 "the highest rates of depressive symptoms": Cathy Schoen et al., The Commonwealth Fund Survey of the Health of Adolescent Girls (The Commonwealth Fund, Nov. 1997), p. 21.

150 "higher levels of stress and anxiety": Desiree Baolian Qin et al., "Parent-Child Relations and Psychological Adjustment Among High-Achieving Chinese and European American Adolescents," *Journal of Adolescence* 35, no. 4 (2012), pp. 863–73; Desiree Baolian Qin et al., "The Other Side of the Model Minority Story: The Familial and Peer Challenges Faced by Chinese American Adolescents," *Youth & Society* 39, no. 4 (2008), pp. 480–506; Carol S. Huntsinger and Paul E. Jose, "A Longitudinal Investigation of Personality and Social Adjustment Among Chinese American and European American Adolescents," *Child Development* 77, no. 5 (2006), pp. 1309–24; Paul E. Jose and Carol S. Huntsinger, "Moderation and Mediation Effects of Coping by Chinese American and European American Adolescents," *Journal of Genetic Psychology* 166, no. 1 (2005), pp. 16–44.

150 study of high-achieving ninth graders: Qin et al., "Parent-Child Relations and Psychological Adjustment Among High-Achieving Chinese and European American Adolescents," p. 870.

150 rates of alcohol abuse and substance dependency: "In 2011, among persons aged 12 or older, the rate of substance dependence or abuse was lower among Asians (3.3 percent) than among other racial/ethnic groups. The rates for the other racial/ethnic groups were 16.8 percent for American Indians or Alaska Natives, 10.6 percent for Native Hawaiians or Other Pacific Islanders, 9.0 percent for persons reporting two or more races, 8.7 percent for Hispanics, 8.2 percent for whites, and 7.2 percent for blacks." Substance Abuse and Mental Health Services Administration, *Results from 2011 National Survey on Drug Use and Health: Summary of National Findings* (2012), http://www.samhsa.gov/data/NSDUH/2k11Results/NSDUHresults2011.htm. Moreover, Asian American rates for alcohol "heavy use" or "binge drinking" are much lower than that of other groups. Ibid.

150 2010 nationwide psychiatric survey: Anu Asnaani et al., "A Cross-Ethnic Comparison of Lifetime Anxiety Disorders," *Journal of Nervous Mental Disorders* 198, no. 8 (2010), pp. 551–5.

150 a 1990s study: David T. Takeuchi et al., "Lifetime and Twelve-Month Prevalence Rates of Major Depressive Episodes and Dysthymia Among Chinese Americans in Los Angeles," *American Journal of Psychiatry* 155, no. 10 (1998), pp. 1407–14; see also Li, *Cultural Foundations of Learning*, pp. 66–7.

150 Asian American suicide rate: All data in the footnote are from the Centers for Disease Control and Prevention, Web-based Statistics Query and Reporting System (WISQARS), Fatal Injury Reports, National and Regional, 1999–2010 (manner of injury: suicide) (years: 2000–2010) (age-adjusted rate), http://webappa.cdc.gov/sasweb/ncipc/mortrate10_us.html. Looking solely at 2010 (the most recent year for which data are available) shows a slightly higher suicide rate for both Asian Americans (6.2 per 100,000) and white Americans (13.6 per 100,000). It has been widely reported that "Asian-American women ages 15–24 have the highest suicide rate of women in any race or ethnic group in that age group." E.g., Elizabeth Cohen, "Push to Achieve Tied to Suicide in Asian-American Women," CNN, May 16, 2007, http://www.cnn.com/2007/HEALTH/05/16/asian.suicides/index.html. As the CDC data indicate, this is not true; nor was it true in 2007, the year of the report just cited. See also American Psychological Association, "Suicide Among Asian-Americans," http://www.apa.org/pi/oema/resources/ethnicity-health/asian-american/suicide.aspx (calling it a "myth" that "[y]oung Asian-American women [aged 15–24] have the highest suicide rates of all racial/ethnic groups"). U.S.-born Asian American women may have rates of suicidal ideation and attempted suicide higher than that of American women in general, see Aileen Alfonso Duldulao et al., "Correlates of Suicidal Behaviors Among Asian Americans," *Archives of Suicide Research* 13, no. 3 (2009), pp. 277–90, but this finding does not appear to apply to Asian American women overall. See Janice Ka Yan Cheng et al., "Lifetime Suicidal Ideation and Suicide Attempts in Asian Americans," *Asian American Journal of Psychology* 1, no. 1 (2010), pp. 18–30; M. Mercedes Perez-Rodriguez et al., "Ethnic Differences in Suicidal Ideation and Attempts," *Primary Psychiatry* 15, no. 2 (2008), pp. 44–53, Table 1.

151 Asian American . . . self-esteem: Douglas S. Massey et al., *The Source of the River: The Social Origins of Freshmen at America's Selective Colleges and Universities* (Princeton, NJ: Princeton University Press, 2003), pp. 118–9, 122; Carl L. Bankston III and Min Zhou, "Being Well vs. Doing Well: Self-Esteem and School Performance Among Immigrant and Nonimmigrant Racial and Ethnic Groups," *International Migration Review* 63, no. 2 (2002), pp. 389–415, especially p. 401.

151 "If you're doing well, you should be feeling good": Stephanie Pappas, "Study: 'Tiger Parenting,' Tough on Kids," *LiveScience*, Jan. 19, 2012, http://www.livescience.com/18023-tiger-parenting-tough-kids.html (quoting Desiree Baolian Qin).

151 Guilt: See Devorah Baum, "Trauma: An Essay on Jewish Guilt," *English Studies in Africa*

52, no. 1 (2009), pp. 15–27; Simon Dein, "The Origins of Jewish Guilt: Psychological, Theological, and Cultural Perspectives," *Journal of Spirituality in Mental Health* 15, no. 2 (2013), pp. 123–37; Joyce Antler, *You Never Call! You Never Write! A History of the Jewish Mother* (New York: Oxford University Press, 2007), pp. 2, 137–8.

151 *naches*: Leo Rosten, *The New Joys of Yiddish* (rev. ed.) (New York: Three Rivers Press, 2001), p. 262 ("Jews use *naches* to describe the glow of pleasure plus pride that only a child can give to its parents").

152 "and my mother's contempt for my father": See Jules Feiffer, *Hold Me! An Entertainment* (New York: Dramatists Play Service, Inc., 1977), p. 44.

152 propensity to challenge authority: See, e.g., Paul Johnson, *A History of the Jews* (New York: Harper & Row, 1987), p. 295 ("it is one of the glories of the Jews that they do not meekly submit to their own appointed authorities. The Jew is the eternal protestant"); Raphael Patai, *The Jewish Mind* (Detroit: Wayne State University Press, 1996), p. 332 (describing Thorstein Veblen's view that Jewish scientists' skepticism gave them "the prerequisite of immunity from the inhibitions of intellectual quietism").

152 *Footnote*: Directed by Joseph Cedar and released in 2011, *Footnote* features a Talmudic scholar who has long labored in obscurity and is mistakenly informed that he has won the prestigious Israeli Prize; the true winner is his son, a much more prominent Talmudic scholar.

152–3 Trilling . . . Bell: Alexander Bloom, *Prodigal Sons: The New York Intellectuals & Their World* (New York: Oxford University Press, 1986), p. 19. Irving Howe describes a "crisis set off in the Jewish family" as a result of the expectations, disappointments, and resentments between traditional immigrant fathers and their more emancipated sons. Irving Howe, *World of Our Fathers* (New York: Harcourt Brace Jovanovich, 1976), p. 254.

152 *The Jazz Singer*: See Ted Merwin, *In Their Own Image: New York Jews in Jazz Age Popular Culture* (New Brunswick, NJ: Rutgers University Press, 2006), p. 152.

153 "a caricature": John H. Davis, *The Guggenheims: An American Epic* (New York: William Morrow and Co., 1978), p. 50.

154 "I have a bit of a phobia": Michael T. Kaufman, *Soros: The Life and Times of a Messianic Billionaire* (New York: Alfred A. Knopf, 2002), p. 5.

154 "My grandparents were Holocaust survivors": E-mail to Amy Chua, Aug. 26, 2012 (on file with authors); see also Helen Epstein, *Children of the Holocaust: Conversations with Sons and Daughters of Survivors* (New York: G. P. Putnam's Sons, 1979), pp. 16, 305–6; Aaron Hass, *In the Shadow of the Holocaust: The Second Generation* (Ithaca, NY, and London: Cornell University Press, 1990), pp. 58, 127–8.

155 "Jews are probably the most insecure group": Dennis Prager, "Explaining Jews, Part 3: A Very Insecure People," WND Commentary, Feb. 21, 2006, http://www.wnd.com/2006/02/34917.

156 "half-savage peoples": Noam Chomsky, *Fateful Triangle: The United States, Israel, and the Palestinians* (Cambridge, MA: South End Press, 1999), pp. 481–3.

156 West Indian . . . immigrants: Mary C. Waters, *Black Identities: West Indian Immigrant Dreams and American Realities* (Cambridge, MA: Harvard University Press, 1999), pp. 64–5.

156 Toni Morrison: Sunaina Marr Maira, *Desis in the House: Indian American Youth Culture in New York City* (Philadelphia: Temple University Press, 2002), p. 72 (alterations to quotation omitted).

156 women are excluded from the priesthood: Claudia L. Bushman, *Contemporary Mormonism: Latter-day Saints in Modern America* (Westport, CT: Praeger, 2006), pp. 31–3, 111–5.

157 "In the beginning": Sheri L. Dew, *Ezra Taft Benson: A Biography* (Salt Lake City: Deseret Book Co., 1987), p. 505.

157 women have been excommunicated: Joanna Brooks, *The Book of Mormon Girl: A Memoir of an American Faith* (New York: Free Press, 2012), pp. 122–31.

157 "Proclamation to the World": LDS Church, "The Family: A Proclamation to the World," www.lds.org/topics/family-proclamation.

157 "For years, I cried": Brooks, *The Book of Mormon Girl*, p. 149.

157 The terms "ward house": C. Mark Hamilton, *Nineteenth-Century Mormon Architecture & City Planning* (New York: Oxford University Press, 1995), p. 165.

158 "sister training leaders": Joseph Walker, "Sister LDS Missionaries Will Have Key Role in New Mission Leadership Council," *Deseret News*, Apr. 5, 2013, http://www.deseretnews.com/article/865577611/Sister-LDS-missionaries-will-have-key-role-in-new-Mission-Leadership-Council.html?pg=all.

158 out-marry themselves into oblivion: See Alan M. Dershowitz, *The Vanishing American Jew: In Search of Jewish Identity for the Next Century* (New York: Little, Brown and Company, 1997), pp. 30–1, 72.

158 strong taboos against marrying outside one's group: See Donald L. Horowitz, *Ethnic Groups in Conflict* (Berkeley, Los Angeles, and London: University of California Press, 1985), p. 62 (noting that "[v]irtually everywhere in Asia . . . endogamy is the norm").

158 inter-Asian marriages: See Rachel L. Swarns, "For Asian-American Couples, A Tie That Binds," *New York Times*, Mar. 30, 2012; see also Richard D. Alba and Victor Nee, *Remaking the American Mainstream: Assimilation and Contemporary Immigration* (Cambridge, MA: Harvard University Press, 2003), p. 265.

159 obvious tokens of success . . . "respectable" careers: see, e.g., Bandana Purkayastha, *Negotiating Ethnicity: Second Generation South Asian Americans Traverse a Transnational World* (New Brunswick, NJ: Rutgers University Press, 2005), p. 91; Mei Tang, Nadya A. Fouad, and Philip L. Smith, "Asian Americans' Career Choices: A Path Model to Examine Factors Influencing Their Career Choices," *Journal of Vocational Behavior* 54 (1999), pp. 142, 142–6; S. Alvin Leung, David Ivey, and Lisa Suzuki, "Factors Affecting the Career Aspirations of Asian Americans," *Journal of Counseling & Development* 72 (March/April 1994), pp. 404, 405; see also Min Zhou, "Assimilation the Asian Way," in Tamar Jacoby, ed., *Reinventing the Melting Pot: The New Immigrants and What It Means to Be American* (New York: Basic Books, 2004), pp. 139, 146–7; Maryam Daha, "Contextual Factors Contributing to Ethnic Identity Development of Second-Generation Iranian American Adolescents," *Journal of Adolescent Research* 26, no. 5 (2011), pp. 560–1.

159 a defensive crouch: We owe the insights in this paragraph to Renagh O'Leary; see also David Brooks, "The Empirical Kids," *New York Times*, Mar. 28, 2013; David Brooks, "The Organization Kid," *The Atlantic*, Apr. 2001.

160 "When I was younger": Amy Tan interview, Academy of Achievement.

160 "[T]hey won't have the guts": Maira, *Desis in the House*, p. 76.

160 "You write, and then you erase": James Atlas, *Bellow: A Biography* (New York: Random House, 2000), pp. 42, 60.

161 "You're only 49": "Ang Lee," *Interview*, http://www.interviewmagazine.com/film/ang-lee (interview by Liev Schreiber).

161 "the only purpose in life": Clifford W. Mills, *Ang Lee* (New York: Chelsea House, 2009), p. 31; see also ibid., pp. 27–9; "Ang Lee—Top 25 Directors," Next Actor, http://www.nextactor.com/ang_lee.html (Lee grew up in a home that "put heavy emphasis on education and the Chinese classics").

161 "I knew I had to please my father": Michael Berry, *Speaking in Images: Interviews with Contemporary Chinese Filmmakers* (New York: Columbia University Press, 2005), p. 329.

161 "devout believer" . . . "She was always nagging him": Atlas, *Bellow*, p. 24.

161 "suffocating orthodoxy": D.J.R. Bruckner, "A Candid Talk with Saul Bellow," *New York Times*, Apr. 15, 1984.

161 got himself fired: Atlas, *Bellow*, p. 3.

162 "[M]y name won't go down": Ibid., p. 219.

162 "I had a lot of guilt": Berry, *Speaking in Images*, p. 329; "Ang Lee—Top 25 Directors."

162 fellow student Spike Lee: Mills, *Ang Lee*, p. 43.

162 "six years of agonizing": Irene Shih, "Ang Lee: A Never-Ending Dream," What Shih Said.com, Feb. 26, 2013, http://whatshihsaid.com/2013/02/26/ang-lee-a-never-ending-dream.

162 "carrying a chip on his shoulder": Atlas, *Bellow*, p. 112 (quoting William Barrett, an editor of the *Partisan Review*).

162 "throwing down the gauntlet": Greg Bellow, *Saul Bellow's Heart: A Son's Memoir* (New York: Bloomsbury, 2013), p. 54.

162 Bellow was famously "disciplined": See Karyl Roosevelt, "Saul Bellow Is Augie, Herzog and Henderson—and of Course the Hero of His Latest Book," *People*, Sept. 8, 1975 (describing Bellow as "highly disciplined" with hard to match "powers of concentration").

162 "insecurity" . . . "first son" . . . "everything rested on my shoulders": Mills, *Ang Lee*, p. 28.

162 "first Jewish-American novelist": Leslie A. Fiedler, *A New Fiedler Reader* (Amherst, NY: Prometheus Books, 1999), p. 110; see also generally Bloom, *Prodigal Sons*, pp. 296–7.

163 sixty-two-week *New York Times* bestseller: Dwight Garner, "Inside the List," Sunday Book Review, *New York Times*, Feb. 11, 2007.

163 "without departing from an American Jewish idiom": Earl Rovit, *Saul Bellow* (St. Paul: University of Minnesota Press, 1967), p. 5.

163 first Chinese to win the Acadamy Award for best director: David Barboza, "The Oscar for Best Banned Picture," *New York Times*, Mar. 12, 2006. Lee's first best director Oscar was for *Brokeback Mountain*; his second was for *Life of Pi*. Nicole Sperling, "Oscars 2013: Ang Lee Wins Best Director for 'Life of Pi,'" *Los Angeles Times*, Feb. 24, 2013.

163 "I never belonged to my own family": Atlas, *Bellow*, p. 8; see also p. 32 ("never felt American").

163 "*métèques*": Saul Bellow, "A Jewish Writer in America," *New York Review of Books*, Oct. 27, 2011, http://www.nybooks.com/articles/archives/2011/oct/27/jewish-writer-america.

164 "I'm a drifter": Rick Groen, "Ang Lee: An Outsider Who Found the Perfect Story for His Gifts in Life of Pi," *The Globe and Mail*, Feb. 23, 2013; see also David Minnihan, "Ang Lee," Senses of Cinema, http://sensesofcinema.com/2008/great-directors/ang-lee.

164 A growing number of Asian college students: See Mitchell J. Chang, Julie J. Park, Monica H. Lin, Oiyan A. Poon, and Don T. Nakanishi, *Beyond Myths: The Growth and Diversity of Asian American College Freshman, 1971–2005* (Los Angeles: UCLA Higher Education Research Institute, 2007), p. 18; Jennie Zhang, "Breaking Out of the Mold: Pursuing Humanities as an APA," *Bamboo Offshoot: USC's Asian Pacific American Magazine*, Jan. 1, 2012, http://bamboooff shoot.com/2012/01/01/breaking-out-of-the-mold-pursuing-the-humanities-as-an-apa.

164 role models in breakout Asian American stars: See, e.g., Caleb Li, "Hikaru Sulu: Hollywood's Asian American Trailblazer," Fresh Patrol, http://freshpatrol.com/hikaru-sulu -hollywoods-asian-american-trailblazer; Louis Peitzman, "'Pitch Perfect' Breakout Utkarsh Ambudkar Takes on 'The Mindy Project,'" BuzzFeed, Jan. 8, 2013, http://www.buzzfeed.com/ louispeitzman/pitch-perfect-breakout-utkarsh-ambudkar-takes-on; Deanna Fei, "The Real Lesson of Linsanity," Huffington Post, Feb. 16, 2012, http://www.huffingtonpost.com/deanna-fei/ jeremy-lin-asian-americans_b_1281916.html.

165 celebrated in the Indian American community: Soon after he graced the cover of *Time*, Bharara was selected as the 2011 *India Abroad* Person of the Year by the most widely circulated *desi* newspaper. See "Preet Bharara, the Man Who Makes Wall Street Tremble, Is India Abroad Person of the Year 2011," Rediff News, June 30, 2012, http://www.rediff.com/news/slide-show/ slide-show-1-preet-bharara-is-india-abroad-person-of-the-year-2011/20120630.htm.

165 consternation and shame: Anita Raghavan, *The Billionaire's Apprentice: The Rise of the Indian-American Elite and the Fall of the Galleon Hedge Fund* (New York and Boston: Business Plus, 2013), pp. 381, 415.

165 over $80 million: Kevin Roose, "Why the Rajat Gupta Trial Is a Big Deal," *New York Magazine*, June 4, 2012.

165 state dinner at the White House . . . "The Court can say without exaggeration": Raghavan, *The Billionaire's Apprentice*, pp. 1–6, 410.

CHAPTER 7: IQ, INSTITUTIONS, AND UPWARD MOBILITY

167 "Horatio Alger Is Dead": David Frum, "Horatio Alger Is Dead," Daily Beast, Feb. 9, 2012 (noting that upward mobility has declined among men but not women).

167 Obituaries reporting the demise of upward mobility: See, e.g., Josh Sanburn, "The Loss of Upward Mobility in the U.S.," *Time*, Jan. 5, 2012; Timothy Egan, "Downton and Downward," *New York Times*, Feb. 14, 2013; Timothy Noah, "The Mobility Myth," *The New Republic*, Feb. 8, 2012; Rana Foroohar, "What Ever Happened to Upward Mobility?," *Time*, Nov. 14, 2011.

167 42 percent of people raised in the lowest economic quintile: Julia B. Isaacs, "Economic Mobility of Families Across Generations," in Julia B. Isaacs, Isabel V. Sawhill, and Ron Haskins, *Getting Ahead or Losing Ground: Economic Mobility in America* (Washington, DC: Economic Mobility Project, The Pew Charitable Trusts, 2008), p. 19; see also Sanburn, "The Loss of Upward Mobility in the U.S."

167 two-thirds of Americans: Bhashkar Mazumder, *Upward Intergenerational Economic Mobility in the United States* (Washington, DC: Economic Mobility Project, The Pew Charitable Trusts, 2008), pp. 7, 11.

167 Rising remains the rule: Isaacs, "Economic Mobility of Families Across Generations," pp. 17–8.

167 cited repeatedly: See, e.g., Egan, "Downton and Downward"; Foroohar, "What Ever Happened to Upward Mobility?"

168 In Denmark: See Julia B. Isaacs, "International Comparisons of Economic Mobility," in Isaacs et al., *Getting Ahead or Losing Ground*, p. 40 and Table 1. Note that international comparisons can be misleading: rising from the bottom to just the *middle* quintile in the U.S. could actually be a greater gain than rising from the bottom to the *top* in a country with less inequality compression. See Jason DeParle, "Harder for Americans to Rise from Lower Rungs," *New York Times*, Jan. 4, 2012 (reporting estimate that "a Danish family can move from the 10th percentile to the 90th percentile with $45,000 of additional earnings, while an American family would need an additional $93,000").

168 "[i]mmigrant families are not included": Isabel V. Sawhil, "Overview," in Isaacs et al., *Getting Ahead or Losing Ground*, p. 6; see also Isaacs, "International Comparisons of Economic Mobility," p. 38. The 2008 Pew Study, like many other American mobility studies, relies on data from the Panel Study of Income Dynamics (PSID), an extraordinary longitudinal data set tracking a sample of families since 1968. Isaacs, "Economic Mobility of Families Across Generations," p. 15. As a result, the Pew sample consisted solely of "individuals who were between the ages of 0 and 18 in 1968," Isaacs et al., *Getting Ahead or Losing Ground*, p. 105, which wholly excludes "the large number of immigrants who have arrived since 1968" and their children. Isaacs, "Economic Mobility of Families Across Generations," p. 22 n. 3. Although the PSID added a sample of immigrant families in the 1990s, to date these immigrant families have still been excluded from upward mobility studies "because they lack historical family and economic data originating with the PSID in 1968." The Pew Charitable Trusts, *Pursuing the American Dream: Economic Mobility Across Generations* (Washington, DC: The Pew Charitable Trusts, July 2012), p. 28.

168 more than 40 million immigrants: U.S. Census, American Community Survey, Table S0201: Selected Population Profile in the United States (2012 3-year dataset); U.S. Census, Table P022: Year of Entry for the Foreign-Born population (2000 SF3 sample data (fewer than 3.3 million of America's foreign-born population as of 2000 had entered the country before 1965).

168 "American dream is alive and well": Isaacs, "Economic Mobility of Families Across Generations," p. 6.

168 experience strong upward mobility: See e.g., Pew Research Center, *Second-Generation Americans: A Portrait of the Adult Children of Immigrants*, (Washington, DC: Pew Research Center, February 7, 2013), p. 7; Ron Haskins, "Immigration: Wages, Education, and Mobility," in Isaacs et al., *Getting Ahead or Losing Ground*, pp. 81–8; Edward E. Telles and Vilma Ortiz, *Generations of Exclusion: Mexican Americans, Assimilation, and Race* (New York: Russell Sage Foundation, 2008), pp. 144–7; Lingxin Hao and Han S. Woo, "Distinct Trajectories in the Transition to

Adulthood: Are Children of Immigrants Advantaged?" *Child Development* 83, no. 5 (2012), pp. 1623, 1635; Rubén G. Rumbaut, "The Coming of the Second Generation: Immigration and Ethnic Mobility in Southern California," *The Annals of the American Academy of Political and Social Science* 620 (November 2008), pp. 196, 205, 219; Rubén G. Rumbaut, "Paradise Shift: Immigration, Mobility, and Inequality in Southern California," Working Paper No. 14 (Vienna: Austrian Academy of Sciences, KMI Working Paper Series, October 2008), pp. 6–9, 30–1; Rubén G. Rumbaut et al., "Immigration and Intergenerational Mobility in Metropolitan Los Angeles (IIMMLA)," Russell Sage Foundation, http://www.russellsage.org/research/Immigration/IIMMLA.

169 "pauper county": Monica Potts, "Pressing on the Upward Way," *The American Prospect*, June 12, 2012.

169 according to some: "America's Poorest County: Proud Appalachians Who Live Without Running Water or Power in Region Where 40% Fall Below Poverty Line," *Daily Mail*, Apr. 23, 2012, http://www.dailymail.co.uk/news/article-2134196/Pictured-The-modern-day-poverty-Kentucky-people-live-running-water-electricity.html.

169 By statute, Stuyvesant High School: Al Baker, "Charges of Bias in Admission Test Policy at Eight Elite Public High Schools," *New York Times*, Sept. 27, 2012 ("In May 1971, after officials began thinking about adding other criteria for admission, protests from many parents, mostly white, persuaded the State Legislature to enshrine the rule in state law").

170 tuition: See the Phillips Exeter website, http://www.exeter.edu/admissions/109_1370.aspx; see also Raquel Laneri, "America's Best Prep Schools," Forbes.com, Apr. 29, 2010, http://www.forbes.com/2010/04/29/best-prep-schools-2010-opinions-private-education.html (reporting that Phillips Exeter sends roughly 29 percent of its graduates to the Ivy League, Stanford, or MIT).

170 Stuyvesant reportedly sends upwards of 25 percent: See Insideschools.Org, http://insideschools.org/high/browse/school/97.

170 the school's new admittees: Thomas Sowell, "Of Stuyvesant, Tests, and Tiger Moms," *New York Post*, Apr. 9, 2013; Warren Kozak, "Call Them Tiger Students. And Get to Work," *Wall Street Journal*, Apr. 4, 2013.

170 Bronx Science: See "Asian Group Critical of Test Policy," *Crain's Insider*, Oct. 4, 2012 ("Asian Americans fill 72 percent of seats at Stuyvesant, 64 percent at Bronx High School of Science").

170 Sunset Park: Beth Fertig, "Around Sunset Park, Tutoring Is Key to Top High Schools," WNYC SchoolBook, Mar. 12, 2013, http://www.schoolbook.org/2013/03/12/around-sunset-park-tutoring-is-key-to-top-high-schools.

170 head start that Asian immigrants bring with them: See Richard Alba and Victor Nee, *Remaking the American Mainstream: Assimilation and Contemporary Immigration* (Cambridge, MA: Harvard University Press, 2003), pp. 174–5, 203–4, 209; Nancy Foner, *From Ellis Island to JFK: New York's Two Great Waves of Immigration* (New Haven, CT: Yale University Press, 2000), p. 163; Haskins, "Immigration: Wages, Education, and Mobility," p. 82.

170 "family reunification": Alba and Nee, *Remaking the American Mainstream*, p. 187; Pew Research Center, *Second-Generation Americans*, p. 35 ("Asian immigrants are more likely than those from other regions to be admitted on employment visas.").

171 cream of the intellectual crop: Thomas L. Friedman, *The World Is Flat: A Brief History of the Twenty-First Century* (New York: Farrar, Straus & Giroux, 2005), pp. 104–5; David Lague, "1977 Exam Opened Escape Route into China's Elite," *New York Times*, Jan. 6, 2008.

171 more likely to be restaurant or factory workers: Fertig, "Around Sunset Park."

171 a majority of Chinese immigrants: In 2010, 54 percent of legal immigrants from mainland China and Hong Kong obtained residence through family-based criteria. Kristen McCabe, "Chinese Immigrants in the United States," Migration Information Source, January 2012, http://migrationinformation.org/USFocus/display.cfm?ID=876#9.

171 "bimodal" . . . exceptional academic success: Rumbaut, "Paradise Shift," p. 12; Rumbaut, "The Coming of the Second Generation," p. 208; see also Renee Reichl Luthra and Roger

Waldinger, "Intergenerational Mobility," in David Card and Steven Raphael, eds., *Immigration, Poverty, and Socioeconomic Inequality* (New York: Russell Sage Foundation, 2013), pp. 169, 192, 196 (second-generation Chinese show upward educational and occupational mobility "regardless of parental educational background"); Jennifer Lee and Min Zhou, "Frames of Achievement and Opportunity Horizons," in Card and Raphael, *Immigration, Poverty, and Socioeconomic Inequality*, pp. 207, 209–11, 216, 221.

171 Chinese Americans' mean IQ is no higher: James R. Flynn, *Asian Americans: Achievement Beyond IQ* (Hillsdale, NJ: Lawrence Erlbaum Associates, 1991), pp. 60, 77. According to one scholar who purports to have found worldwide racial intelligence differences, Chinese IQ in some countries is significantly higher than white IQ in the United States. See Richard Lynn, "Race Differences in Intelligence: A Global Perspective," *Mankind Quarterly* 31 (1991), pp. 264–5 (reporting a median Chinese IQ of 101 in China, but 110 in Singapore and 116 in Hong Kong). But even if Lynn's findings are credible, his own estimate of the median IQ of East Asians *in North America* is 103, almost indistinguishable from white Americans (101–2). Ibid. Moreover, as has been acknowledged even by IQ-proponents Charles Murray and Richard Herrnstein, Lynn's higher Chinese IQ numbers did not incorporate certain important corrections for temporal shifts; when "such corrections were made," Lynn's own data showed Chinese IQ to be generally comparable to North American white IQ. Richard J. Herrnstein and Charles Murray, *The Bell Curve: Intelligence and Class Structure in American Life* (New York: The Free Press, 1994), pp. 272–73 (citing Lynn, "Race Differences in Intelligence"). Reevaluating Lynn's data and other sources, Flynn found that Chinese Americans' mean IQ appeared to be slightly below white Americans'. Flynn, *Asian Americans*, p. 1.

171 more bang for their intelligence buck: Richard E. Nisbett, *Intelligence and How to Get It: Why Schools and Culture Count* (New York: W. W. Norton & Co., 2009), p. 157; Roy F. Baumeister and John Tierney, *Willpower: Rediscovering the Greatest Human Strength* (New York: Penguin Press, 2011), p. 195.

172 "If Asian students": Laurence Steinberg, *Beyond the Classroom: Why School Reform Has Failed and What Parents Need to Do* (New York: Simon & Schuster, 1996), p. 87.

172 "probably 95 out of 100 Chinese students": Fertig, "Around Sunset Park"; see also Nisbett, *Intelligence and How to Get It*, p. 158 ("Asian and Asian American achievement is not mysterious. It happens by working harder").

172 NAACP Legal Defense Fund: Baker, "Charges of Bias in Admission Test at Eight Elite Public High Schools"; Kyle Spencer, "For Asians, School Tests Are Vital Steppingstones," *New York Times*, Oct. 26, 2012.

173 $5,000 a year: Fertig, "Around Sunset Park."

173 free tutoring . . . study "excessively" . . . "This is the easy part": Spencer, "For Asians, School Tests Are Vital Steppingstones."

173 "Most of our parents": Ibid.

174 about half of Kentucky and Tennessee: For the Appalachian Regional Commission's boundary definition of the region, see http://www.appalachiancommunityfund.org/html/where wefund.html.

174 "[G]et out, stay out of people's lives" . . . "[Y]our elite group": See Roger Catlin, "What's on Tonight: Diane Sawyer in Appalachia; 'Dollhouse,'" *Hartford Courant*, Courant .com, Feb. 13, 2009, http://blogs.courant.com/roger_catlin_tv_eye/2009/02/whats-on-tonight -diane-sawyer.html (comments).

174 "non-welfare drawing, non–Mountain Dew guzzling": "Does Diane Sawyer Get Appalachia?," The Revivalist, Apr. 1, 2010, http://therevivalist.info/does-diane-sawyer-get-appalachia (quoted text slightly edited by authors).

174 contradictory impressions: See, e.g., Anthony Harkins, *Hillbilly: A Cultural History of an American Icon* (New York: Oxford University Press, 2004), pp. 6–7; Kai T. Erikson, *Everything in Its Path: Destruction of Community in the Buffalo Creek Flood* (New York: Simon & Schuster, 1976), pp. 84–9; Silas House and Jason Howard, *Something's Rising: Appalachians Fighting Mountaintop Removal* (Lexington: University Press of Kentucky, 2009), pp. 1, 59, 133–5; John O'Brien, *At Home in the Heart of Appalachia* (New York: Alfred A. Knopf, 2001), pp. 3–5.

175 "southern mountain folk": Harkins, *Hillbilly*, p. 4.

175 crystal meth addicts: Zhiwei Zhang et al., *An Analysis of Mental Health and Substance Abuse Disparities and Access to Treatment Services in the Appalachian Region* (Appalachian Regional Commission and the National Opinion Research Center, August 2008), p. 2.

175 Rates of cancer . . . fewer than 12 percent: Appalachian Regional Commission, *Economic Overview of Appalachia–2011*, http://www.arc.gov/images/appregion/Sept2011/EconomicOver viewSept2011.pdf.

175 42 percent rural: Appalachian Regional Commission, "The Appalachian Region," http://www.arc.gov/appalachian_region/TheAppalachianRegion.asp.

175 neighboring Kentucky counties: See Potts, "Pressing on the Upward Way."

175 America's one hundred lowest median-income counties: U.S. Census Bureau, "Small Area Poverty and Income Estimates" (2010 U.S. and all States and Counties data file), http://www.census.gov/did/www/saipe/data/statecounty/data/2010.html; Appalachian Regional Commission, "Counties in Appalachia," http://www.arc.gov/counties; see also Housing Assistance Council, "Central Appalachia," http://www.ruralhome.org/storage/documents/appalov.pdf, p. 58 ("Over 43 percent of Central Appalachia's counties experienced poverty rates of 20 percent or more in 1960, 1970, 1980, 1990, and 2000").

176 far more socially acceptable: See Jim Goad, *The Redneck Manifesto: How Hillbillies, Hicks, and White Trash Became America's Scapegoats* (New York: Simon & Schuster, 1997), p. 15; Harkins, *Hillbilly*, p. 8.

176 "Our magazines and sitcoms": Goad, *The Redneck Manifesto*, pp. 15, 100; see also Harkins, *Hillbilly*, especially chaps. 1 and 2; Anne Shelby, "The 'R' Word: What's So Funny (and Not So Funny) About Redneck Jokes," in Dwight B. Billings, Gurney Norman, and Katherine Ledford, eds., *Back Talk from an American Region: Confronting Appalachian Stereotypes* (Lexington: University Press of Kentucky, 1999), pp. 153–4.

176 "just to go along" . . . "anger": Shelby, "The 'R' Word," p. 154.

176–7 "counter the 'dumb hillbilly' stereotype": Phillip J. Obermiller, "Paving the Way: Urban Organizations and the Image of Appalachians," in Billings et al., *Back Talk from an American Region*, pp. 251, 258.

177 perhaps contributing to the stereotypes: See Dwight B. Billings and Kathleen M. Blee, *The Road to Poverty: The Making of Wealth and Hardship in Appalachia* (Cambridge, UK: Cambridge University Press, 2000), pp. 13–4 (critiquing "culture-of-poverty" theories for "blatant stereotyping and victim blaming" and noting that even richer theories embrace "perjorative views about mountain people").

177 "defeatism," "dejection": Harry M. Caudill, *Night Comes to the Cumberlands: A Biography of a Depressed Area* (Boston: Little, Brown and Company, 1962), pp. 79, 346, 392; see also Jack E. Weller, *Yesterday's People: Life in Contemporary Appalachia* (Lexington: University of Kentucky Press, 1965), pp. 2, 20, 37.

177 "Appalachian fatalism": O'Brien, *At Home in the Heart of Appalachia*, p. 24.

177 "We're a religious bunch": House and Howard, *Something's Rising*, pp. 52, 59.

177 "helpless before the God": Erikson, *Everything in Its Path*, p. 85; see Weller, *Yesterday's People*, p. 104.

177 working two and three jobs: Potts, "Pressing on the Upward Way."

177 undermined by government welfare: Charles Murray, *Coming Apart: The State of White America, 1960–2010* (New York: Crown Forum, 2012), pp. 170–81, 216–9; see Caudill, *Night Comes to the Cumberlands*, pp. 275–6.

177 Obesity is common: Centers for Disease Control and Prevention, "Estimated County-Level Prevalence of Diabetes and Obesity—United States, 2007," Morbidity and Mortality

Weekly Report 58, no. 45 (2009), p. 1259 (reporting over 30 percent obesity rates in West Virginia and Appalachian Kentucky and Tennessee).

177 "As kids, we never learned": J. D. Vance, manuscript on file with authors (New York: HarperCollins, forthcoming 2014).

178 abuse rates of prescription opioid painkillers: See "Prescription Drug Abuse in Appalachia," http://www.nytimes.com/slideshow/2011/04/03/us/DRUGS.html; Appalachian Regional Commission, "ARC Study: Disproportionately High Rates of Substance Abuse in Appalachia," August 2008, http://www.arc.gov/news/article.asp?ARTICLE_ID=113; Lisa King, "Oxycontin Is the Drug of Choice of Appalachian Addicts, *The Washington Times*, June 9, 2012.

178 "pillbillies": "Editorial: 'Pillbilly' Addicts," *The Charleston Gazette*, Oct. 10, 2012.

178 1 in 10 newborns tested positive: Sabrina Tavernise, "Ohio County Losing Its Young to Painkillers' Grip," *New York Times*, Apr. 19, 2011.

178 highest teen pregnancy rates: Commonwealth of Kentucky, Department for Public Health, Division of Women's Health, *Teen Pregnancy Prevention Strategic Plan* (Cabinet for Health and Family Services, Commonwealth of Kentucky), p. 1. The New Hampshire rate is from U.S. Department of Health and Human Services, Centers for Disease Control and Prevention, "Births: Final Data for 2010," *National Vital Statistics Reports* 61, no. 1 (Aug. 28, 2012), p. 7, Table B.

178 a version of the "resource curse": The "resource curse"—a term coined by Richard Auty in 1993—refers to the idea that societies with too much oil, gold, or other extremely valuable resources typically end up mired in poverty. Subsequent research has broadly confirmed Auty's thesis. See Richard M. Auty, *Sustaining Development in Mineral Economies: The Resource Curse Thesis* (New York: Routledge, 1993); Macartan Humphreys, Jeffrey Sachs, Joseph E. Stiglitz et al., *Escaping the Resource Curse* (New York: Columbia University Press, 2007); Thomas L. Friedman, "The First Law of Petropolitics," *Foreign Policy*, May/June 2006. For empirical studies finding evidence of the resource curse thesis as applied to Appalachian poverty, see Annie Walker, "An Empirical Analysis of Resource Curse Channels in the Appalachian Region," Feb. 19, 2013, http://www.be.wvu.edu/econ_seminar/documents/12-13/walker.pdf; Mark D. Partridge, Michael R. Betz, and Linda Lobao, "Natural Resource Curse and Poverty in Appalachian America," *American Journal of Agricultural Economics* 95, no. 2 (2013), pp. 449–56.

178 Traditional industry . . . Salt and timber . . . mechanized: Billings and Blee, *The Road to Poverty*, pp. 243, 264–9; Housing Assistance Council, "Central Appalachia," p. 58; Christopher Price, "The Impact of the Mechanization of the Coal Mining Industry on the Population and Economy of Twentieth Century West Virginia," *West Virginia Historical Society Quarterly* 22, no. 3 (2008), pp. 2–3; Amanda Paulson, "In Coal Country, Heat Rises over Latest Method of Mining," *The Christian Science Monitor*, Jan. 3, 2006, p. 2.

178 Mountaintop removal: See, e.g., House and Howard, *Something's Rising*, pp. 1–2, 12–3; Billings and Blee, *The Road to Poverty*, p. 243; Paulson, "In Coal Country, Heat Rises over Latest Method of Mining," p. 2.

179 "double jeopardy": Christopher Bollinger, James P. Ziliak, and Kenneth R. Troske, "Down from the Mountain: Skill Upgrading and Wages in Appalachia," *Journal of Labor Economics* 29, no. 4, October 2011, p. 843.

179 catastrophic industrial accidents: House and Howard, *Something's Rising*, p. 9; Erikson, *Everything in Its Path*, p. i (describing a 1972 industrially caused "avalanche of black water and mine waste" in West Virginia and its aftermath). Some of the Buffalo Creek victims did eventually win a lawsuit and receive damages. Ibid., pp. 247–8.

179 marshmallow test: See Walter Mischel, Ebbe B. Ebbeson, and Antonette Raskoff Zeiss, "Cognitive and Attentional Mechanisms in Delay of Gratification," *Journal of Personality and Social Psychology* 21, no. 2 (1972), pp. 204–18.

179 reran the test with a new wrinkle: Celeste Kidd et al., "Rational Snacking: Young Children's Decision-Making on the Marshmallow Task Is Moderated by Beliefs About Environmental Reliability," *Cognition* 126 (2013), pp. 109–14.

180 childhood poverty and abuse: See, e.g., W. R. Lovallo et al., "Early Life Adversity Contributes to Impaired Cognition and Impulsive Behavior: Studies from the Oklahoma Family Health Patterns Project," *Alcoholism: Clinical and Experimental Research* 37, no. 4 (2013), pp. 616–23. One researcher theorizes that this result is due to long-term change in brain functioning. William R. Lovallo, "Early Life Adversity Reduces Stress Reactivity and Enhances Impulsive Behavior: Implications for Health Behaviors," *International Journal of Psychophysiology* (in press), abstract available at http://www.ncbi.nlm.nih.gov/pubmed/23085387.

180 Girls who become pregnant: Rumbaut, "Paradise Shift," pp. 23–4, 33–4.

180 "growing concern": Cathy Brownfield, "Abuse Is Devastating in Appalachia," *Salem News*, Mar. 11, 2012, http://www.salemnews.net/page/content.detail/id/552029.

180 roughly two hundred thousand: Joe Mackall, *Plain Secrets: An Outsider Among the Amish* (Boston: Beacon Press, 2007), pp. 5, 7.

181 Renno Amish: Richard A. Stevick, *Growing Up Amish: The Teenage Years* (Baltimore: The Johns Hopkins University Press, 2007), pp. 10–1, 43; "Indiana Amish," Amish America, http://amishamerica.com/indiana-amish ("Swiss Amish only travel by open buggy").

181 Beachy Amish . . . Swartzentruber Amish: Mackall, *Plain Secrets*, pp. xxi–xxii.

181 "when they are in diapers": Stevick, *Growing Up Amish*, pp. 41, 47, 55, 62, 81, 106–7, 112–3.

181 unconditional compliance: John A. Hostetler, *Amish Society* (4th ed.) (Baltimore and London: The Johns Hopkins University Press, 1993), pp. 77.

181 Fairy tales, science fiction: Stevick, *Growing Up Amish*, p. 72.

181 conducted in "high" German: Ibid., p. 22.

181 "usually keep to the task": Ibid., p. 106.

181 extreme humility . . . "Do not be haughty": Hostetler, *Amish Society*, pp. 77, 247.

181 taught to abhor any effort by one individual to rise: Mackall, *Plain Secrets*, pp. 40–1; Stevick, *Growing Up Amish*, p. 43.

181–2 "*high* school and *higher* education produce *Hochmut*": Stevick, *Growing Up Amish*, p. 61.

182 Nietzsche . . . Christianity . . . reverse superiority: Friedrich Nietzsche, *On the Genealogy of Morality*, trans. Maudemarie Clark and Alan J. Swensen (Indianapolis, IN: Hackett Publishing Co., 1998) (1887), pp. 16–7 (describing Christianity as motivated by a "desire for revenge" and as consummating Judaism's "slave morality," which proclaims that "the poor, powerless, lowly alone are the good . . . whereas you, you noble and powerful ones, you are in all eternity the evil"); see also James Q. Whitman, *Harsh Justice: Criminal Punishment and the Widening Divide Between America and Europe* (New York: Oxford University Press, 2005), p. 21 (according to Nietzsche, the horrific punishments in the afterlife contemplated by Christianity offered people "the immense satisfaction of witnessing the torments of the damned, and being thereby confirmed in their own blessed superiority").

182 "not highly ethnocentric" . . . accept other people: Hostetler, *Amish Society*, p. 77.

182 don't even believe they are saved: Mackall, *Plain Secrets*, p. 8.

182 "contaminat[ed]": Stevick, *Growing Up Amish*, p. 63; see also Hostetler, *Amish Society*, pp. 75–6.

182 "Letting children go unsupervised": Stevick, *Growing Up Amish*, p. 109.

183 "to be separate from the world": Ibid., pp. 7–8.

183 try to suppress the kind of thinking: Ibid., pp. 43–4.

183 "spirit of competition" . . . "when batting and catching": Ibid., p. 112.

183 the Amish have their worries: See, e.g., Hostetler, *Amish Society*, pp. 119–20.

183 aren't worried about proving themselves in America's eyes: See, e.g., Stevick, *Growing Up Amish*, pp. 55–6.

184 "love of money" . . . "last shall be" . . . "camel": Timothy 6:10; Matthew 20:16; Matthew 19:24.

184 "remarkable" fact: Max Weber, *The Protestant Ethic and the Spirit of Capitalism* (New York: Routledge, 1992), p. 35.

185 Calvinist doctrine taught: Ibid., pp. 93–106; Guy Oakes, "The Thing That Would Not Die: Notes on Refutation," in Hartmut Lehmann and Guenther Roth, eds., *Weber's Protestant Ethic: Origins, Evidence, Contexts* (Washington, DC: Cambridge University Press, German Historical Institute, 1993), p. 286.

185 "The question, *Am I one of the elect*": Weber, *The Protestant Ethic and the Spirit of Capitalism*, p. 110 (italics added).

185 "doctrine of proof": Ibid., p. 122.

185 "calling": Ibid., pp. 160–2.

185 "could measure his worth": Max Weber, "Anticritical Last Word on the Spirit of Capitalism," *American Journal of Sociology* 83, no. 5 (1978), p. 1124, quoted in Malcolm H. MacKinnon, "The Longevity of the Thesis: A Critique of the Critics," in Lehmann and Roth, *Weber's Protestant Ethic*, pp. 211, 224; see also Weber, *The Protestant Ethic and the Spirit of Capitalism*, pp. 176–7; Oakes, "The Thing That Would Not Die," pp. 285, 286–9.

185 They had to show everyone: Oakes, "The Thing That Would Not Die," pp. 287–8.

186 Inherited wealth . . . national differences . . . minorities or majorities: Weber, *The Protestant Ethic and the Spirit of Capitalism*, pp. 35–46.

186 geography can shape: See Jared Diamond, *Guns, Germs and Steel: The Fates of Human Societies* (New York and London: W. W. Norton & Co., 1999).

186 success is a "sign of divine favor": James Carroll, "The Mormon Arrival," *Boston Globe*, Aug. 7, 2011.

187 Word of Wisdom: Matthew Bowman, *The Mormon People: The Making of an American Faith* (New York: Random House, 2012), pp. 169–70.

187 "Family Home Evening": Ibid., pp. 168–9.

187 Missionary work increased: The Church of Jesus Christ of Latter-day Saints, "History of Missionary Work in the Church," Jun. 25, 2007, http://www.mormonnewsroom.org/article/history-of-missionary-work-in-the-church. The over 650,000 post-1990 estimate stated in the text is an extrapolation from the figure given by the Church for 1990–2007, which was approximately 535,000. Ibid.

187 banned polygamy, in 1904: Bowman, *The Mormon People*, pp. 159–60.

187 Mormons turned outward: Ibid., pp. 151, 152–3, 217; Jan Shipps, *Mormonism: The Story of a New Religious Tradition* (Chicago: University of Illinois Press, 1985), pp. 114–6. On the separatist Utah phase of Mormon history, see Franklin D. Daines, "Separatism in Utah, 1847–1870," *Annual Report of the American Historical Association for the Year 1917* (Washington, DC, 1920), pp. 331–43.

188 refused to go along with the renunciation: Bowman, *The Mormon People*, pp. 178–9.

188 Fundamentalist Church of Jesus Christ of Latter-Day Saints: Jon Krakauer, *Under the Banner of Heaven: A Story of Violent Faith* (New York: Doubleday, 2003), pp. 10–5, 167.

188 don't refrain from alcohol, cigarettes, or coffee: Nate Carlisle, "Alcohol, Coffee and Why the FLDS Drink Them," *Salt Lake Tribune*, Mar. 1, 2013, http://www.sltrib.com/sltrib/blog spolygblog/55924890-185/flds-coffee-church-sltrib.html.csp.

188 Colorado City is one of the poorest: U.S. Census, American Community Survey, Table DP03: Selected Economic Characteristics (2010 5-year dataset) (geographical designations: United States; Colorado City, Arizona) (showing a 36.5 percent poverty rate for all Colorado City families, compared with 10 percent for the country overall).

188 post-traumatic stress . . . genetic scars: Jeffrey Kluger, "Genetic Scars of the Holocaust: Children Suffer Too," *Time*, Sept. 9, 2010; Harvey A. Barocas and Carol B. Barocas,

"Separation-Individuation Conflicts in Children of Holocaust Survivors," *Journal of Contemporary Psychology* 11, no. 1 (1998), pp. 6–14.

189 outperform other groups economically: Aaron Hass, *In the Shadow of the Holocaust: The Second Generation* (Ithaca, NY, and London: Cornell University Press, 1996), p. 44; Ephraim Yuchtman Yaar and Gila Menachem, "Socioeconomic Achievements of Holocaust Survivors in Israel: The First and Second Generation," *Contemporary Jewry* 13 (1992), pp. 95–123; Morton Weinfeld and John J. Sigal, "Educational and Occupational Achievement of Adult Children of Holocaust Survivors," in Usial Oskar Schmelz and Sergio Della Pergola, eds., *Papers in Jewish Demography 1985: Proceedings of the Demographic Sessions Held at the 9th World Congress of Jewish Studies* (Jerusalem: Hebrew University of Jerusalem, 1985), p. 359.

189 "I remember as a child always worrying": Fran Klein-Parker, "Dominant Attitudes of Adult Children of Holocaust Survivors Toward their Parents," in John P. Wilson, Zev Harel, and Boaz Kahana, eds., *Human Adaptation to Extreme Stress: From the Holocaust to Vietnam* (New York: Plenum, 1988), pp. 207–8.

189–90 "'For this I survived'" . . . "I, my needs" . . . "Acts of rebellion": Hass, *In The Shadow of the Holocaust*, pp. 51, 53. For additional quotations from other survivors' children, see ibid., pp. 60, 62–3; Helen Epstein, *Children of the Holocaust: Conversations with Sons and Daughters of Survivors* (New York: G. P. Putnam's Sons, 1979), p. 37.

190 "inhibited from making noise": Barocas and Barocas, "Separation-Individuation Conflicts in Children of Holocaust Survivors," p. 11.

190 "My parents always said": Epstein, *Children of the Holocaust*, pp. 23, 27.

190 "When college ended": Sonia Taitz, *The Watchmaker's Daughter: A Memoir* (New York: McWitty Press, 2012), p. 152.

191 "My mother raised us": Hass, *In the Shadow of the Holocaust*, p. 58. For an article discussing the prevalence of anxiety and feelings of insecurity among second generation survivors, see Miri Scharf, "Long-term Effects of Trauma: Psychosocial Functioning of the Second and Third Generation of Holocaust Survivors," *Development and Psychopathology* 19 (2007), pp. 604–5, 617.

191 "They felt it was up to them": Klein-Parker, "Dominant Attitudes," p. 206; see also Leslie Gilbert-Lurie, *Bending Toward the Sun: A Mother and Daughter Memoir* (New York: Harper Perennial, 2010).

191 "need to resurrect their lost families": Barocas and Barocas, "Separation-Individuation Conflicts in Children of Holocaust Survivors," pp. 8–9, 12.

191 "I was not David Greber": Natan P. F. Kellerman, *Holocaust Trauma: Psychological Effects and Treatment* (Bloomington, IN: iUniverse Books, 2009), p. 73.

192 "I wondered what I could do in my life": Epstein, *Children of the Holocaust*, pp. 169–70.

192 "I make a differentiation": Hass, *In the Shadow of the Holocaust*, p. 122.

192 "I may have said": Epstein, *Children of the Holocaust*, p. 19.

193 "My father and mother were both": Taitz, *The Watchmaker's Daughter*, p. 9.

193 "sense of superiority": Hass, *In the Shadow of the Holocaust*, p. 108.

193 "lingering insecurities": Weinfeld and Sigal, "Educational and Occupational Achievement of Adult Children of Holocaust Survivors," p. 363.

193 "Permeated by an intense drive": "Conspiracy of Silence: A Conversation with Yael Danieli," Reform Judaism Online (2009), http://reformjudaismmag.org/Articles/index.cfm?id=1530.

193 Ron Unz: Ron Unz, "The Myth of American Meritocracy," *The American Conservative*, Nov. 28, 2012.

194 many have criticized: See, e.g., Janet Mertz, "Janet Mertz on Ron Unz's 'Meritocracy' Article," andrewgelman.com/wp-content/uploads/2013/03/Mertz-on-Unz-Meritocracy-Article.pdf?a43d93 (suggesting that the true figure for Jewish top scorers in the Math Olympiad more like 25 to 30 percent in the 1970s, not over 40 percent, and 13 percent since 2000, not 2.5 percent).

194 a mean ranging from 108 to 115: See Steven Pinker, "The Lessons of the Ashkenazim: Groups and Genes," *The New Republic*, June 26, 2006.

194 U.S. mean of 100: In most IQ studies, the average American IQ is set at 100. See James R. Flynn and Lawrence G. Weiss, "American IQ Gains from 1932 to 2002: The WISC Subtests and Educational Progress," *International Journal of Testing* 7, no. 2 (2007), p. 210.

194 theory most in the news . . . "essentially cannot do": Gregory Cochran, Jason Hardy, and Henry Harpending, "Natural History of Ashkenazi Intelligence," *Journal of Biosocial Science* 38 (2006), pp. 659, 662–5, 674–9.

194 "bad genetics" . . . "bullshit": Jennifer Senior, "Are Jews Smarter?," *New York Magazine*, Oct. 24, 2005 (quoting Harry Ostrer and Sander Gilman). For work building on the Cochran/Hardy/Harpending thesis, see, e.g., Hanna David and Richard Lynn, "Intelligence Differences Between European and Oriental Jews in Israel," *Journal of Biosocial Science* 39 (2007), pp. 465–73.

195 IQ is not a complete predictor of success: See, e.g., Steinberg, *Beyond the Classroom*, pp. 59, 64–5; Flynn, *Asian Americans*, p. 77 (discussing the "IQ/Achievement Gap"); Harold W. Stevenson et al., "Mathematics Achievement of Chinese, Japanese, and American Children," *Science* 231 (1986), pp. 693–9; Harold W. Stevenson et al., "Cognitive Performance and Academic Achievement of Japanese, Chinese, and American Children," *Child Development* 56, no. 3 (1985), pp. 733–4. See also the sources cited in chapter 5 on the superior predictive power of impulse control over IQ in numerous studies of academic and economic outcomes.

195 "first- and second-generation Asian": Suet-ling Pong et al., "The Roles of Parenting Styles and Social Capital in the School Performance of Immigrant Asian and Hispanic Adolescents," *Social Science Quarterly* 86 (2005), pp. 928, 946–7; see also, e.g., Hao and Woo, "Distinct Trajectories in the Transition to Adulthood," pp. 1623, 1635; Grace Kao and Marta Tienda, "Optimism and Achievement: The Educational Performance of Immigrant Youth," *Social Science Quarterly* 76 (1995), pp. 9–17; Rumbaut, "The Coming of the Second Generation," pp. 196, 205, 219.

195 third-, fourth-, or fifth-generation: Jewish Virtual Library, "Family, American Jewish," http://www.jewishvirtuallibrary.org/jsource/judaica/ejud_0002_0006_0_06269.html.

195 secure . . . economically: Paul Burstein, "Jewish Educational and Economic Success in the United States: A Search for Explanations," *Sociological Perspectives* 50, no. 2 (2007), pp. 209–14; Barry R. Chiswick, "The Postwar Economy of American Jews," in Jeffrey S. Gurock, ed., *American Jewish History* (New York: Routledge, 1998), pp. 93–98 (discussing the "dramatic increase in the professionalization of the Jewish labor force" in the second half of the twentieth century and the disproportionately high mean incomes of American Jews).

195 and in their identity as Americans: See Alan M. Dershowitz, *The Vanishing American Jew: In Search of Jewish Identity for the Next Century* (New York: Touchstone, 1997) pp. 1–2.

195 Bellow and Roth . . . wrote "Jewish American" literature: Ethan Goffman, "The Golden Age of Jewish American Literature," *Proquest Discovery Guides*, March 2010, pp. 1, 5–8, http://www.csa.com/discoveryguides/jewish/review.pdf.

195 Matthew Weiner: Celia Walden, "Matthew Weiner: The Man Behind Mad Men," *The Telegraph*, Mar. 18, 2012.

195 not unlike the mid-twentieth-century WASP establishment: On WASP decline, see, e.g., Robert C. Christopher, *Crashing the Gates: The De-WASPing of America's Power Elite* (New York: Simon & Schuster, 1989); Eli Wald, "The Rise and Fall of the WASP and Jewish Law Firms," *Stanford Law Review* 60 (2008), pp. 1828–58; James D. Davidson et al., "Persistence and Change in the Protestant Establishment, 1930–1992," *Social Forces* 74, no. 1 (1995), pp. 157–75; E. Digby Baltzell, "The Protestant Establishment Revisited," *The American Scholar* 45, no. 4 (1976), pp. 499–518; Robert Frank, "That Bright, Dying Star, the American WASP," *The Wall Street Journal*, May 15, 2010; Noah Feldman, "The Triumphant Decline of the WASP," *New York Times*, June 27, 2010.

196 but the memory yet lingers: See Charles E. Silberman, *A Certain People: American Jews and Their Lives Today* (New York: Summit Books, 1985), pp. 327–31.

196 an almost ethnic identification: See Theodore Sasson et al., *Still Connected: American*

Jewish Attitudes About Israel (Waltham, MA: Brandeis University, Marice and Marilyn Cohen Center for Modern Jewish Studies, 2010), pp. 1, 9–13.

196 220 million people: See Amy Chua, *World on Fire: How Exporting Free Market Democracy Breeds Ethnic Hatred and Global Instability* (New York: Doubleday, 2003), pp. 212, 223–4. Population estimates for the Middle East vary considerably. Our rough figure for the region's total population is a conservative one, based on the 1990 estimate reported in Youssef M. Choueiri, *Arab Nationalism: A History* (Oxford, UK: Blackwell Publishers, 2000), p. vii.

196 *Cheerful Money*: See Tad Friend, *Cheerful Money: Me, My Family, and the Last Days of Wasp Splendor* (New York: Hachette Book Group, 2009).

196 Jews are insecure about losing their insecurity and as anxious as ever: See, e.g., Dershowitz, *The Vanishing American Jew*, pp. 1–18; Lee Siegel, "Rise of the Tiger Nation," *Wall Street Journal*, Oct. 27, 2012.

196 Jews are also among America's preeminent poets . . . opinion leaders: See Steven L. Pease, *The Golden Age of Jewish Achievement* (Sonoma, CA: Deucalion, 2009), pp. viii–ix, 76–7, 103–4, 126–7, 156–7, 388–9.

197 Mormons . . . bridling at the culture of conformity: Bowman, *The Mormon People*, pp. 234–45.

197 "speak up, stand out": Buck Gee and Wes Hom, "The Failure of Asian Success in the Bay Area: Asians as Corporate Executive Leaders" (Corporate Executive Initiative, March 28, 2009), p. 5; see also Wesley Yang, "Paper Tigers," *New York Magazine*, May 8, 2011.

197 "no longer confined": Anita Raghavan, *The Billionaire's Apprentice: The Rise of the Indian-American Elite and the Fall of the Galleon Hedge Fund* (New York and Boston: Business Plus Books, 2013), p. 416.

197 much less observant: Fewer than 30 percent, and perhaps as few as 16 percent, of American Jews attend synagogue even once a month; 80 percent or more no longer keep kosher. See Humphrey Taylor, "While Most Americans Believe in God, Only 36% Attend a Religious Service Once a Month or More Often," Harris Poll No. 59, Oct. 15, 2003, p. 5, Table 3; Sue Fishkoff, *Kosher Nation* (New York: Schocken Books, 2010), p. 330 n. 29.

197 often described as "permissive": See, e.g., Elliott J. Rosen and Susan F. Weltman, "Jewish Families: An Overview," in Monica McGoldrick, Joe Giordano, and Nydia Garcia-Preto, *Ethnicity & Family Therapy* (3d ed.) (New York: The Guilford Press, 2005), pp. 667, 675 ("Jewish parents have tended to be permissive"); Lila Corwin Berman, "Blame, Boundaries, and Birthrights: Jewish Intermarriage in Midcentury America," in Susan A. Glenn and Naomi B. Sokoloff, *Boundaries of Jewish Identity* (Seattle: University of Washington Press, 2010), pp. 100–2 (describing turn to "permissive parenting" among American Jews in the 1950s and 60s).

198 drawn to American attitudes: Institute for Jewish & Community Research, "Projects: Ethnic and Racial Diversity: The Be'chol Lashon Initiative," http://www.jewishresearch.org/projects_diversity.htm (discussing how "Jews have become so integrated into American society that they tend to mirror American culture in many ways").

CHAPTER 8: AMERICA

200 "all the animals are much smaller": See Dwight Boehm and Edward Schwartz, "Jefferson and the Theory of Degeneracy," *American Quarterly* 9, no. 4 (1957), pp. 448–53; Lee Alan Dugatkin, *Mr. Jefferson and the Giant Moose: Natural History in Early America* (Chicago: University of Chicago Press, 2009), p. ix. The Buffon quotes are from Count de Buffon, *Natural History: General and Particular,* trans. William Smellie (3rd ed.) (A. Strahan and T. Cadell, 1791), vol. 5, p. 115; see also ibid., pp. 124–53.

201 No wonder, concluded Raynal: The Raynal quote is from Kevin J. Hayes, ed., *The Oxford Handbook of Early American Literature* (New York: Oxford University Press, 2008), p. 599.

201 American "exceptionalism": See Deborah L. Madsen, *American Exceptionalism* (Edinburgh: Edinburgh University Press, 1998), p. 2 ("America and Americans are special, exceptional because they are charged with saving the world from itself. . . . America must be as 'a city upon a hill' exposed to the eyes of the world"); Kammen, "The Problem of American Exceptionalism: A Reconsideration," *American Quarterly* 45 (1993), pp. 1, 6 ("American exceptionalism is as old as the nation itself . . . and has played an integral part in the society's sense of its own identity"). There is a large literature critiquing, or pronouncing the death of, American exceptionalism. For a small sampling, see Godfrey Hodgson, *The Myth of American Exceptionalism* (New Haven, CT: Yale University Press, 2010); Andrew J. Bacevich, *The Limits of Power: The End of American Exceptionalism* (New York: Henry Holt & Co., 2009); Michael Ignatieff, ed., *American Exceptionalism and Human Rights* (Princeton, NJ: Princeton University Press, 2005); Harold Hongju Koh, "On American Exceptionalism," *Stanford Law Review* 55 (2003); Daniel Bell, "The End of American Exceptionalism," *The Public Interest* 41 (1975), pp. 193–234.

201 "City upon a Hill": George McKenna, *The Puritan Origins of American Patriotism* (New Haven, CT, and London: Yale University Press, 2007), pp. 6–7.

201 "reserved to the people of this country": Alexander Hamilton, "Federalist 1" (1787), in Jack N. Rakove, ed., *The Federalist by Alexander Hamilton, James Madison, and John Jay: The Essential Essays* (Boston: Bedford/St. Martin's Press, 2003), p. 36.

201 "the Israel of our time": Herman Melville, *White-Jacket, or The World in a Man-of-War*, in *Redburn, White-Jacket, Moby-Dick* (New York: Library of America, 1983), p. 506.

201 exhorting his hunter friends: Dugatkin, *Mr. Jefferson and the Giant Moose*, pp. ix–xii; see also Mark V. Barrow Jr., *Nature's Ghosts: Confronting Extinction from the Age of Jefferson to the Age of Ecology* (Chicago: University of Chicago Press, 2009), pp. 15–9; Boehm and Schwartz, "Jefferson and the Theory of Degeneracy."

201 sent one proof after another: See Howard C. Rice Jr., "Jefferson's Gift of Fossils to the Museum of Natural History in Paris," *Proceedings of the American Philosophical Survey* 95, no. 6 (1951), pp. 597–627.

201 *Notes on Virginia*: Thomas Jefferson, "Notes on Virginia," in Paul Leicester Ford, ed., *The Works of Thomas Jefferson* (New York: G. P. Putnam's Sons, 1904), vol. 3, pp. 411-5 (discussing the "tremendous" mammoth); pp. 422–4 (chart comparing animals in Europe and America).

202 wrote an entire book: Richard L. Bushman, "The Romance of Andrew Carnegie," *American Studies* 6, no. 1 (1965), pp. 36–7; Andrew Carnegie, *Triumphant Democracy* (New York: J. Little & Co. 1886).

202 "the most self-conscious people": Henry James, *Hawthorne* (1887) (New York: AMS Press, 1968), p. 153.

202 no "stations": Alexis de Tocqueville, *Democracy in America*, trans. Henry Reeve (London: Longman, Green, Longman, and Roberts, 1862), vol. 2, part III, chap. 14, p. 262; see also ibid., chap. 21, pp. 301–2 (noting that in America, nothing keeps men "settled in their station," that "every man, finding himself possessed of some education and some resources, may choose his own path, and proceed apart from all his fellow-men," and that "there is no longer a race of poor men . . . [or] a race of rich men").

203 "longing to rise": Alexis de Tocqueville, *Democracy in America*, trans. George Lawrence, ed. J. P. Mayer and Max Lerner, vol. 2, part III, chap. 19 (New York: Harper & Row, 1966), p. 603.

203 Franklin's parents were Puritan: See Walter Isaacson, *Benjamin Franklin: An American Life* (New York: Simon & Schuster, 2003) pp. 5–15, 526, n. 35.

203 "Industry, Perseverance & Frugality": "Poor Richard for 1744," in Paul Leicester Ford, ed., *The Prefaces, Proverbs, and Poems of Benjamin Franklin Originally Printed in Poor Richard's Almanacs for 1733–1758* (New York and London: G. P. Putnam's Sons), p. 147.

203 "Dost thou love Life" . . . **"There are no Gains"** . . . **"Leisure is Time":** "Poor Richard for 1758: Preface," in Ford, *The Prefaces*, pp. 270, 271, 273.

203 "No man e'er was glorious": "Poor Richard for 1734," in Ford, *The Prefaces*, p. 38.

203 "Be at War with your Vices": "Poor Richard for 1755," in Ford, *The Prefaces*, p. 245.

203 "To lengthen thy life": "Poor Richard for 1733," in Ford, *The Prefaces*, p. 28.

203 "He that can have patience" . . . **"Diligence is the mother of good luck":** "Poor Richard for 1736," in Ford, *The Prefaces*, pp. 61, 62.

204 antiauthoritarian ferment: See, e.g., Akhil Reed Amar, *The Bill of Rights: Creation and Reconstruction* (New Haven, CT, and London: Yale University Press, 1998), pp. 21, 23, 46, 79; see generally Gordon S. Wood, *The Radicalism of the American Revolution* (New York: Alfred Knopf, 1991).

204 "loosened the bonds": Claude S. Fischer, *Made in America: A Social History of American Culture and Character* (Chicago: University of Chicago Press, 2010), p. 110 (quoting John Adams).

204 More wives . . . College students . . . Young couples: Ibid., pp. 111–2.

204 A third of the brides: Ibid., p. 113.

204 "a country of beginnings": Ralph Waldo Emerson, *Nature: Addresses and Lectures* (Boston: Houghton Mifflin and Company, 1903), p. 371.

204 Alexander Hamilton: Ashamed of his illegitimate origins, Hamilton "decided to cut himself off from the past and forge a new identity. He would find a home where he would be accepted for what he did, not for who he was." Ron Chernow, *Alexander Hamilton* (New York: The Penguin Press, 2004), p. 40.

204 "With the past": Ralph Waldo Emerson, *The Journals and Miscellaneous Notebooks of Ralph Waldo Emerson, 1838–1842* (Boston: Harvard University Press, 1969) p. 241.

205 "[T]he earth belongs to the living": Letter from Thomas Jefferson to James Madison (Sept. 6, 1789), in Julian P. Boyd, ed., *The Papers of Thomas Jefferson, Volume 15, 27 March 1789 to 30 November 1789* (Princeton, NJ: Princeton University Press, 1958), p. 396; Jed Rubenfeld, *Freedom and Time* (New Haven, CT, and London: Yale University Press, 2001), pp. 18–21.

205 unprecedented experiment, which many expected to fail: Gordon S. Wood, *The American Revolution: A History* (New York: Modern Library, 2002), pp. 139–46; see also Michael J. G. Cain and Keith L. Dougherty, "Suppressing Shays' Rebellion: Collective Action and Constitutional Design Under the Articles of Confederation," *Journal of Theoretical Politics* 11, no. 2 (1999), pp. 233–60 (discussing the "constitutional failures associated with the Articles of Confederation"—especially the "collective action problem"—which were "solved by the new Constitution").

205 full of lawlessness, ineffectual government: Gordon S. Wood, *The Creation of the American Republic 1776–1787* (Chapel Hill: University of North Carolina Press, 1969), pp. 367–9; Cain and Dougherty, "Suppressing Shay's Rebellion," pp. 233–4.

205 close to anarchy: See Alexander Hamilton, "Federalist 15" (1787), in Rakove, *The Federalist*, p. 65 ("It must in truth be acknowledged that . . . there are material imperfections in our national system, and that something is necessary to be done to rescue us from impending anarchy."); Richard Sylla, *The Rise of Securities Markets: What Can Government Do?* (The World Bank: November 1995), pp. 6–7 (describing the disarray when the Articles of Confederation governed).

205 But if the majority themselves turned tyrant: See, e.g., James Madison, "Note to His Speech on the Right of Suffrage" (1821), in Max Farrand, ed., *The Records of the Federal Convention of 1787, Volume III* (New Haven, CT: Yale University Press, 1911), pp. 450, 452 (warning of the "danger" to "the rights of property" posed by universal suffrage "whenever the Majority shall be without landed . . . property"); Wood, *The Creation of the American Republic*, pp. 409–11; Morton J. Horwitz, "Tocqueville and the Tyranny of the Majority," *The Review of Politics* 28, no. 3 (1966), p. 293 ("the problem of tyranny of the majority had dominated the political thought of no other nation as it had that of America"); cf. Charles A. Beard, *An Economic Interpretation of the Constitution of the United States* (New York: MacMillan Co., 1952), pp. 15–18 (speculating that the Constitution's Framers may have been chiefly concerned with protecting their own property from majoritarian appropriation).

206 "passions . . . of the public": James Madison, "Federalist 49" (1788), in Rakove, *The Federalist*, pp. 130–3.

206 "Constitutions are chains": John E. Finn, *Constitutions in Crisis: Political Violence and the Rule of Law* (New York: Oxford University Press, 1991), p. 5 (quoting U.S. Senator John Potter Stockton in debates over the Ku Klux Klan Act of 1871); see Abraham Lincoln, First Inaugural Address, Mar. 4, 1861 ("A majority held in restraint by constitutional checks and limitations . . .

is the only true sovereign of a free people"); David J. Brewer, "An Independent Judiciary as the Salvation of the Nation" (1893), reprinted in *The Annals of America: Agrarianism and Urbanization 1884–1894* (1968), vol. 11, pp. 423, 428 ("Constitutions . . . are rules proscribed by Philip sober to control Philip drunk"); Thomas M. Cooley, *A Treatise on the Constitutional Limitations Which Rest upon the Legislative Power of the States of the American Union* (Boston: Little, Brown, and Co., 1868), pp. 54–5 (Constitution protects against "the danger that the legislature will be influenced by temporary excitements and passions among the people"); Stephen Holmes, *Passions and Constraint: On the Theory of Liberal Democracy* (Chicago: University of Chicago Press, 1995), chap. 5; cf. Akhil Reed Amar, "The Central Meaning of Republican Government: Popular Sovereignty, Majority Rule, and the Denominator Problem," *University of Colorado Law Review* 65 (1994), p. 761 (noting the widely held view that "the Bill of Rights was designed to inhibit majority tyranny and limit popular passion" but pointing out that the Constitution was also designed to restrict wayward legislators and presidents).

206 structure and restraint: William J. Brennan Jr., "Construing the Constitution," *University of California, Davis Law Review* 19, p. 6 ("It is the very purpose of a Constitution—and particularly of the Bill of Rights—to declare certain values transcendent, beyond the reach of temporary political majorities. The majoritarian process cannot be expected to rectify claims of minority right that arise as a response to the outcomes of that very majoritarian process"); Wood, *The Creation of the American Republic*, p. 453.

206 powerful national government: Wood, *The American Revolution*, p. 151 (noting that the Constitution created an "extraordinarily powerful national government" that "possessed far more than the additional congressional powers that were required to solve the United States' difficulties in credit, commerce, and foreign affairs"); Wood, *The Creation of the American Republic*, p. 467; see generally Akhil Reed Amar, *America's Constitution: A Biography* (New York: Random House, 2005).

206 tyranny: See Hannah Arendt, *On Revolution* (London: Penguin, 1965), p. 233 (describing Jefferson's view).

206 "a more perfect Union": United States Constitution, preamble.

207 A great many Americans believed: See Rogers M. Smith, "Beyond Tocqueville, Myrdal, and Hartz: The Multiple Traditions in America," *American Political Science Review* 87, no. 3 (1993), p. 549 ("For over 80% of U.S. history, its laws declared most of the world's population to be ineligible for American citizenship solely because of their race, original nationality, or gender. . . . [At the time of the Founding,] [m]en were thought naturally suited to rule over women. . . . White northern Europeans were thought superior culturally—and probably biologically—to black Africans, bronze Native Americans, and indeed all other races and civilizations. Many British Americans also treated religions as an inherited condition and regarded Protestants as created by God to be morally and politically, as well as theologically, superior to Catholics, Jews, Muslims, and others"); Thurgood Marshall, "Reflections on the Bicentennial of the United States Constitution," *Harvard Law Review* 101 (1987), pp. 1–5; Ellen Carol DuBois, "Outgrowing the Compact of the Fathers: Equal Rights, Woman Suffrage, and the United States Constitution, 1820–1878," *The Journal of American History* 74 (1987), pp. 836–62.

207 stains on American history: See, e.g., Mary Frances Berry, *Black Resistance/White Law: A History of Constitutional Racism in America* (New York: Penguin Books, 1994) (discussing discrimination and racial violence that has persisted over the past century and a half, afflicting blacks as well as Asian Americans, Latinos, and Native Americans); Robert A. Williams Jr., *Like a Loaded Weapon: The Rehnquist Court, Indian Rights, and the Legal History of Racism in America* (Minneapolis: University of Minnesota Press, 2005) (criticizing the racist roots of federal Indian law); Jose Monsivais, "A Glimmer of Hope: A Proposal to Keep the Indian Child Welfare Act of 1978 Intact," *American Indian Law Review* 22 (1997), p. 2 ("When Andrew Jackson became President of the United States in 1829, the policy of removal began. The Indian Removal Act of

1830 was enacted to relocate most of the eastern tribes west of the Mississippi river. . . . Vast numbers of American Indians were marched westward onto lands considered unfit for human life"); Andrew Gyory, *Closing the Gate: Race, Politics, and the Chinese Exclusion Act* (Chapel Hill, NC, and London: University of North Carolina Press, 1998); William D. Carrigan, "The Lynching of Persons of Mexican Origin or Descent in the United States, 1848 to 1928," *Journal of Social History* 37, no. 2 (2003), pp. 411-38.

208 "History is more or less bunk": Charles N. Wheeler, "Fight to Disarm His Life's Work, Henry Ford Vows," *Chicago Tribune*, May 25, 1916, p. 10.

209 "disintegration of the work ethic": Robert Rector and Jennifer A. Marshall, "The Unfinished Work of Welfare Reform," *National Affairs* (Winter 2013).

209 "There is still today a frontier": President Franklin D. Roosevelt, "Radio Address on the Third Anniversary of the Social Security Act," August 15, 1938, The American Presidency Project, http://www.presidency.ucsb.edu/ws/index.php?pid=15523.

209 "Second Bill of Rights" . . . "inadequate": President Franklin D. Roosevelt, "State of the Union Message to Congress," January 11, 1944, The American Presidency Project, http://www.presidency.ucsb.edu/ws/index.php?pid=16518.

209 "freedom from insecurity": President Franklin D. Roosevelt, "Greeting to the Economics Club of New York," Dec. 2, 1940, The American Presidency Project, http://www.presidency.ucsb.edu/ws/index.php?pid=15907; see also David M. Kennedy, *Freedom from Fear: The American People in Depression and War, 1929–1945* (New York: Oxford University Press, 1999), p. 247. The term "social security" is believed to have been coined by the economist Abraham Epstein, who in 1933 wrote, "Ever since Adam and Eve . . . insecurity has been the bane of mankind." Abraham Epstein, *Insecurity: A Challenge to America* (New York: H. Smith & R. Haas, 1933), p. 1.

209 chances of graduating: Greg Toppo, "Big-City Schools Struggle with Graduation Rates," *USA Today*, June 20, 2006.

209 $130,000 a year selling drugs: Steven D. Levitt and Sudhir Alladi Venkatesh, "An Economic Analysis of a Drug-Selling Gang's Finances," *The Quarterly Journal of Economics* (August 2000), pp. 755, 770.

210 at the average age of twenty: See Leon Bing, *Do or Die* (New York: Harper Perennial, 1992), p. 268 (estimating the life expectancy of active gang members in South Central Los Angeles at nineteen years); William J. Harness, Chief of Police, *Gang Facts and Myths: A Guide for School Administrators* (Conroe ISD Police Department, 1994–2006), p. 21 (stating that the average life expectancy of an active gang member is 20 years, 5 months).

210 everything that's important about the big picture: See, e.g., William Julius Wilson, *When Work Disappears: The World of the New Urban Poor* (New York: Alfred A. Knopf, 1996), pp. 52–3 ("Neighborhoods that offer few legitimate employment opportunities, inadequate job information networks, and poor schools lead to the disappearance of work . . . many people eventually lose their feeling of connectedness to work in the formal economy . . . . These circumstances also increase the likelihood that the residents will rely on illegitimate sources of income"), p. 107 (Where "young people have little reason to believe that they have a promising future," there tends to be "an explosion of single-parent families"); see also Thomas J. Sugrue, *The Origins of the Urban Crisis: Race and Inequality in Post-War Detroit* (Princeton, NJ: Princeton University Press, 1996); Bruce Western, *Punishment and Inequality in America* (New York: Russell Sage Foundation, 2006); Alex Kotlowitz, *There Are No Children Here: The Story of Two Boys Growing Up in the Other America* (New York: Anchor, 1991).

210 apply at McDonald's: See Paul Krugman, *End This Depression Now!* (New York and London: W. W. Norton & Co., 2012), p. 7; see also Levitt and Venkatesh, "An Economic Analysis of a Drug-Selling Gang's Finances," p. 771 (noting that gang members often also hold low-paying jobs in shopping malls and fast-food restaurants).

210 wiped out all the sectors: See, e.g., Sugrue, *The Origins of the Urban Crisis*, p. 6; Dwight

B. Billings and Kathleen M. Blee, *The Road to Poverty: The Making of Wealth and Hardship in Appalachia* (Cambridge, UK: Cambridge University Press, 2000), pp. 243, 264–9; Christopher Price, "The Impact of the Mechanization of the Coal Mining Industry on the Population and Economy of Twentieth Century West Virginia," *West Virginia Historical Society Quarterly* 22, no. 3 (2008), pp. 2–3.

210 rise in American income and standards of living: Claude S. Fischer and Michael Hout, *Century of Difference: How America Changed in the Last One Hundred Years* (New York: Russell Sage Foundation, 2006), pp. 139–40 (stating that "Americans easily quadrupled their real earnings" over the course of the twentieth century); U.S. Congress, "The U.S. Economy at the Beginning and End of the 20th Century," usinfo.org/enus/economy/overview/docs/century.pdf ("Today, the average full-time employee works about 40 hours per week rather than 60, and the average family spends just 15 percent of its income on food today, compared to 44 percent in 1900"); Raghuram G. Rajan, *Fault Lines: How Hidden Fractures Still Threaten the World Economy* (Princeton, NJ: Princeton University Press, 2010), p. 31 (emphasizing U.S. policies that widened availability of credit to ordinary Americans starting in the early 1980s).

210 greatest wealth explosions: See, e.g., Niall Ferguson, *Colossus: The Price of America's Empire* (New York: HarperCollins, 2004), pp. 18–9; John Steele Gordon, *An Empire of Wealth: The Epic History of American Economic Power* (New York: HarperPerennial, 2004), pp. 416–8; Michael Lewis, *The New New Thing: A Silicon Valley Story* (New York: W. W. Norton & Company, 2000), p. 30 (quoting venture capitalist John Doerr as describing Silicon Valley in the 1990s as "the greatest legal creation of wealth in the history of the planet").

210 entire decade came to be symbolized: Tom Wolfe, *Bonfire of the Vanities* (New York: Farrar, Straus Giroux, 1987); see Jessica Winter, "Greed Is Bad. Bad!" Slate.com, Sept. 25, 2007, http://www.slate.com/articles/arts/dvdextras/2007/09/greed_is_bad_bad.html.

210 1990s were astronomically even richer: For graphs dramatically illustrating the record highs in stock prices, corporate earnings, and home prices, see Robert J. Shiller, *Irrational Exuberance* (2d ed.) (Princeton, NJ: Princeton University Press, 2005), pp. 3–6, 13; see also Bill Hutchinson, "He Brakes for Cash: Day-Trader Cabbie Winning Wall St. Game," *New York Daily News*, Aug. 19, 1999, p. 3.

210 multimillionaires overnight: Eryn Brown, "Valley of the Dollars: The Young, Wealthy Netheads of San Francisco and Silicon Valley Protest That It's Not About the Money. Give Us a Break," CNN.com, Sept. 27, 1999, http://money.cnn.com/magazines/fortune/fortune_archive/1999/09/27/266205; Ilana DeBare, "Young, Rich, Now What?: Tech Millionaires Face the Rest of Their Lives," *S.F. Chronicle*, June 4, 1999, http://www.sfgate.com/business/article/Young-Rich-Now-What-Tech-millionaires-face-2927263.php; Eryn Brown, "So Rich, So Young, but Are They Really Happy?," CNNMoney.com, Sept. 18, 2000, http://money.cnn.com/magazines/fortune/fortune_archive/2000/09/18/287692.

210 stock market tripled: Shiller, *Irrational Exuberance*, p. 13.

210 sixty-four new millionaires per day: Rusty Dornin, "New Anxieties Can Accompany Silicon Valley's New Money," CNN.com, Feb. 26, 2000, http://archives.cnn.com/2000/US/02/26/sudden.wealth.syndrome.

210 Corporate compensation soared: Nouriel Roubini and Stephen Mihm, *Crisis Economics: A Crash Course in the Future of Finance* (New York: The Penguin Press, 2010), pp. 68–9; Krugman, *End This Depression Now!*, p. 78.

210 few complained: Shiller, *Irrational Exuberance*, p. 3 (chart showing soaring stock prices in the 1990s), p. 213.

210 family net worth climbed: Felix Salmon, "Chart of the Day: Median Net Worth, 1962–2010," Reuters, June 12, 2012, http://blogs.reuters.com/felix-salmon/2012/06/12/chart-of-the-day-median-net-worth-1962-2010.

211 the top 1 percent of U.S. earners: Congressional Budget Office, *Trends in the Distribu-*

tion of Household Income Between 1979 and 2007 (Washington, DC: Congressional Budget Office, 2011), pp. ix–xi.

211 vast majority of Americans: Ibid., p. ix (reporting an approximate 65 percent income gain for Americans in the top quintile from 1979 to 2007 and 40 percent gain for those in the second through fourth quintiles). For other estimates of historical American income growth broken down by quintile, see U.S. Census Bureau, Historical Income Tables: Households, Table H-3: Mean Household Income Received by Each Fifth and Top 5 Percent (All Races), www .census.gov/hhes/www/income/income/data/historical/household; but cf. Russell Sage Foundation, "Chartbook of Social Inequality," www.russellsage.org/sites/all/files/chartbook/Income %20and%20 Earnings.pdf.

211 As economist Robert Shiller observes: Shiller, *Irrational Exuberance*, p. 213.

211 globalization seemed to herald: See Amy Chua, *World on Fire: How Exporting Free Market Democracy Breeds Ethnic Hatred and Global Instability* (New York: Doubleday, 2003), p. 233.

211 law students at Georgetown: Brooke Masters, "GU Legal Eagles Flying to Estonia's Aid," *Washington Post*, June 18, 1992.

211 "end point of mankind's ideological evolution" . . . "final form": Francis Fukuyama, *The End of History and the Last Man* (New York: Avon Books, 1992), p. xi.

211 started U.S.-style stock exchanges: See Klaus Weber, Gerald F. Davis, and Michael Lounsbury, "Policy as Myth and Ceremony? The Global Spread of Stock Exchanges, 1980–2005," *Academy of Management Journal* 52, no. 6 (2009), pp. 1319, 1320; Kathryn C. Lavelle, *The Politics of Equity Finance in Emerging Markets* (New York: Oxford University Press, 2004), pp. 19–22.

211 "hyperpower": See Amy Chua, *Day of Empire: How Hyperpowers Rise to Global Dominance—and Why They Fall* (New York: Doubleday, 2007), pp. 259–61.

212 "I cannot think of a single psychological problem": Nathaniel Branden, *The Six Pillars of Self-Esteem* (New York: Bantam Books, 1994), p. xv.

212 a bestseller: See Nathaniel Branden, *My Years with Ayn Rand* (San Francisco: Jossey-Bass, 1999), p. 368 (claiming about a million copies sold worldwide).

212 "virtually every *social* problem": Roy F. Baumeister and John Tierney, *Willpower. Rediscovering the Greatest Human Strength* (New York: Penguin Press, 2011), pp. 188–9 (italics added) (quoting the chairman of California's task force on self-esteem).

212 "[M]any, if not most": Ibid., p. 189 (quoting sociologist Neil Smelser).

212 "self-esteem was the most important thing": Carol S. Dweck, "Mindsets: How Praise Is Harming Youth and What Can Be Done About It," *School Library Media Activities Monthly* 24, no. 5 (2008), p. 55.

213 accomplishment remained central: See William James, *The Principles of Psychology* (New York: Henry Holt, 1890), vol. 1, pp. 309–11. James observes that a person has to succeed not in everything, but in what's important to him, in order to sustain his self-esteem: "I, who for the time have staked my all on being a psychologist, am mortified if others know much more psychology than I. But I am contented to wallow in the grossest ignorance of Greek." *Id.* at 310.

213 severed self-esteem from esteem-worthy conduct: This severance can be seen in the Rosenberg Self-Esteem Scale, the standard measure of self-esteem used by psychologists everywhere today. See, e.g., University of Maryland, Department of Sociology, "Rosenberg-Self-Esteem Scale," www.socy.umd.edu/quick-links/rosenberg-self-esteem-scale ("The Rosenberg Self-Esteem Scale is perhaps the most widely-used self-esteem measure in social science research"). Created by Morris Rosenberg in 1965, the scale is a question-and-answer survey instrument, entirely attitudinal, asking individuals in a variety of ways how good they feel about themselves. With the arguable exception of asking respondents how strongly they agree with the statement, "I am able to do things as well as most other people," the Rosenberg test asks no questions about actual conduct, performance, successes, failures, etc.

213 feel good about themselves: See Lori Gottlieb, "How to Land Your Kid in Therapy," *Atlantic Monthly* (July/August 2011) (questioning "self-esteem" that "comes from constant accommodation and praise rather than earned accomplishment").

213 much more satisfied with themselves: Jean M. Twenge and W. Keith Campbell, *The Narcissism Epidemic: Living in the Age of Entitlement* (New York: Free Press, 2009), p. 13 ("[s]elf-esteem is at an all-time high in most groups").

213 Asian American students: Douglas S. Massey et al., *The Source of the River: The Social Origins of Freshmen at America's Selective Colleges and Universities* (Princeton, NJ: Princeton University Press, 2003), pp. 120–1.

213 world's leaders in self-esteem . . . among the lower-scoring: See, e.g., Tom Loveless, *The 2006 Brown Center Report on American Education: How Well Are American Students Learning* (Washington, DC: Brookings Institution, 2006), pp. 13–20.

213 controlled experiment: See Donelson R. Forsyth, Natalie K. Lawrence, Jeni L. Burnette, and Roy F. Baumeister, "Attempting to Improve the Academic Performance of Struggling College Students by Bolstering Their Self-Esteem: An Intervention That Backfired," *Journal of Social and Clinical Psychology* 26 (2007), pp. 447–59; see also Baumeister and Tierney, *Willpower*, pp. 190–1.

213 another study: Carol S. Dweck, *Mindset: The New Psychology of Success* (New York: Random House, 2006), pp. 71–3.

213 "secretly feel bad about themselves": Lauren Slater, "The Trouble with Self-Esteem," *New York Times*, Feb. 3, 2002 (quoting Nicholas Emler).

213 Serial rapists: Baumeister and Tierney, *Willpower*, p. 191; see Betsy Hart, *It Takes a Parent: How the Culture of Pushover Parenting Is Hurting Our Kids* (New York: Penguin Group, 2005), p. 93.

213 depression and anxiety . . . narcissism: Gottlieb, "How to Land Your Kid in Therapy."

214 "primary task of psychotherapy": Nathaniel Branden, *The Psychology of Self-Esteem: A Revolutionary Approach to Self-Understanding that Launched a New Era in Modern Psychology* (San Francisco: Jossey-Bass, 2001), p. 251.

214 "so much red on the page": Gottlieb, "How to Land Your Kid in Therapy."

214 the self-esteem movement erodes: Baumeister and Tierney, *Willpower*, pp. 188–97; Gottlieb, "How to Land Your Kid in Therapy."

214 "People with incredibly positive": David Dent, "Bursting the Self-Esteem Bubble," *Psychology Today*, Mar. 1, 2002 (quoting Nicholas Emler); Baumeister and Tierney, *Willpower*, p. 192.

214 not taught to endure hardship: See Caitlin Flanagan, "The Ivy Delusion," *The Atlantic*, Feb. 24, 2011 (Today's "good mothers" desperately want their children to discover their "natural talent" and to achieve "effortless success" without experiencing any stress or pain); Elizabeth Kolbert, "Spoiled Rotten," *The New Yorker*, July 2, 2012; see also Sally Koslow, *Slouching Toward Adulthood: Observations from the Not-So Empty Nest* (New York: Viking, 2012).

214 monitor their every move . . . "special": Madeline Levine, *Teach Your Children Well: Parenting for Authentic Success* (New York: HarperCollins, 2012), p. 29.

214 anti-inhibition, live-in-the-moment decade: Ron Chepesiuk, *Sixties Radicals, Then and Now: Candid Conversations with Those Who Shaped the Era* (Jefferson, NC: McFarland & Co., 1995), p. 29 (quoting Paul Krassner) ("The ultimate credo of the sixties was to live in the present"); Myron Magnet, *The Dream and the Nightmare: The Sixties' Legacy to the Underclass* (New York: William Morrow, 1993), p. 35 (describing Norman Mailer's call to "give up 'the sophisticated inhibitions of civilization,' to live in the moment, to follow the body and not the mind"); Harvey C. Mansfield, "The Legacy of the Late Sixties," in Stephen Macedo, ed., *Reassessing the Sixties: Debating the Political and Cultural Legacy* (New York: W. W. Norton, 1997), p. 24.

215 hardly the first to call attention to immediate gratification: See, e.g., Georgie Anne Geyer, *Americans No More: The Death of Citizenship* (New York: Atlantic Monthly Press, 1996), p. 134 (describing polling data showing America shifting between 1950 and 1990 "from future gratification to immediate gratification"); J. Eric Oliver, *Fat Politics: The Real Story Behind America's Obesity Epidemic* (New York: Oxford University Press, 2006), p. 10 ("the prevalence of such chronic diseases is also the by-product of a fast-paced culture of instant gratification and individual license"); Michael S. Rothberg, *American Greed: A Personal and Professional Look at How Greed Caused the Great Recession of 2008* (Bloomington, IN: AuthorHouse, 2010), p. 112 ("America is an instant gratification society"); Christopher Muther, "Instant Gratification Is Making Us Perpetually Impatient," *Boston Globe*, Feb. 2, 2013; Jules Lobel, "America's Penchant for Instant Gratification," *Christian Science Monitor*, Nov. 18, 2003. A "main finding" in a 2011 Pew survey on the future of the Internet was that "[n]egative effects include a need for instant gratification [and] loss of patience." Pew Research Center, *Millennials Will Benefit and Suffer Due to Their Hyperconnected Lives* (Washington, DC: Pew Research Center's Internet & American Life Project, 2012), p. 11.

215 U.S. public debt: See Congressional Budget Office, *Federal Debt and Interest Costs* (September 1984), p. 2, Table 1; "Federal Debt: Total Public Debt as Percent of Gross Domestic Product," Federal Reserve Bank of St. Louis, Economic Research, http://research.stlouisfed.org/fred2/series/GFDEGDQ188S; Matt Phillips, "The Long Story of U.S. Debt, from 1790 to 2011, in 1 Little Chart," *The Atlantic*, Nov. 13, 2012; see also Niall Ferguson, *The Great Degeneration: How Institutions Decay and Economies Die* (London: Allen Lane, 2012), pp. 41–2 (noting that the "statistics commonly cited as government debt" do not "include the often far larger unfunded liabilities of welfare schemes" like "Medicare, Medicaid and Social Security").

215 America's infrastructure: American Society of Civil Engineers, *2013 Report Card for America's Infrastructure*, http://www.infrastructurereportcard.org.

215 investment declined . . . even during the . . . 1980s and '90s: Samuel Sherraden, "The Infrastructure Deficit," *New America Foundation*, Feb. 3, 2011, http://newamerica.net/publica tions/policy/the_infrastructure_deficit. See also Robert B. Reich, *Beyond Outrage: What Has Gone Wrong with Our Economy and Our Democracy and How to Fix It* (New York: Vintage, 2012), p. 45 ("[t]he puzzle is why so little was done during those years" of "continued gains from economic growth").

215 research and development spending has dropped by over 50 percent: Thomas L. Friedman and Michael Mandelbaum, *That Used to Be Us : How America Fell Behind in the World It Invented and How We Can Come Back* (New York: Farrar, Straus and Giroux, 2011), pp. 229 31.

215 spending more on potato chips: National Academy of Sciences, National Academy of Engineering, Institute of Medicine, *Rising Above the Gathering Storm, Revisited: Rapidly Approaching Category Five* (Washington, DC: National Academies Press, 2010), pp. 6, 12 n. 13; Friedman and Mandelbaum, *That Used to Be Us*, p. 246.

215 personal savings rate: "Personal Savings Rate," Federal Reserve Bank of St. Louis, Economic Research, http://research.stlouisfed.org/fred2/series/PSAVERT (charting history of U.S. savings rate); "How Household Savings Stack Up in Asia, the West, and Latin America," *Bloomberg Business Week*, June 10, 2010 (reporting 38 percent Chinese household savings rate).

215 Gambling has skyrocketed too: Shiller, *Irrational Exuberance*, p. 53; see generally Sam Skolnik, *High Stakes: The Rising Cost of America's Gambling Addiction* (Boston: Beacon Press, 2011).

215 almost one in five high schoolers: *National Survey of American Attitudes on Substance Abuse XVII: Teens* (New York: The National Center on Addiction and Substance Abuse at Columbia University, August 2012), p. 2. See also *Adolescent Substance Use: America's #1 Public Health Problem* (New York: The National Center on Addiction and Substance Abuse at Columbia University, June 2011), p. 1.

216 Over half the students at America's private high schools: *National Survey of American Attitudes on Substance Abuse XVII: Teens*, p. 3.

216 study in California: "Study Finds Rich Kids More Likely to Use Drugs Than Poor," The Partnership for a Drugfree America, Feb. 20, 2007, http://www.drugfree.org/join-together/drugs/study-finds-rich-kids-more; see generally *California Healthy Kids Survey*, http://chks.wested.org.

216 adolescents in a suburb where the average family income was over $120,000: Amy Novotner, "The Price of Affluence," *American Psychological Association Monitor on Psychology* 40, no. 1

(2009), p. 50 (describing a series of studies conducted by Suniya Luthar with a cohort of suburban adolescents); see, e.g., Suniya S. Luthar and Adam S. Goldstein, "Substance Use and Related Behaviors Among Suburban Late Adolescents: The Importance of Perceived Parent Containment," *Development and Psychopathology* 20 (2008), pp. 591–614.

216 "Avoiding discipline is endemic to affluent parents": Dan Kindlon, *Too Much of a Good Thing: Raising Children of Character in an Indulgent Age* (New York: Hyperion, 2001), p. 15; Gottlieb "How to Land Your Kid in Therapy."

216 explosion of narcissism: Twenge and Campbell, *The Narcissism Epidemic*, pp. 1–2; Jean M. Twenge, *Generation Me: Why Today's Young Americans Are More Confident, Assertive, Entitled— and More Miserable Than Ever Before* (New York: Simon & Schuster, 2007), pp. 64–9.

216 millennials . . . "CEO tomorrow": Ron Alsop, "The 'Trophy' Kids Go to Work," *Wall Street Journal*, Oct. 21, 2008.

216 children of well-off white baby boomers: Eric Hoover, "The Millennial Muddle: How Stereotyping Students Became a Thriving Industry and a Bundle of Contradictions," *The Chronicle of Higher Education*, Oct. 11, 2009 ("commentators have tended to slap the Millennial label on white, affluent teenagers").

216 "Millennials don't always want to work": Dan Schawbel, "Reviving Work Ethic in America," Forbes.com, Dec. 21, 2011 (quoting Eric Chester).

217 teaser rates and easy credit: See Robert J. Shiller, *The Subprime Solution: How Today's Global Financial Crisis Happened, and What to Do About It* (Princeton, NJ: Princeton University Press, 2008), pp. 6–7; Ben Steverman and David Bogoslaw, "The Financial Crisis Blame Game," *Business Week*, Oct. 18, 2008.

217 bought $500,000 houses with . . . loans they couldn't afford: Steverman and Bogoslaw, "The Financial Crisis Blame Game"; see also Shiller, *The Subprime Solution*, p. 45 (2005 survey showed that San Francisco home buyers on average expected a 14% price increase per year and that "a third of the respondents reported truly extravagant expectations – occasionally over 50% a year").

217 banks offered mortgages: Nouriel Roubini and Stephen Mihm, *Crisis Economics: A Crash Course in the Future of Finance* (New York: Penguin Press, 2010), p. 65 (in the 1990s and 2000s both "ordinary banks" and investment banks "no longer subjected would-be borrowers to careful scrutiny. So-called liar loans became increasingly common, as borrowers fibbed about their income and failed to provide written confirmation of their salar[ies]"); Shiller, *The Subprime Solution*, p. 6 ("Mortgage originators, who planned to sell off the mortgages to securitizers, stopped worrying about repayment risk").

217 AAA grades: Roubini and Mihm, *Crisis Economics*, pp. 33, 66–7.

218 "They could explode a day later": Steverman and Bogoslaw, "The Financial Crisis Blame Game"; see also Roubini and Mihm, *Crisis Economics*, p. 65 (the problem with these "newfangled" financial products was that "the bank or firm originating the securities had little incentive to conduct the oversight and due diligence necessary"; "a bad mortgage is passed down the line like a hot potato").

218 a contagious "excessively optimistic" conviction: See Shiller, *Irrational Exuberance*, p. 1; Shiller, *The Subprime Solution*, p. 41 ("social contagion"), pp. 46–7 ("excessively optimistic"), p. 52 (describing mind-set that a catastrophic downturn would simply never happen); Roubini and Mihm, *Crisis Economics*, p. 18 ("boom will never end"), p. 88 ("blind faith that asset prices would only continue to rise").

218 Cassandras: Roubini and Mihm, *Crisis Economics*, p. 88; Shiller, *The Subprime Solution*, p. 52.

219 Tony Blair once called: Jemima Lewis, "The Greatest Nation on Earth? I Don't Think So," *The Independent*, May 12, 2007; "A Blessed Nation?, *"The Guardian*, May 10, 2007 (asking readers, "What do you make of [Blair's] description" of Britain as "the greatest nation on Earth"?).

219 "I suspect that the Brits believe in British exceptionalism": Seema Mehta, "Romney, Obama and God: Who Sees America as More Divine?," *Los Angeles Times*, Apr. 13, 2012 (quoting Barack Obama and describing reactions to his quote).

220 "humiliation" . . . "sharp goad": Orville Schell and John DeLury, *Wealth and Power: China's Long March to the Twenty-First Century* (New York: Random House, 2013), p. 7; see also Suisheng Zhao, "'We Are Patriots First and Democrats Second': The Rise of Chinese Nationalism in the 1990s," in Edward Friedman and Barrett L. McCormick, eds., *What If China Doesn't Democratize? Implications for War and Peace* (New York: M. E. Sharpe, 2000), p. 23 ("[T]here is a rallying cry for Chinese everywhere . . . that after a century of humiliation" the time has come for China to "rise in the world to the place it deserves") (quoting James Lilley, former U.S. ambassador to China and Taiwan).

221 Her father was an alcoholic, and her mother's "way of coping": Sonia Sotomayor, *My Beloved World* (New York: Alfred A. Knopf, 2013), pp. 11–4.

221 gave herself painful insulin shots: Ibid., pp. 3–4, 9.

221 "fragile world" . . . "blessed": Ibid., p. 11.

221 "decided to approach one of the smartest girls in the class": Ibid., p. 72; see also pp. 117–8, 143.

222 "can make an enormous difference": Ibid., p. 16.

222 "acting white": See John H. McWhorter, *Losing the Race: Self-Sabotage in Black America* (New York: HarperCollins, 2001); John McWhorter, "Guilt Trip," *The New Republic*, June 24, 2010; John U. Ogbu and Herbert D. Simons, "Voluntary and Involuntary Minorities: A Cultural Ecological Theory of School Performance with Some Implications for Education," *Anthropology & Educational Quarterly* 29, no. 2 (1998), pp. 155, 161. In a 2004 speech, Henry Louis Gates said:

> I read the results of a poll from the *Washington Post* recently that interviewed inner-city black kids, and it said, 'List things white.' You know what they said? The three most prevalent answers: getting straight A's in school, speaking standard English, and visiting the Smithsonian. Had anybody said anything like this when we were growing up, they would have smacked you upside your head and checked you into an insane asylum. Somehow, we have internalized our own oppression.

Henry Louis Gates, "America Beyond the Color Line," in Catherine Ellis and Stephen Drury Smith, eds., *Say It Loud! Great Speeches on Civil Rights and African American Identity* (New York: The New Press, 2010), p. 235.

222 phenomenon did not exist at all-black schools: See Roland E. Fryer, "'Acting White': The Social Price Paid by the Best and Brightest Minority Students," *Education Next* 6, no. 1 (2006).

222 "Remember that Bill Gates": Anne-Marie Slaughter, "Rebellion of an Innovation Mom," CNN World, June 5, 2011, http://globalpublicsquare.blogs.cnn.com/2011/06/05/rebellion-of-the-innovation-mom.

223 on entertainment media: Friedman and Mandelbaum, *That Used to Be Us*, p. 128; see also Mark Bauerlein, *The Dumbest Generation: How the Digital Age Stupefies Young Americans and Jeopardizes Our Future* (New York: Jeremy P. Tarcher, 2008), chap. 3; Nicholas Carr, *The Shallows: What the Internet Is Doing to Our Brains* (New York: W. W. Norton & Co., 2010).

223 25 percent more time watching television: See Nancy Zuckerbroad and Melissa Trujillo, "U.S. Schools Weigh Extending Hours, Year," Associated Press, Feb. 25, 2007 (citing a study finding that the average school day is 6.5 hours and the average school year is 180 days); The Nielsen Co., "TV Viewing Among Kids at an Eight-Year High," Oct. 26, 2009 (estimating that kids age 6–11 spend 28 hours per week watching TV, which comes to 1,456 hours/year); "Children and Watching TV," The American Academy of Child and Adolescent Psychology No. 54, December 2011, http://www.aacap.org/AACAP/Families_and_Youth/Facts_for_Families/Facts_for_Families_Pages/Children_And_Wat_54.aspx (estimating that kids watch 3–4 hours of television each day, which comes to 1,100–1,400 hours per year).

223 "Discipline. Patience. Perseverance": "Khaled Hosseini: By the Book," *New York Times*, June 6, 2013.

223 Google, Facebook, or the iPod: Amy Chua, "Tiger Mom's Long Distance Cub," *Wall Street Journal*, Dec. 24, 2011.

224 Jeff Bezos founded Amazon when he was "dead broke": Robert Spector, *Amazon.com: Get Big Fast* (New York: HarperCollins, 2000), p. 84.

224 The present moment by itself is too small: See Rubenfeld, *Freedom and Time*, p. 16; Jed Rubenfeld, *Revolution by Judiciary: The Structure of American Constitutional Law* (Cambridge, MA: Harvard University Press, 2005), p. 91 ("[A] person's freedom . . . is bound up with his capacity to give his life purposes of his own making and to pursue those purposes over time").

224 Happiness . . . "cannot be pursued": Mihaly Csikszentmihalyi, *Flow: The Psychology of Optimal Experience* (New York: Harper Perennial, 1991), p. 2 (quoting Victor Frankl).

INDEX

academic achievement, *see* education and
    academic achievement
Academy Awards, 52, 163
Acemoglu, Daron, 17*n*
Achebe, Chinua, 81
Adams, James Truslow, 18
Adler, Alfred, 59*n*
Adventures of Augie March, The
    (Bellow), 162, 163
African Americans, 2, 100, 156
    elite schools and, 172
    equality and, 76–77, 80–81
    at Harvard University, 6, 41–42
    income of, 43–44
    Mormons and, 30–31
    "oppositional" urban culture and, 222
    poverty and, 44, 74–75
    self-esteem of, 112
    stereotype threat and, 78, 80, 81
    superiority complex and, 72–78
    at Yale Law School, 42
    see also African immigrants; blacks
African immigrants, 105, 156, 194
    Nigerian, *see* Nigerian Americans
    at universities, 41–42
    upward mobility and, 168
alcohol abuse, 150, 216
Alexander, Clifford, 43
Alexander, Elizabeth, 73
Alexander the Great, 90–91
Allen, Woody, 52, 108
Amanat, Abbas, 92
Amazon, 224
America, 26–27
    Constitution, 2, 62, 204–6, 209
    debt in, 215
    Declaration of Independence, 204–6

exceptionalism of, 201, 219
    impulse control and, 142–44, 197, 200,
        203–6, 208–9, 214–18, 221–24
    infrastructure in, 215
    insecurity and, 200–203, 208–12, 214,
        215, 218, 220–21
    live-in-the-moment message and, 1,
        2, 10, 27, 143–44, 204, 205, 207–9,
        214, 224
    rebelliousness of, 204, 205, 208
    research and development in, 215
    Revolution, 204, 205
    rise in standards of living in, 210–11
    savings in, 215
    superiority complex and, 203, 207, 209,
        211–12, 219–20, 225
    Triple Package and, 199–225
American Conservative, 193
American Dream, 5–6, 18, 168, 173, 208
    see also upward mobility
American Express, 5, 32, 135
American Motors, 32
Amish, 119, 180–84, 187
    ball games and, 183
    education and, 119, 181
    impulse control and, 119, 180–81
    insecurity and, 182–83
    superiority complex and, 181–82
Ansari, Aziz, 164
anti-Semitism, 12, 15, 54–55, 61,
        138, 141–42, 154, 194*n*
Appalachia, 174–80, 210
    impulse control and, 177–80
    poverty in, 169, 174, 175, 178
    substance abuse in, 175, 178, 180
Arabs, 156
Aronson, Joshua, 78

Ashton, Alan, 32
Asian Americans, 2, 13, 45, 52, 151,
    194, 196
    depression among, 104, 150
    education and accomplishment among,
        24, 45–48, 78, 79, 110–11, 131,
        151, 170, 172, 173, 194–95, 213*n*
    entry visas of, 170
    family and parenting among, 13,
        110–11, 147–51
    impulse control and, 133, 173
    insecurity and, 13, 110–11, 173
    musical training and, 46–47, 126–28,
        129, 164
    opposition to stereotypes among, 164
    self-esteem of, 111–12, 151, 213*n*
    stereotype boost and, 79
    stereotype threat and, 78
    suicide among, 150
    superiority complex and, 13, 173
    tutoring and, 173
    upward mobility and, 168
    see also East Asian Americans; Chinese
        Americans; Indian Americans
assimilation, 20, 83, 195, 196
AT&T Mobility, 38
At Home in the Heart of Appalachia
    (O'Brien), 177
Atlas, James, 161

Bacardi family, 37, 71
Bain Capital, 31
Bar Kokhba revolt, 12
battle, life as, 15
Baughman, Gary, 32
Baumeister, Roy, 117
Beck, Glenn, 33, 34
Bell, Daniel, 153
Bellow, Greg, 162
Bellow, Saul, 12, 160–64, 195
Benedict, Jeff, 34
Beneficial Life, 35
Benny, Jack, 52
Benson, Ezra Taft, 67–68, 157
Bezos, Jeff, 224
Bharara, Preet, 49, 131, 165
Billete, Pepe, 68–69, 71–72

Black & Decker, 5, 32
blacks, *see* African Americans
black schools, 77, 222
Blaine, David, 119–20
Blair, Tony, 219
Bloom, Harold, 137
Bloomberg, Michael, 52
Boggs, Lilburn, 64
Bonfire of the Vanities (Wolfe), 210
Book of Mormon, 25, 30
Book of Mormon Girl, The (Brooks), 157
Borushek, Grisha, 161
Bose, Amar Gopal, 49
Bose Corporation, 49
Bowman, Matthew, 65
Brandeis, Louis, 62
Branden, Nathaniel, 212, 214
Breaking Bad, 143
Brigham Young University, 25,
    32–33, 35, 157
Britain, 168, 219
Bronx High School of Science, 170,
    172, 173
Brooks, David, 54
Brooks, Joanna, 157
Buber, Martin, 63
Buffalo Creek flood, 179
Buffon, George-Louis Leclerc,
    Comte de, 200–201
Bush, George H. W., 46
Bush, George W., 31
Bushman, Claudia, 66, 67

Calvinism, 137, 184–85
Carmichael, Stokely, 43
Carnegie, Andrew, 202
Carroll, James, 186
Carter, Jimmy, 31
Castro, Fidel, 36, 37, 69, 88
Catherine of Valois, 122
Catholics, 35–36, 53, 184, 185
Catmull, Edwin, 32
Chabon, Michael, 63
Chaplin, Charlie, 52
Checketts, Dave, 32, 137
Chen, Steve, 48
Chernin, Peter, 54

Chester, Eric, 216
chi ku, 125
childhood, 11, 146–47
 Confucian approach to, 147–48
China, 120–22, 215, 223
 Confucian tradition of, *see* Confucian
 principles
 decline of, 123
 impulse control and, 220
 Japan and, 121, 122, 123, 158
 Ming Dynasty, 121–22
 rising power of, 124, 220
Chinese Americans, 7, 8, 45–51, 56, 57–58
 academic achievement and, 13, 123–24,
 126–31, 142, 171–73
 bimodal communities of, 171
 and breaking out of Triple Package, 197
 chi ku and, 125
 discrimination against, 104, 124
 family honor and, 110
 immigrant selection criteria and, 170–71
 impulse control and, 120, 125–31,
 132, 142
 insecurity and, 50, 123–24
 IQ and, 171
 parenting among, 126–29, 132, 142,
 147, 148, 150
 psychological symptoms among, 150
 in Sunset Park, 170, 171, 172
 superiority complex and, 124
 tutoring and, 172–73
 upward mobility of, 171
 see also Taiwanese Americans
Chinese Exclusion Act, 207
Chinese superiority complex, 72, 120–23,
 130–31, 156, 220
chip on the shoulder, 11–12, 18, 21, 26,
 114, 159, 183
 America and, 200, 201
 Jobs and, 22
 Mormons and, 137
Chisholm, Shirley, 43
Cho, John, 164
Chomsky, Noam, 156
Christensen, Clayton, 32
Christianity, 60, 98, 182, 184, 207
 Amish and, 180, 181, 183–84

Catholicism, 35–36, 53, 184, 185
 Lebanese and, 112–13
 Mormonism and, 64–65, 136
 Protestantism, 8, 31, 53, 136,
 184–86, 207
Churchill, Winston, 23
Church of England, 35
Church of Jesus Christ of Latter-day
 Saints (LDS Church), 5, 30–31,
 34–36, 134, 136–37, 156–58
 see also Mormons
Chyao, Amy, 46
Cicero, 112
Cincinnati, Ohio, 176
Citigroup, 5, 32
Clark, Kim, 32
class rigidity, 169
coal mining, 178
Coca-Cola, 36
Cohen, Steve, 52
Colorado City, Ariz., 188
Columbus, Christopher, 122
Combs, Sean "Diddy," 73–74, 75
Confucian principles, 103, 110, 142,
 151, 161, 220
 childhood and, 147–48
 learning and, 125–26, 147
Congressional Budget Office, 211
Congreve, William, 87
Constitution, 2, 62, 204–6, 209
Cooper, Helene, 80, 82–83
Covey, Stephen, 32
Crash, 92
creative destruction, 208
creativity, 146, 223
Crittendon, Gary, 32, 135
Cromwell, Oliver, 60
Cruz, Ted, 39
crystal meth, 175
Cuba, 69–71
Cuban Americans, 36n, 37,
 39–40, 109
 Golden Exiles, 36n, 87
 Hispanic label and, 68–69, 156
 Marielitos, 36n, 40
 New Cubans, 36n, 40
 superiority complex and, 68–72, 156

Cuban Exiles, 6, 12, 36–41, 58, 71–72, 87–89, 105
  insecurity and, 87–89
  status loss among, 88–89, 92
cultural groups, 30
culture, definitions of, 30*n*

Daily Beast, 167
Davis, John, 153–54
Davis-Blake, Alison, 32
Death of a Salesman (Miller), 108
Declaration of Independence, 204–6
De La Torre, Miguel, 71
De la Vega, Ralph, 38
Dell, 5, 32
Deloitte, 5, 32
Delury, John, 220
democracy, 62, 182, 206, 211, 217
Democracy in America (Tocqueville), 202
Denmark, 168
depression, 212, 213*n*
  in Asian Americans, 104, 150
Diamond, Jared, 17*n*, 186
Diaz, Cameron, 39
Diller, Barry, 52
discipline, 11, 163
  see also impulse control
Disraeli, Benjamin, 15
drive, 15, 17, 84, 89, 114–15, 151, 163, 194, 200, 209
  pathology of, 165–66
drugs, *see* substance abuse
D'Souza, Dinesh, 44
Duckworth, Angela, 117
Dweck, Carol, 117, 212

East Asian Americans, 50
  academic achievement and, 7, 13, 196
  bullying of, 103–4
  family and parenting among, 147–48, 151, 152
  out-marrying and, 158
Eckstein, Susan, 36*n*, 39
education and academic achievement, 24–26
  Amish and, 119, 181

Asian Americans and, 24, 45–48, 78, 79, 110–11, 131, 151, 170, 172, 173, 194–95, 213*n*
  black schools and, 77, 222
  Chinese Americans and, 13, 123–24, 126–31, 142, 171–73
  conventional success and, 159
  East Asian Americans and, 7, 13, 196
  expectations and, 13–14
  insecurity and, 105–6
  Iranian Americans and, 56, 95
  Jews and, 13, 20, 24–26, 193–96
  Korean Americans and, 13, 110, 131
  self-esteem and, 111–12, 213*n*
  tutoring and, 172–73
Einstein, Albert, 55
Eire, Carlos, 70–71
Eisen, Arnold, 62
Emerson, Ralph Waldo, 204
Emler, Nicholas, 213*n*
Emory University, 36
End of History, The (Fukuyama), 211
enkrateia, 138
entrepreneurialism, 223–24
Epstein, Helen, 190, 192
Epstein, Lawrence J., 107
equality, 20, 62, 76–77, 80–81, 83, 144, 207, 219–20, 221, 225
Erikson, Kai, 177
Estefan, Gloria, 39
Ethiopia, 6
ethnic armor, 22–23, 82–83
ethnocentrism, 15, 59
Exxon Valdez, 179

failure, 117, 159, 214
Fairbanks, John K., 121
family, parents, 86, 106–14, 131–33, 115, 143, 146–50
  Asian Americans and, 13, 110–11, 147–51
  basing life decisions on expectations of, 160
  Chinese Americans and, 126–29, 132, 142, 147, 148, 150
  East Asian Americans and, 147–48, 151, 152

Indian Americans and, 101–2
Jews and, 106–8, 109, 140–41, 142, 151–53
Korean Americans and, 131, 132
Lebanese Americans and, 113–14
parental sacrifice, 109–10
rebellion against, 161–63
self-esteem parenting, 214, 216
Fanjul, Alfonso, 37
Fanjul Corporation, 37
Farrakhan, Louis, 78
fear, 86, 104
in Jews, 104–6, 153–55, 195
Federalist Papers, 201, 206
Feiffer, Jules, 151–52
Fiddler on the Roof, 152
Fiedler, Leslie, 162
financial collapse of 2008, 215, 217–18, 220
Fischer, Claude, 204
Fisher-Price, 5, 32
Flynn, James, 171
Footnote, 152
Forbes, 8
Forbes 400, 32, 52
Ford, Henry, 208
Fortune, 114
Fortune 500 companies, 33, 50
Forward, 140–41
Frankfort, Lew, 51
Frankl, Victor, 224
Franklin, Benjamin, 203
Freud, Sigmund, 10, 62, 152, 153
Friedman, Thomas, 54
Friend, Tad, 87–88
Fryer, Roland, 222
Fukuyama, Francis, 211
Fundamentalist Church of Jesus Christ of Latter-Day Saints (FLDS), 188

Galleon Group, 165
gambling, 215
Gandhi, Indira, 96
Gandhi, Mohandas, 96
García, Andy, 39
Gardner, Howard, 23
Garvey, Marcus, 43

Gates, Bill, 222, 223
Gates, Henry Louis, Jr., 41–42
Gawande, Atul, 49–50
Gehry, Frank, 53
Ghana, 6
Giardina, Denise, 177
Glazer, Nathan, 25
globalization, 211
Goad, Jim, 176
Goizueta, Roberto, 36, 38, 71
Goizueta Business School, 36
Golden Exiles, 36n, 87
Goldman Sachs, 33, 34, 42
Good as Gold (Heller), 152
Graham, Robert, 91–92
Greek Americans, 57
Greeks, ancient, 89, 112, 121
Greenberg, Clement, 12
Grenier, Guillermo, 69
Grey, Brad, 54
Grimm, Thomas, 32
Grinberg, Gedalio, 38
group generalizations, 14
Grove, Andy, 32
Guggenheim, Meyer, 153–54
Guinier, Lani, 41–42
Guitar Hero, 48–49
Gupta, Ashwini, 100
Gupta, Rajat, 100, 165–66
Gupta, Sanjay, 49
Gutierrez, Carlos, 38

Hadrian, 12
Haiti, 6
Haley, Nikki, 49
Hamilton, Alexander, 204
happiness, 11, 27, 146, 147, 206–7, 209, 224
hardship endurance, 16–17, 21, 119–20, 125
self-esteem movement and, 214
Harkins, Anthony, 175
Harold & Kumar, 164
Harvard Business School, 32, 42
Harvard University, 47, 110
black students at, 6, 41–42
Hass, Aaron, 189, 190, 192

Hatch, Orrin, 31, 136
Hebrew Union College, 139
Hedâyat, Sâdeq, 91
Heder, Jon, 33
Heller, Joseph, 152
Hemingway, Ernest, 162, 163
Henry V, King, 121, 122
Heritage Foundation, 209
Hinduism, 102
Hispanic and Latino Americans, 2, 6, 39,
    40, 58, 68–69, 79, 87, 100, 194
  Cuban, *see* Cuban Americans; Cuban
    Exiles
  Cubans' sense of distinctness from,
    68–69, 156
  elite schools and, 172
  self-esteem of, 112
  upward mobility and, 168
Hitler, Adolf, 154
Hollywood, 54
Holmes, Oliver Wendell, Sr., 7
Holocaust, 63, 154
  children of survivors of, 188–93
Hosseini, Khaled, 223
Hostetler, John, 182
House at Sugar Beach, The (Cooper), 80
House of Sand and Fog, 92
housing bubble, 217–18
Hsieh, Tony, 48, 127–28
Huang, Kai, 49
Huang, Sirena, 47
Huguenots, 20, 186
human capital, 87
humility, 181–82
Huntsman, Jon, Jr., 31
Huntsman, Jon, Sr., 32
Huntsman Corporation, 5

Icahn, Carl, 52
Igbo people, 9, 81
Iger, Robert, 51, 54
immigrant selection, 169, 170–71
impulse control, 10–11, 23, 117–44
  America and, 142–44, 197, 200,
    203–6, 208–9, 214–18, 221–24
  Amish and, 119, 180–81
  Appalachia and, 177–80

Asian Americans and, 133, 173
Chinese Americans and, 120, 125–31,
    132, 142
Chinese tradition of, 220
Holocaust survivors' children and,
    189–90
Indian Americans and, 131, 132
Jews and, 20, 138–42, 197
live-in-the-moment message and, 1,
    2, 10, 27, 143–44, 204, 205, 207–9,
    214, 224
marshmallow test and, 23, 118, 179, 218
Mormons and, 134–37, 186–87
research on, 23, 117–19, 179
self-esteem movement and, 214
in Stoicism, 120
superiority complex and, 15–16, 17, 120
transference to other areas of life,
    133–34
underside of, 17, 18, 147–51
India:
  Hinduism in, 102
  Institutes of Technology in, 170
  social hierarchy in, 95–98
Indian Americans, 7, 9, 14, 45, 47–51, 53,
    57–58, 95*n*–96*n*
  and breaking out of Triple Package, 197
  caste and, 9, 96–97, 99, 101
  Hinduism and, 102
  immigrant selection criteria and, 170–71
  impulse control and, 131, 132
  insecurity and, 50, 95–102
  out-marrying and, 158
  prejudice against, 99
  superiority complex and, 95, 97–99,
    101–2
individualism, 202, 203
industrial accidents, 179
innovation, 222–24
Innovator's Dilemma, The (Christensen), 32
insecurity, 9–10, 20, 85–115, 159, 163,
    182, 195
  America and, 200–203, 208–12, 214,
    215, 218, 220–21
  in Americans, Tocqueville's observation
    of, 27, 85–86, 202–3
  Amish and, 182–83

Appalachia and, 175
Asian Americans and, 13, 110–11, 173
of capitalism, 202
Chinese Americans and, 50, 123–24
Cuban Exiles and, 87–89
drive and, 15, 17
family and, 86, 106–114, 115
fear and, 86, 104–6, 153–55, 195
Holocaust survivors' children and,
    190–91, 193
Indian Americans and, 50, 95–102
of individualism, 202
Iranian Americans and, 89–95
Jews and, 104–8, 109, 153–55,
    195–96, 197
Lebanese Americans and, 112–14
Mormons and, 187
Protestantism and, 184–85
scorn-based, 86, 87–104, 114
self-esteem movement and, 213
studies on, 23
underside of, 17, 18, 151–55
superiority complex and, 1, 10, 11–14,
    15, 17, 114, 124
welfare and, 209
Intel Science Talent Search, 46, 193
intermarriage, 20, 158, 196
International Tchaikovsky
    Competition, 46
IQ, 169, 171–72, 186, 194, 291n
    Jews and, 194
Iran, 90, 91, 94
Iranian Americans, 7, 12, 55–58
    academic achievement and, 56, 95
    insecurity and, 89–95
    Persian culture and, 94–95
    prejudice and animus against, 92–93
    status and, 92, 93–94
Iranian superiority complex, 72, 90–92
Islamic Revolution, 92, 93
Israel, 154, 195
Ivy League, 7, 41, 47–48, 170

Jackson, Jesse, 77
Jackson, Samuel L., 77
Jacobs, Joseph L., 112–14
Jamaica, 6, 43

James, Henry, 202
James, Josh, 31
James, William, 212–13
Japan:
    China and, 121, 122, 123, 158
    imperial, 219
Japanese Americans, 57
    family honor and, 110
Jazz Singer, The, 152–53
Jefferson, Thomas, 200, 201–2, 203, 205,
    206, 208
Jeffs, Rulon, 34
Jennings, Ken, 33
Jeopardy!, 33
JetBlue, 5, 31, 137
"Jewish Genius" (Murray), 63
Jews, 7, 17n, 20, 50, 51–55, 57–58, 72, 78,
    88, 193–97
    academic achievement and, 13, 20, 24–
        26, 193–96
    anti-Semitism and, 12, 15, 54–55, 61,
        138, 141–42, 154, 194n
    Ashkenazi, 194n
    authority questioned by, 151, 152
    bar and bat mitzvahs of, 139, 142
    chip on the shoulder and, 12
    Disraeli on, 15
    family and parenting among, 106–8,
        109, 140–41, 142, 151–53
    Holocaust and, *see* Holocaust
    impulse control and, 20, 138–42, 197
    income of, 7–8, 53
    insecurity and, 104–8, 109, 153–55,
        195–96, 197
    IQ and, 194
    Israel and, 154, 195
    music practice and, 141, 161
    out-marrying and, 158, 196
    poverty and, 53
    Sabbath and, 139–40
    superiority complex and, 9, 60–64, 194,
        195, 197
    Syrian, 24
    Triple Package pathologies of, 151–55
Jindal, Bobby, 49
Jobs, Steve, 22, 223
Jordan, Michael, 117

Judaism, 62, 64, 138, 151
  orthodox, 139, 142
  Reform, 139
Juilliard School of Music, 46, 127, 129
Jung, Andrea, 123

Kafka, Franz, 105, 119–20
Kahn, Louis, 53
Kaling, Mindy, 164
Kant, Immanuel, 138, 142
Kazin, Alfred, 12, 13
Keister, Lisa, 33
Kellogg, 38
Kennedy, Paul, 123
Khosla, Vinod, 49
Kimball, Spencer, 25
King, Martin Luther, Jr., 77, 78
Kiryas Joel, 26
Kleiner Perkins, 49
Kodak, 32
Koh, Jennifer, 47
Korean Americans, 7
  academic achievement and, 13, 110, 131
  family honor and, 110
  parenting among, 131, 132
Kovner, Bruce, 52
Krugman, Paul, 54
Kwon, Yul, 104

Lahiri, Jhumpa, 98
Lala, Nupur, 47
Landers, Ann, 53
Latino Americans, see Hispanic and
  Latino Americans
Lauren, Ralph, 51
Lebanese Americans, 7, 20, 55–58
  insecurity and, 112–14
Lebanon, 113
Lee, Ang, 160–64
Lee, Jennifer, 79
Lee, Noah, 47
Lee, Spike, 77, 162
Leonidas I, 89
Lewis, Jerry, 52
Lewis and Clark expedition, 201
Li, Jin, 147
liberal principles, 9, 11

Liberia, 6, 80, 82
Liebeskind, Daniel, 53
Lim, Phillip, 148
Lin, Alfred, 49
Lin, Jeremy, 103, 124–25, 164
Lord & Taylor, 32
Los Angeles, Calif., 171
Los Angeles Times, 54
Louie, Vivian, 110, 128, 129
Lufthansa, 32
Lynton, Michael, 54

Madison, James, 206
Madison Square Garden, 5, 32, 137
Mad Men, 195
Madoff, Bernie, 165
Mailer, Norman, 12, 163
Majd, Hooman, 90–91
Malcolm X, 43, 77
Mao Zedong, 123
Marielitos, 36n, 40
Maronites, 112–13
marriage out of the group, 20, 158, 196
Marriott, J. W., 31–32
Marriott International, 5, 32
Marriott School of Management, 33
Marshall, Thurgood, 77
marshmallow test, 23, 118, 179, 218
Martin County sludge spill, 179
Marx Brothers, 52
Massad, Alex, 113–14
Massey, Douglas, 41
Math Olympiad, 193
Mauss, Armand, 136–37
May, Elaine, 106
Mehta, Ved, 97
Meier, Richard, 53
Melville, Herman, 201
Mendelssohn, Felix, 61
Mendelssohn, Moses, 61
Mendes, Eva, 39
Menendez, Robert, 39
methamphetamine addiction, 175
Mexican Americans, 170, 207
Meyer, Barry, 54
Meyer, Stephenie (author of Twilight
  series), 33, 137

Miami, Fla., 6, 37, 39, 71, 88
Midler, Bette, 52
millennials, 216
Miller, Arthur, 108
Min, Anchee, 125
mining, 178
Mischel, Walter, 118
Miss Manners, 53
Mobasher, Mohsen, 93
Mobil Oil, 114
Mondrian, Piet, 223
Mongols, 121
Montgomery Ward, 114
Moonves, Leslie, 54
Mormons, 5, 14, 30–36, 52, 105, 169,
    181, 186–88
  afterlife and, 66
  Book of Mormon and, 25, 30
  and breaking out of Triple
    Package, 197
  Christianity and, 64–65, 136
  education and, 25
  Family Home Evening and, 187
  fundamentalist, 34–35, 188
  impulse control and, 134–37,
    186–87
  insecurity and, 187
  LDS Church, 5, 30–31, 34–36, 134,
    136–37, 156–58
  missions of, 16, 66, 134–35, 158,
    186, 187
  morality and, 66–67
  original sin rejected by, 66
  persecution of, 64, 136
  polygamy and, 67, 136, 187, 188
  separatist Utah phase of, 187
  superiority complex and, 8, 16, 64–68,
    136, 156–58, 186–87
  women's status among, 156–58
  Word of Wisdom and, 187, 188
Morrison, Toni, 77, 156
Mosaic law, 138
Movado Group, 38
Mukherjee, Siddhartha, 49–50, 97
Murray, Charles, 63
Murillo, Bartolomé Esteban, 71
musical training:

Asian Americans and, 46–47, 126–28,
    129, 164
  Jews and, 141, 161
My Beloved World (Sotomayor), 221

NAACP Legal Defense Fund, 172
Naked and the Dead, The (Mailer), 163
Napoleon Dynamite, 33
narcissism, 213*n*, 216
national security, 220
National Study of Youth and
    Religion, 134
Native Americans, 207
Nazi Germany, 9, 63, 189, 219
  see also Holocaust
Neeleman, David, 31, 137
Nehru, Jawaharlal, 96
New Cubans, 36*n*, 40
Newsweek International, 49
New York, 102
New Yorker, 32, 49
New York Knicks, 32, 103
New York Times, 46, 53–54, 163
Nichols, Mike, 106
Nietzsche, Friedrich, 12, 182
Nigerian Americans, 6–7, 9, 42–45, 58
  parenting among, 132
  superiority complex and, 81–82
Niña, 122
9/11 attacks, 93, 99, 101, 220
Nixon, Richard, 31
Nobel Prize, 7, 54, 159–60, 162, 196
nobility, English, 83
Nooyi, Indra, 49
North Korea, 17*n*
Notes on Virginia (Jefferson), 201–2
Notorious B.I.G., 73
Novick, Peter, 63

Obama, Barack, 219
Obama, Michelle, 164
O'Brien, John, 177
Obstacles Welcome (de la Vega), 38
Odyssey (Homer), 112
Oppenheimer, Robert, 55
Oxycontin, 178
Owsley County, Ky., 169, 175

Palestine, 156
parents, *see* family, parents
Passage from Home (Rosenfeld), 108
Penn, Kal, 164
People Magazine, 104
PepsiCo, 49
Pérez, Lisandro, 69
Persia, Persians, 8, 89–91, 94–95,
   121, 156
Pew Research Center, 6*n*, 45, 124,
   129–30, 167, 168
Phillips Exeter Academy, 170
Philo, 138
Phoenicians, 112
Physics Olympiad, 193
Picasso, Pablo, 223
Pixar Studios, 32
Podhoretz, Norman, 12
politics, 216–17
Pollack, Kenneth, 90
Poor Richard's Almanack (Franklin), 203
Portnoy's Complaint (Roth), 107
poverty, 18–19, 169
  in Appalachia, 169, 174, 175, 178
  black, 44, 74–75
  escape from, 167–68, 173; *see also*
    upward mobility
  Jewish, 53
Powell, Colin, 43
Powers, Nicholas, 74–75
Prager, Dennis, 104–5, 155
PricewaterhouseCoopers, 32
Princeton University, 47, 110, 222
Protestant Ethic and the Spirit of Capitalism,
  The (Weber), 8, 184
Protestants, 8, 31, 53, 136,
  184–86, 207
Psychology of Self-Esteem, The
  (Branden), 212
public-interest careers, 164–65
Pulitzer Prize, 7, 52, 162
Puritans, 26, 137, 184, 186, 201, 203,
  204, 207
Purkayastha, Bandana, 100

Raghavan, Anita, 100
Rand, Ayn, 212

Rape of Nanjing, 158
Ravitz, Martha, 107
Raynal, Guillaume, 201
Redneck Manifesto (Goad), 176
Reformation, 184
Reid, Harry, 31
resentment, 12, 87, 89
resource curse, 178
Rise of Asian Americans, The, 45
Roberts, B. H., 66
Robinson, James, 17*n*
Rollins, Kevin, 32
Romans, 112
Romney, George, 31
Romney, Mitt, 31, 136
Roosevelt, Franklin D., 209
Rosenfeld, Isaac, 108
Ross School of Business, 32
Roth, Philip, 64, 107, 108, 195
Roubini, Nouriel, 218
Roy, Arundhati, 97
Rubio, Marco, 39, 109

Sacks, Lord, 63
Samburg, Andy, 52
Sam's Club, 5, 32
Sandler, Adam, 52
Saramago, José, 61
Sawyer, Diane, 174
Sayles, Peter, 88
Schachter, Stanley, 25
Schell, Orville, 220
Schumpeter, Joseph, 208
Schwartz, Delmore, 12
scorn, 86, 87–104, 114
  Cuban Exiles and, 87–89
  East Asian Americans and,
    103–4
  Indian Americans and, 95–102
  Iranian Americans and, 89–95
SCP Worldwide, 32
Sears, Roebuck, 5, 32
security:
  national, 220
  see also insecurity
Seidenberg, Ivan, 51
Seinfeld, Jerry, 52

self-esteem, 10, 112, 114, 115,
212–14, 220
  academic performance and,
  111–12, 213*n*
  of Asian Americans, 111–12, 151, 213*n*
self-worth, external measures of, 160
Sen, Amartya, 97
*Seven Habits of Highly Effective People,
The* (Covey), 32
Shaheen, Bill, 57
Shaheen, Jeanne, 57
Shahs of Sunset, 55–56, 94
Shelby, Ann, 176
Shic, Emily, 47
Shiller, Robert, 211, 218
Shulchan Aruch, 138, 139
Silicon Valley, 49, 210
Silverman, Sarah, 52
Simons, Jim, 52
Skullcandy, 5, 32
SkyWest Airlines, 32
slavery, 17, 19, 73, 74, 75, 76, 202, 207
Slim, Carlos, 56
Sloan, Harry, 54
Smalls, Biggie, 73
Smith, Daniel, 105
Smith, Joseph, 30, 64–65, 67, 136
social capital, 87
Social Security, 209
Socrates, 7
Solomon, Robert, 89
Somalia, 43
Soros, George, 154
Sotomayor, Sonia, 221–22
South Korea, 17*n*
Spellbound, 47
spelling bees, 47
Spinoza, Baruch, 61
stamina, 133
Stanford University, 47
status loss, 12
  Cuban Exiles and, 88–89, 92
  Iranians and, 92
Steele, Claude, 78
Stein, Joel, 54
stereotype boost, 22, 79–80, 83–84
stereotypes, 2, 14

stereotype threat, 22, 78–80, 83–84
  blacks and, 78, 80, 81
Stewart, Jon, 52
Stoicism, 120, 138
Stone, Oliver, 210
Stuyvesant High School, 169–72
substance abuse, 150, 215, 216
  in Appalachia, 175, 178, 180
success, 15, 20, 21, 117, 147, 151, 195, 196
  conventional, 145, 159–60, 196
  defining, 7, 159–60
  external measures of, 18, 159–60
Sudan, 43
suicide, 150
Sumner, William Graham, 59
Sun Microsystems, 49
Sunset Park, 170, 171, 172
Sununu, John E., 57
Sununu, John H., 57
superiority complex, 8–9, 15, 20–22,
59–84
  America and, 203, 207, 209, 211–12,
  219–20, 225
  Amish and, 181–82
  Appalachia and, 175
  Asian Americans and, 13, 173
  assimilation and, 83
  blacks and, 72–78
  Chinese, 72, 120–23, 130–31, 156, 220
  Chinese Americans and, 124
  Cuban Americans and, 68–72, 156
  drive and, 15, 17, 84
  Holocaust survivors' children and, 193
  humility and, 182
  impulse control and, 15–16, 17, 120
  Indian Americans and, 95, 97–99, 101–2
  insecurity and, 1, 10, 11–14, 15,
  17, 114, 124
  intolerance and, 158, 225
  Iranians and, 72, 90–92
  Jews and, 9, 60–64, 194, 195, 197
  Mormons and, 8, 16, 64–68, 136,
  156–58, 186–87
  Nigerian Americans and, 81–82
  Protestantism and, 184, 185
  stereotype boost and, 22, 79–80
  stereotype threat and, 22, 78–80

superiority complex (*cont.*)
  Stoicism and, 120
  underside of, 17, 155–58
  use of term, 59*n*
Survivor, 104
Syrian Jews, 24

Tagore, Rabindranath, 96, 98
Taitz, Sonia, 190, 192–93
Taiwanese Americans, 48, 124
  family honor and, 13, 110
Talmud, 138, 151
Tan, Amy, 149, 160
teen pregnancy, 178, 180
Teller, Edward, 55
Ten Commandments, 138
Theodorakis, Mikis, 61
300, 89–90
Three Stooges, 52
Time, 49
Tocqueville, Alexis de, 27, 85–86, 202–3
Tony Awards, 7, 52
Trilling, Lionel, 152
Triple Package, 5–27
  America and, 199–225
  breaking out of, 160–64, 196–97
  underside of, 145–66
  see also impulse control; insecurity;
    superiority complex
20/20, 174
Twilight series (Meyer), 33, 137

Udall, Morris, 31
United States Presidential Scholars, 47
University of Michigan, 32
University of Pennsylvania, 33
Unz, Ron, 193–94
upward mobility, 5–6, 11, 147, 167–69,
    173–74
  immigrant selection criteria and, 171

Van Buren, Abigail, 53
Vance, J. D., 177–78
Vietnamese Americans, 7
Voltaire, 87
Von Fürstenberg, Diane, 52
Von Neumann, John, 55

Walker, Alice, 77
Wall Street, 210
Walt Disney Animation Studios, 32
Wang, Vera, 148
WASPs, 20, 87, 112, 113, 195
Watchmaker's Daughter, The
    (Taitz), 190
wealth, 169
Weber, Max, 8, 137, 184, 185
Weiner, Matthew, 195
welfare, 177, 209, 210
West Indian Americans, 41,
    43, 156
Wharton School, 33
What Makes Sammy Run?
    (Schulberg), 140
white supremacy, 72–73
Wilde, Oscar, 146
Williams, Ted, 21
Winthrop, John, 201
Wire, The, 143
Wittgenstein, Ludwig, 146
Wolfe, Tom, 210
Wong, Freddie, 164
WordPerfect Corporation, 32
work ethic, 26, 87, 129, 137,
    143, 173
  self-esteem movement and, 216
  welfare and, 177, 209
Wu, Jason, 164
Wu, Tim, 103

Yahoo, 48
Yale Law School, 42
Yale University, 25, 47, 110, 111
Yang, Jerry, 48
Yang, Wesley, 14, 102–3
Yoruba people, 81
Young, Brigham, 64, 67
youth culture, 11, 146
YouTube, 48

Zakaria, Fareed, 49
Zappos, 48, 127
Zhou, Min, 79
Zucker, Jeff, 54
Zuckerberg, Mark, 222, 223